T0340065

Time in Variance

The Study of Time

Founding Editor

Julius T. Fraser†

VOLUME 17

The titles published in this series are listed at *brill.com/stim*

Time in Variance

Edited by

Arkadiusz Misztal
Paul A. Harris
Jo Alyson Parker

BRILL

LEIDEN | BOSTON

Cover illustration: The Universe Cascade, The Garden of Cosmic Speculation. Photograph by Paul A. Harris.

Library of Congress Cataloging-in-Publication Data

Names: Misztal, Arkadiusz, editor. | Harris, Paul (Paul A.), editor. |
    Parker, Jo Alyson, 1954- editor.
Title: Time in variance / edited by Arkadiusz Misztal, Paul A. Harris, Jo
    Alyson Parker.
Description: Leiden ; Boston : Brill, [2021] | Series: The study of time,
    0170-9704 ; volume 17 | Includes bibliographical references and index.
Identifiers: LCCN 2021034478 (print) | LCCN 2021034479 (ebook) | ISBN
    9789004470163 (hardback) | ISBN 9789004470170 (ebook)
Subjects: LCSH: Time. | Time in mass media.
Classification: LCC BD638 .T5645 2021 (print) | LCC BD638 (ebook) | DDC
    115—dc23
LC record available at https://lccn.loc.gov/2021034478
LC ebook record available at https://lccn.loc.gov/2021034479

Typeface for the Latin, Greek, and Cyrillic scripts: "Brill". See and download: brill.com/brill-typeface.

ISSN 0170-9704
ISBN 978-90-04-47016-3 (hardback)
ISBN 978-90-04-47017-0 (e-book)

To the memory of Marlene Pilarcik Soulsby,
dear friend and long-time ISST colleague

∴

Contents

Acknowledgements

The efforts of many people have made this volume possible. We wish to begin by thanking those who helped plan and facilitate the 2019 "International Study of Time Triennial Conference: Time in Variance" from which the contributions in this volume derived. The ISST Conference Committee – Emily DiCarlo, Carol Fischer, Raji Steineck, Daniela Tan, and Tom Weissert – helped facilitate the work of Conference Coordinator Paul Harris. The conference would not have been possible without the generosity of many academic units at Loyola Marymount University, including the President's Office (Timothy Snyder, President), Hannon Library (Kristine Brancolini, Dean), Bellarmine College of Liberal Arts (Robbin Crabtree, Dean), and the Department of English (Barbara Rico, Chair). The Conference Committee is especially grateful for the tireless work of Elmo Johnson and Maria Jackson in making the conference a success.

Although nameless in the pages of this volume, the anonymous reviewers generously gave of their time and expertise. Their often extensive suggestions for revision and recommendations for further reading bespoke their commitment to their disciplines and helped ensure that the contributions herein not only adhered to disciplinary standards but brought forward original insights.

Since 2001, Brill Publishing has demonstrated a strong commitment to the Study of Time series. Acquisitions Editor Joed Elich has championed the series for the last two decades, and he has also been a most welcome participant in our conferences.

Jo Alyson Parker would like to thank the Saint Joseph's University Board on Faculty Research and Development for granting her a sabbatical for fall 2019, which helped support her work on this volume.

Illustrations

Figures

Table

Notes on Contributors

Vroni Ammann
is a PhD candidate in Japanese Studies at the University of Zurich. She is currently part of the ERC Advanced Grant Project titled "Time in Medieval Japan" and working on her dissertation, tentatively titled "Scent of Time: Incense between Marketplace and Time in Medieval Japan." In 2017 her MA thesis "Das Tierschutzgesetz Japans: Die Reform von 2012 und ihre Rezeption in der Asahi Shinbun," about the reformation of the Protection of Animals Act in Japan in 2012, was published.

Adam Barrows
is a Professor of English at Carleton University in Ottawa, Ontario. He has written two books on time and literature: *The Cosmic Time of Empire* (2011) and *Time, Literature, and Cartography After the Spatial Turn* (2016).

Lucía Cash Beare
has a PhD in Spanish from University of California, Irvine. Her areas of interest include environmental critical theory, embodiment, molecularity, and Latin American literature.

Raphaëlle Costa de Beauregard
is Emeritus Professor at the Université de Toulouse Jean Jaurès. She has specialized in Art and Film Anglo-Saxon Studies since 1970. In 1993 she founded the SERCIA (Société d'Etudes et de Recherches sur le Cinéma Anglosaxon): www.sercia.net. Her book publications are *Nicholas Hilliard et l'imaginaire élisabéthain* (Paris: CNRS, 1992) and *Silent Elizabethans-The Language of Colour of Two Miniaturists* (Montpellier: CERRA, 2000). She edited the collections *Le Cinéma et ses objets – Objects in Film* (Poitiers: La Licorne, 1997) and *Cinéma et Couleur – Film and Colour* (Paris: Michel Houdiard, 2009). She has published many articles on Elizabethan Portrait Miniatures and on Shakespeare's plays, as well as on Anglo-American cinema. Her present research is about phenomenology and film, with a focus on early cinema.

Emily DiCarlo
is an artist and writer whose interdisciplinary work applies methodologies that often produce collaborative, site-specific projects. Evidenced through video, performance and installation, her research connects the infrastructure of time with the intimacy of duration. A recent graduate of the Master of

Visual Studies at the University of Toronto, she is the recipient of the Canadian Social Sciences and Humanities Research Council grant (SSHRC). Since 2007, her work has been shown both locally and abroad, with most recent exhibitions at the Art Museum in Toronto and at SÍM Gallery in Reykjavik, Iceland, as part of their artist-in-residence program. Since 2013, she has served as ISST's Communications Officer. She currently lives and works in Toronto (Tkaronto), Canada. www.emilydicarlo.com.

Sonia Front

is Assistant Professor in the Institute of Literary Studies, the University of Silesia in Katowice, Poland. Her research interests include time and temporality as well as representations of consciousness in twenty-first-century literature, film and television. Her most recent book is the monograph *Shapes of Time in British Twenty-First Century Quantum Fiction* (Cambridge Scholars, 2015).

Paul Harris

is a Professor of English at Loyola Marymount University. He served as President of the International Society for the Study of Time 2004–2013, and he is co-editor of the literary theory journal *SubStance*. His work under the appellation "The Petriverse of Pierre Jardin" encompasses a rock garden, blog, and an open-access e-book in the experimental digital theory series SubStance@Work. He has designed "slow time gardens" at LMU and exhibited stone installations at the National Gallery of Denmark, Chapman University, and the Arizona State Art Museum Project Space. His recent collaborations include a story written with novelist David Mitchell and *Contemporary Viewing Stone Display* (VSANA, 2020), a book co-authored with Richard Turner and Thomas S. Elias.

David Harris-Birtill

is a Lecturer in Computer Science at the University of St. Andrews, UK, where he is Director of the Digital Health Masters program and Director of the Human Computer Interaction group. He has led projects funded by the Digital Health and Care Institute, Innovate UK, and industrial partners. David leads research projects in the fields of Artificial Intelligence (machine learning) and medical imaging, and he lectures on machine learning, signal processing, and human computer interaction. David has a background in Medical Physics and holds a PhD in Physics from the Institute of Cancer Research; he was also previously a Post-Doctoral Researcher at Imperial College, London. David is the Founder and Director of Beyond Medics, a company established to create medical imaging and sensing technology for patient benefit, where he is currently leading development of a system that measures people's vital signs at a distance using video camera footage.

Rose Harris-Birtill

is an Honorary Research and Teaching Fellow at the School of English at the University of St. Andrews and serves as Managing Editor at the Open Library of Humanities, Birkbeck, University of London. Her academic monograph, *David Mitchell's Post-Secular World: Buddhism, Belief and the Urgency of Compassion* (Bloomsbury Academic, 2019) brings together post-secular literary fiction and Buddhist philosophies, investigating the cross-temporal redeployment of Buddhist influences across the complete fictions of contemporary author David Mitchell. Rose holds the International Society for the Study of Time Founder's Prize for New Scholars, the Frank Muir Prize for Writing, and a McCall MacBain Teaching Excellence Award, and she is an Associate Fellow of the Higher Education Academy. Rose is also the Secretary for the British Association for Contemporary Literary Studies, and her research interests include contemporary narratives of time, digital storytelling, science and speculative fiction, post-secular literature and theory, globalization, and critical and cultural theory.

Arkadiusz Misztal

is Assistant Professor in American Studies at the University of Gdańsk, Poland. His research and teaching interests focus on contemporary American fiction, narrative theory, and the philosophy of time. His most recent book is *Time and Vision Machines in Thomas Pynchon's Novels* (Peter Lang, 2019).

Steve Ostovich

is a past member of the ISST Council and a former Managing Editor of *Krono-Scope*. With Sabine Gross, he so-edited *Time and Trace: Multidisciplinary Investigations of Temporality: The Study of Time XV* (2016). He is Professor (emeritus) of Philosophy at the College of St. Scholastica in Minnesota, USA. Recent publications include "Dangerous Memories: Nostalgia and the Historical Sublime" in *Nostalgia Now: Cross-Disciplinary Perspectives on the Past in the Present*, edited by Michael Hviid Jacobsen (Routledge, 2020); and "Claude Lanzmann's *Shoah* and the Historical Sublime" in *The Cinematic Sublime: Negative Pleasures, Structuring Absences*, edited by Nathan Carroll (Intellect, 2020).

Jo Alyson Parker

is a Professor Emerita of English at Saint Joseph's University in Philadelphia. Her publications include *The Author's Inheritance: Henry Fielding, Jane Austen, and the Establishment of the Novel*; *Narrative Form and Chaos Theory in Sterne, Proust, Woolf, and Faulkner*; and articles on Kate Atkinson, Ted Chiang, David Mitchell, and narrative and time. With Michael Crawford and Paul Harris, she co-edited *Time and Memory: The Study of Time XII*, and, with Paul Harris and Christian Steineck, she co-edited *Time: Limits and Constraints: The Study of Time XIII*. From 2014–2018, she was Managing Editor of *KronoScope: Journal for the Study of Time*.

Sue Scheibler

When she is not playing *Skyrim*, Sue is busy teaching film, TV, and video game studies at Loyola Marymount University. She puts her graduate degrees in Biblical Studies, Philosophy of Religion, and Critical Studies to good use in both her scholarly and pedagogical work, all of which emphasize the intersections of philosophy, theology, critical theory, and disabilities studies with those of film, TV, and video games. Sue's current work includes representations of neurodivergence in TV and video games; the meditative gaze in film, TV, and video games; and a Levinasian reading of video games.

Martin Scheuregger

is a Senior Lecturer in Music in the School of Fine and Performing Arts at the University of Lincoln (UK), where he leads the BA (Hons) Music program. He takes an interdisciplinary approach to research, combining musical analysis and composition as he explores notions of time and brevity in music. His current research is expanding on areas of his PhD thesis, with an extensive study of the music of George Benjamin. His research into issues of temporality continue to develop, with current work focusing on approaches to musical stasis in analysis and composition. He completed a PhD at the University of York in 2015, funded by the Arts and Humanities Research Council (AHRC) and supervised by Professor Tim Howell and Professor Thomas Simaku. Scheuregger's music has been performed in the UK, Australia, Hong Kong, Germany, and Holland, and in 2015 a new work was premiered at the Huddersfield Contemporary Music Festival, supported by Sound and Music. His music has been recorded by Dark Inventions and Percussing, and it has been published by University of York Music Press.

Raji C. Steineck

is Professor of Japanology at University of Zurich (UZH), president of ISST, and principal investigator of the European Research Council's Advanced Grant project "Time in Medieval Japan" (TIMEJ). His research interests combine the history of ideas in Japan, the theory of symbolic forms, and the philosophy of time. Recent time-related publications include *Critique of Symbolic Forms II* (in German, 2017, on mythological time), "Chronographical Analysis" (*KronoScope* 18/2, 2018), and "Should we give up 'Time'?" (*Time's Urgency: The Study of Time XVI*, 2019).

Thomas Weissert

is Director of Technology at the Agnes Irwin School. His publications include *The Genesis of Simulation in Dynamics: Pursuing the Fermi-Pasta-Ulam Problem* (Springer) and articles on tsunami travel-time charts, dynamical systems theory, dynamics and narrative, Jorge-Luis Borges, and Stanislaw Lem. He served as Executive Secretary of the International Society for the Study of Time from 1998–2016.

David Wood

has taught environmental and continental philosophy at the University of Warwick (UK) and Vanderbilt University. His recent books include *Eco-Deconstruction: Derrida and Environmental Philosophy* (ed. with Fritsch and Lynes); *Reoccupy Earth: Toward an Other Beginning*; *Deep Time: Dark Times*; and *Thinking Plant Animal Human*. He is currently director of YellowBird Artscape in rural Tennessee.

Introduction

Paul A. Harris, Arkadiusz Misztal, and Jo Alyson Parker

This volume features selected work presented at Time in Variance, the 17th triennial conference of the International Society for the Study of Time (ISST), held on the campus of Loyola Marymount University in Los Angeles, California, in June, 2019. The active engagement of nearly one hundred participants from six continents made for a deeply collaborative, convivial, and rewarding conference. In addition to the 70 plenary papers presented over four days, conferees were treated to an evening of improvisational jazz by the David Ornette Cherry Trio and a dance-theatre performance by Susan Banyas entitled *Kundalini History/Voices from the Great Serpent*, featuring Frederick Turner reading and David Ornette Cherry's live accompaniment. The ISST was also delighted to collaborate with new colleagues from the Japanese Society for Time Studies, and members look forward to continuing this cooperation in the future. For conference coordinator Paul Harris, the event had special significance, as it marked his 30 years as a member of the ISST and of the English faculty at Loyola Marymount University.

The theme of Time in Variance resonates with the core principles of J. T. Fraser's hierarchical theory of time. "What we unproblematically call 'time,'" Fraser writes, "though a single and coherent feature of the world, is, in fact, a hierarchy of distinct temporalities corresponding to certain semiautonomous integrative levels of nature" (1975, 435). Over the course of his voluminous corpus, Fraser elucidated the six temporalities he designated as "levels" in his hierarchical theory, drawing deeply on different disciplinary discourses to describe each one in detail. In summarizing the propositions of his theory, Fraser posits that "processes characteristic of each of these levels function with different types of causation and must be described in different languages" (1999, 26). In Fraser's hierarchy, it is "variance all the way down." Not only does time encompass various temporal levels; each level is axiomatically at variance with itself. Each temporal level is fundamentally incomplete, in the sense of Gödel's theorem: "*Incompleteness in nature* consists of the impossibility of determining the boundaries of the laws of an integrative level, using only the language of that level" (2007b, 61). For Fraser, the incompleteness of each level is not only a question of description, of needing a metalanguage to provide a consistent account of a level. It is, rather, that temporal levels are at variance with themselves, in Gödel's sense of inconsistency: "*Inconsistency in nature*

© KONINKLIJKE BRILL NV, LEIDEN, 2021 | DOI:10.1163/9789004470170_002

may be recognized in the existence upon each integrative level of certain contradictions or unresolvable conflicts which define that level" (2007b, 61). The
overarching implication is that, because variance permeates time across all
levels and within each level of reality, "the foundational dynamics of nature is
not Platonic harmony but a nested hierarchy of conflicts" (2007b, 61). Fraser's
hierarchical theory remains conceptually robust, precisely because it provides
a framework for thinking about different ways that temporalities function,
evolve, and emerge.

The initial part of essays in this volume, entitled "Variations on J. T. Fraser's
Hierarchical Theory of Time," provide a series of critical perspectives on
Fraser's work in the context of the conference theme. In his President's
Address that opened the conference, Raji Steineck surveys the different meanings of variance and their rich resonances with the study of time, opening up
avenues for exploration taken up by subsequent papers. Steineck uses a cita-
tion from Fraser's dissertation to show that the creative conflicts that drive
time and human life were integral to the ISST's Founder from the outset of
his career. In "Out of Plato's Cave," Steve Ostovich repurposes the title of a
1980 Fraser essay to clarify the arc and underlying sense of the hierarchical
theory. Ostovich underscores Fraser's reversal of the Platonic path from the
dark cave to the light of truth, which moves from time towards the timeless.
Instead, Fraser's work descends from the timeless to immerse itself in the full
complexity of time, where truth is a process and evolution and human lives are
shaped by conflict. Citing the lasting influence of Fraser's personal and intellectual encounter with Joseph Needham on his work, Ostovich illustrates how
Fraser's framework allowed him to parse tangled questions such as the relation
between time and entropy.

The remaining papers in this part sound out speculative intersections
between ecological temporalities and Fraser's hierarchical theory and envision
ways that its scheme of temporal levels might be extended or impacted by such
engagements. Lucia Cash Beare reflects on ecological temporality to analyze
Atilio Caballero's novel La última playa, arguing that narrative threads individual lives into ecological assemblages. Embeddedness in these assemblages
entangles humans with other organisms and inhuman matter in processes
that call for an "ecotemporal" mode of understanding the distributed interactions of agents bound by a common material context and fate. In the Founder's
Lecture, David Wood delves deeply into the question of how impending catastrophic climate change disrupts received models of history and time, including Fraser's hierarchical theory. In his intricate paper, Wood argues that, while
change is central to time, catastrophic climate change produces a variation
in the very shape of change and therefore must mark a new "threshold" in

Fraser's sense. If imminent climate change leaves us passive spectators to radical transformation, Wood asks, how will a shrinking future horizon impact our identity or any hope of progress? While speculating on the arrival of future states entangles us in strange loops that signal the breakdown of the inductive assurances that the past traditionally supplied, Wood contends, he concludes on a note of hope, that the necessity of extending Fraser's scheme opens the possibility of "a new threshold, a new temporal dispensation." In his exploration of "Slow Time," Paul A. Harris argues that the Anthropocene demands what David Wood calls "temporal phronesis," the ability to negotiate and reconcile radically variant timescales, and that this challenge is most effectively met through a contemplative mode of thinking. Harris's essay imaginatively walks the reader through a series of art installations (created for the conference) and two landscape art gardens that address relations between human and ecological, geologic, and cosmic timescales. In the process, Harris recapitulates his proposal (made in 1999) that Fraser's theory would be complicated by an emerging "Gaiatemporality," constituted by the conflicts created by the intrusion of humanity into earth processes.

The second set of essays explores "Variant Narratives" – narratives in fiction, video games, film, and graphic novels that play with the notions of temporal variance and invariance.

Adam Barrows examines the figuration of the Disabled Time Child as a prominent trope within the cultural histories of childhood and disability by focusing on the trope's rendition/translation into the genres of science fiction, horror, and fantasy. Drawing on the work of Michael Bérubé and others, Barrows argues that the disability chronotrope reveals culturally inflected dynamics of temporal otherness as a gateway to non-human and ecologically oriented forms of being. Disability in narrative, seen as a departure from adult human rational temporality, offers an abnormal perspective on time's passage, its form and texture. This perspective moves beyond narrow definitions of human temporality by opening narrative sensibilities to disjunctive temporalities, temporal indigenity, and ecological modes of futurity.

In a similar vein, Sue Scheibler addresses temporal otherness by approaching the Role Playing Game (RPG) *Elder Scrolls V: Skyrim* through the lens of an autistic player with ADD. Scheibler examines the games mechanics, design, and narrative architecture as constitutive elements of *Skyrim*'s sophisticated interplay of seriality and simultaneity that creates immersive and interactive temporalities mapped onto time in the real world. By providing the player with new modes of temporal control, the game offers a richly rewarding experience inducive of "stimming" that helps autistic people counter a temporal cacophony and sensory onslaught in their lives. Sonia Front's chapter in turn probes critical points and pressures of the contemporaneity and its new temporal regime in twenty-first century cinema and television narratives. The narrative as a mediator between manifold forms and shapes of time reflects, Front contends, personal and cultural awareness of imminent crisis, brought about by abuses of technology and science, environmental exploitation, and various zero-events marked by the turn of the new millennium and 9/11 terrorist attacks. The series *FlashForward* and the film *Annihilation* reveal the contemporaneity as comprising temporal rifts, clashing temporalities, and disjunctive timescales and remind us of the fragility of our human time when exposed to such radical ruptures and displacements.

Jo Alyson Parker and Thomas Weissert contribute to the study of variation in narrative temporalities by tracing connections among temporal awareness, self-narration, and consciousness in the television science fiction series *Westworld*. The series features android/gynoids characters whose lives are determined by apparently unvarying narrative loops, made up of backstories and implanted memories that trap the characters in a meaningless existence of an endless repetition. And yet, as Parker and Weissert argue, out of repetition can come a meaningful variation that makes possible to break the scripted temporal invariance of the loops and develop a narrative sense of self. This self-making is not merely confined to the fictional world of *Westworld* but in a more general sense demonstrates the importance of narrative processes as necessary for and constitutive of emergence and the continuity of human consciousness over time. Narrative articulations of in/variance are also a focal point of Arkadiusz Misztal's contribution devoted to the interplay of consistencies and variations in the game-changing graphic novel *Here* by Richard McGuire. Making full use of narrative techniques and the logic of the visual, the novel, Misztal argues, explores the peculiar human constructions of hereness as intrinsic parts of the changing timescapes by framing both a stasis and an unfolding of time. Through its appeal to the productive dynamics of imagining, *Here* injects a heightened sense of temporal fluidity and plasticity into the static architecture of the page and encourages the reader to transcend panels' spatial juxtaposition and temporal succession. In doing so, McGuire's work brings to the fore the coexistence of temporal tensions and constants in our constructions of places and invites us to negotiate the former without undermining the latter.

The final part of the volume addresses "Measuring Time's Variance." Commenting on Immanual Kant's awe over "the starry heaven above me and the moral law within me," J. T. Fraser notes, "This tie between man and the cosmos paved the way for the employment of time measurements for a better understanding of the universe" (2007a, 113). The essays here explore the variances that can both complicate and facilitate our "understanding of the universe."

The part opens with two essays dealing directly with measuring time's passage – whether in centuries past or in the modern age. Exploring the varying morphologies of time in medieval Japan, Vroni Ammann focuses on the incense clock: as a clock, it reflects a cyclic understanding of time, but, as a commodity, it participates in trading networks based on a linear conception of time and indeed reinforces the idea of time itself as a commodity. David Harris-Birtill and Rose Harris-Birtill's contribution takes on a contemporary problem of time measurement – the varying times that it takes a computer to run an algorithm because of differences in computer hardware. After providing a thorough analysis of the problem, the writers submit a set of recommended best practices for alleviating it, and they include helpful templates for recording computation time.

The last three essays explore artistic creations that highlight measuring temporal variance. With Raphaëlle Costa de Beauregard's contribution, we return to the clock – in this case, the big clock that looms over the action and provides the title for the film *The Big Clock*. Closely examining key scenes and motifs in the film, Costa de Beauregard contends that the invariant march of time represented by the clock is in tension with variant contingent situations occurring in the film and that, as the film progresses, invisible time is made visible. With her discussion of her recent video project *I Need to Be Closer to You*, Emily DiCarlo draws our attention to an anomaly of the Canadian Daylight Saving Time system whereby a one-hour gap between time zones extends to two, thus impacting the relationship and collaboration between two partners. Locating the video project within a Romantic Conceptualist framework and emphasizing its dependence on time-specificity, DiCarlo explains how the telepresence that the project brings about enables the partners to transcend the temporal distance occasioned by time's variance. In his essay, Martin Scheuregger also deals with different "timezones" – specifically those achieved through the compositional devices employed by George Benjamin in his orchestral composition *Sudden Time*. Through his thorough analysis of these devices – his measurements of them – and of the way that *Sudden Time* negotiates these different timezones, Scheuregger demonstrates how Benjamin brings "time front and center" into the composition and how it is best understood through a temporal lens.

In an essay on "critical interdisciplinarity," J. T. Fraser points out the importance of the interdisciplinary nature of time studies:

A uniform method of testing for truth or the use of a single jargon or a single personality of knowledge would prohibit the division of labor that made the growth of civilizations possible. These pluralities should, therefore, not be judged as inhibiting the development of an integrated study of time but rather as guiding it. An integrated theory of time should not be required to eliminate the differences among ways of knowing but facilitate communication among them.

2005, 168

As the following essays demonstrate, the ongoing mission of the ISST continues – embracing disciplinary pluralities, facilitating communication among them, enabling the growth of knowledge.

References

Fraser, J. T. 1975. *Of Time. Passion, and Knowledge: Reflections on the Strategy of Existence.* New York: Braziller.

Fraser, J. T. 1999. *Time. Conflict, and Human Values.* Urbana and Chicago: University of Illinois Press.

Fraser, J. T. 2005. "Space-Time in the Study of Time: An Exercise in Critical Interdisciplinarity." *KronoScope* 5 (2): 151–175.

Fraser, J. T. 2007a. "How to Use a Clock." In *Time and Time Again: Reports from a Boundary of the Universe,* 113–114. Leiden and Boston: Brill.

Fraser, J. T. 2007b. "Mathematics and Time." In *Time and Time Again: Reports from a Boundary of the Universe,* 53–64. Leiden and Boston: Brill.

PART 1

Variations on J. T. Fraser's Hierarchical Theory of Time

∵

CHAPTER 1

President's Address: Time in Variance

Raji C. Steineck

The theme of our conference, "Time in Variance," reportedly emerged while three friends were riding a car from Hadrian's Wall through Yorkshire. The friends were our previous president, Paul Harris, our long time *KronoScope* chief editor and incoming vice president, Jo Alyson Parker, and our esteemed former executive secretary, Tom Weissert. By virtue of its origin in a car ride, "Time in Variance" is an apt theme for a conference held in a country that invented the road movie, in the city that brought this genre to perfection. By the same token, it is an apt theme for our society, since it was born from on-the-road conversations between a physicist and two literature scholars, one of whom is also a gardener. Most importantly however, "Time in Variance" elegantly expresses a fundamental trait of what ISST is all about: the manifold aspects of time.

At the beginning of our last conference in Edinburgh, I was so bold as to present you with my own take on the problem of time. I remember unnerving some of you by suggesting that time, at least "Time" with a capital "T," best be regarded as a concept.[1] Today, I promise to do nothing of that sort – not because I want to retract on that statement, but because our conference theme resonates with the old aesthetic principle *variatio delectat*: there is pleasure in variation. Consequently, I will do something different than the last time around and simply contemplate the definitions of variance and how they relate to the subject of time.

The Merriam-Webster Dictionary of US American English lists five meanings of "variance."[2] The first and most general one is "the fact, quality, or state of being variable or variant: difference, variation." The sample expression is "yearly variance in crops," and it seems no mere coincidence that this sample invokes time. There may be a great variance in our beliefs about the nature of time. But we will probably agree that one of the most important reasons for invoking time is our experience of the same as different. Agriculturalists plant the same soil but with different seeds and thus realize a "yearly variance in crops." They learned – and forgot again – this principle of crop rotation

---

1 Steineck, "Presidential Address," 49–53.
2 https://www.merriam-webster.com/dictionary/variance, accessed Oct. 17, 2019.

© KONINKLIJKE BRILL NV, LEIDEN, 2021 | DOI:10.1163/9789004470170_003

because they have planted the same crops on the same soil and have as a consequence experienced a negative variance in harvests over time. Again, a difference of the same, or, to put it paradoxically: the variance of an invariant. And how is this paradoxical variance of an invariant possible? Because of time: we identify an object, such as a field, as the same, and distinguish its variant states over time.

Time is invoked by variance, but variance also shapes the image of time. Time takes on different forms according to the variance in question: there is linear time, as evoked by diminishing harvests or, inversely, economic growth. There is oscillating time, as in the alternation of day and night, working and leisure time. There is cyclical time, as in the course of the seasons. These different morphologies, succinctly described by Maki Yūsuke in his *Comparative Sociology of Time*,[3] create what you might call a second order of variance. Here comes spring, or the ISST conference, again, but with each cycle, we have grown older. Depending on age, conduct, and luck, our growth in age may translate into greater maturity or decay. We feel the contrast between chronology, a purely linear progression of time, the arc of life, and the cycle of recurring events. Perhaps even more poignantly, we miss our friends and colleagues we used to meet at ISST, but who, for some reason or other in their personal timelines are no longer present on this occasion.

As if to account for such contrasts between linearity and cyclicity and for the sadness and sometimes outright anger we feel when they hit home, Merriam-Webster's second definition of "variance" is "the fact or state of being in disagreement: dissension, dispute." "Time," as I just pointed out, very often is experienced as such disagreement. The disagreement may be described as one inherent to its own form – because, even if we believe, as the moderns tend to believe, that the fundamental or objective schema of time is that of a straight, continuous line that conforms to a homogeneous metric, it can only be measured as such in relation to an invariant. In the absence of a substantial invariant, such as God, or the Chinese way of the heavens, invariance is sought in uniform recurrence – that is, cyclicity. Time is thus literally, and of the essence, "in variance"; it is fundamentally defined by "a state of being in disagreement."

Remember that, having edited the classic *Voices of Time* in 1966, which assembled variant views of time in the major disciplines of the sciences and the humanities, J. T. Fraser went on to develop his theory of "time as a hierarchy of creative conflicts." Here is my triennial Fraser quote, this time from the abstract of his PhD dissertation, submitted 50 years ago, in 1969:

---

3   Maki 真木, *Jikan No Hikaku Shakaigaku* 時間の比較社会学, 195. See, too, Steineck, "Time in Old Japan," 24–28.

Observation and introspection suggest the existence in nature of certain conflicts which appear to be unresolvable on their indigenous levels of occurrence. These conflicts display irreversibilities of various richness with respect to their origins and indeterminacies of varying degrees with respect to their possible resolutions. Thus, they show two characteristic attributes of temporality, namely the irreversibility of the past and the indeterminate nature of the future. The conditions surrounding these conflicts seem to permit their partial solutions through the emergence of new levels of existents. Critical examination of the new existents shows, however, that for them the earlier arguments of irreversibility, indeterminism and unresolvability of conflicts once again hold. The conflicts and their partial resolutions may be arranged in a hierarchy of increasing complexity and axiological weight. The thesis of this paper is that the idea of time comprises such a hierarchy of conflicts.[4]

The creative conflict J. T. Fraser found at the heart of human existence is the one between the knowledge that our life has a limited timeline and the desire to negate that knowledge. While in his dissertation, he conceived of this conflict as a "conflict between the time-knowing and time-ignorant strata of the mind,"[5] we might more cautiously understand it as a conflict between two temporal morphologies in which we, as humans, both participate: that of directed, linear time, and that of cyclical, recurring time. The conflict remains as dramatic as ever because each morphology is tied to different "carriers" of invariance. Linear time is "monadic," related to a personal and irreplaceable "I"; cyclical time is partly organic, partly inorganic, and in any case, impersonal. We can opt out of finite, linear time and into the "oceanic consciousness"; we can feel we take part in the undulating rhythms of living and dead matter because that is what we also are. But we can only do so at the price of losing the monadic aspect of ourselves; and that aspect, or so I would argue with Fraser, carries a heavier axiological weight because it is a necessary condition of participating in the symbolic domains of art, science, religion, law, and so forth. To reduce ourselves to the "oceanic" realities within and around ourselves means a flattening of the horizon.[6] If our world is to retain the totality of its dimensions, we will have to live with the conflicts and disagreements inherent to time in variance.

---

4 Fraser, "Time as a Hierarchy of Creative Conflicts," 598.
5 Fraser, 599.
6 Fraser, 655–656.

And lest we forget: Even descending into oceanic consciousness would not mean the end of variance or conflict. Even on the level of inanimate matter, time is tied to a variance between the world of pure electromagnetic radiation and the gravitational universe of ponderable mass. As Fraser states in his dissertation, without that variance, time would have no "physical meaning"; it would be "operationally undefinable."[7]

If we think of nature in terms of "laws of nature," this finding of Fraser's could be easily re-stated in terms of the third of Merriam-Webster's definitions of variance: "a disagreement between two parts of the same legal proceeding that must be consonant." The search for a "theory of everything" in physics conforms to the idea that the fundamental laws of physics must be in agreement. However, for the time being, the fundamental variance stands. In terms of the theory of time as a hierarchy of creative conflicts, it is a variance that is constitutive with respect to time. And this is only the primordial level in a hierarchy of ever more complex arrays sustained by a balancing of opposing forces.

The creative aspect of variance, as disagreement, harkens back to the old Heraclitean statement "Πόλεμος πάντων μὲν πατήρ ἐστι, πάντων δὲ βασιλεύς. [Conflict is the father and king of all.]"[8] I quote this, with due apologies for its patriarchal ring, only to remind us that much of what we cherish in this world and explore as scientists, humanists, or artists in ISST, can only exist in variance.

If variance is the rule, or even the ruler in Heraclitean diction, in another sense it refers to exception: Definition number 4 points to variance as "a license to do some act contrary to the usual rule," such as "a zoning variance." In terms of time, invocations of "eternity," be it in law or religion, constitute such "zoning variances" with respect to different aspects of human life. The questions here are: Who commands the authority to grant such license? On what grounds? And what does it mean for our understanding of time? Several papers during this conference return to these questions.

With this, and in view of time, I turn to the fifth and last definition of variance: "the square of the standard deviation." All I can say to this is: it is important to recognize that deviation can be squared, and that, conversely, the squares may be deviant. This observation may even be applied to the theory of time – but I promised not to bother you with my own thoughts today, and I am square enough to keep that promise. Thank you for your attention, and I wish us all a stimulating conference.

---

7 Fraser, 627.
8 Heraclitus, Diels and Kranz 22 B 53.

References

Diels, Hermann, and Walther Kranz. *Die Fragmente der Vorsokratiker: griechisch und deutsch.* 7th ed. Berlin: Weidmannsche Verlagsbuchhandlung, 1954.

Fraser, J. T. "Time as a Hierarchy of Creative Conflicts." *Studium Generale; Zeitschrift für die Einheit der Wissenschaften im Zusammenhang ihrer Begriffsbildungen und Forschungsmethoden* 23, no. 7 (1970): 597–689.

Maki Yūsuke 真木悠介. *Jikan no hikaku shakaigaku* 時間の比較社会学 [*Comparative Sociology of Time*]. Tōkyō: Iwanami Shoten, 2003.

Steineck, Raji C. "Time in Old Japan: In Search of a Paradigm." *KronoScope* 17, no. 1 (2017): 16–36.

Steineck, Raji C. "Presidential Address: Should We Give Up 'Time'?" In *Time's Urgency*, edited by Carlos Montemayor and Robert Daniel, 44–56. Leiden: Brill, 2019.

CHAPTER 2

Out of Plato's Cave

Steve Ostovich

Abstract

The title is lifted from an essay by J. T. Fraser in his book *Time and Time Again* (2007). It conveys Fraser's conviction, a conviction shared here, that understanding time and reality requires us to redirect our thinking process. Plato describes a path out of the dark cave of confusion into the realm of truth and light, that is, from time towards the timeless. But we should "reverse course" along this path and move from the timeless into the complexity of time. Time is not one thing foundational to reality; reality rather is a series of temporal levels developed through evolution and related in a nested hierarchy driven by conflict and towards increasing complexity. This theory makes possible critical and fruitful reflection on issues like entropy, indeterminacy, and mind/body dualism. It entails embracing our position as knowers in time and the complexity of truth as temporal rather than timeless.

Keywords

J. T. Fraser – hierarchical theory of time – theory of time's conflicts – complexity – evolution of time – entropy – Joseph Needham

J. T. Fraser was the founder of the International Society for the Study of Time (ISST) and a friend, mentor and inspiration to many of us concerned with the nature of time and temporal existence.[1] He was a scholar of time or, his preferred term, "timesmith," whose theory of time provides "a framework within which diverse intellectual traditions could converse productively" (Steineck and Clausius 2013, ix). Founding the ISST was one way Fraser facilitated multidisciplinary conversation about time.[2] The ISST embodies Fraser's conviction that the perspectives of scholars socialized into multiple disciplines would be

---

1　This article was published in *Kronoscope* 20 (2020), 121–134.
2　Fraser described his work as "interdisciplinary," but the word used here is "multidisciplinary." The former describes the goal of this work as opening up a new place for thinking between

the most fruitful way to approach the study of time (see Fraser 1981, 590; in Parker 2013, 272).

Studying time is neither easy nor straightforward, as "time" is a notoriously difficult concept to pin down. Augustine realized: "What then *is* time? If no one asks me, I know; if I want to explain it to a questioner, I do not know" (Augustine 2006, 242). Answering this question is made more difficult by issues of competency: scientists and scholars tend to hold "dogmatic and often contradictory" views regarding "which domains of knowledge are equipped to deal with this question" (Fraser 1981, xix). Fraser himself studied physics and was an engineer, and it might have been tempting for him to follow our usual course of turning to physics as the discipline assumed to be most attuned to the foundations of reality to sort out the nature of time. "But no universal theory is within sight to combine with ease and self-evidence – the hallmarks of a good theory – the physics of light, of particles, and of the universe on equal terms" (Fraser 1982, 28). Instead, Fraser proposes a theory of time that assumes the "diversity of time, that is the existence of its multiple manifestations," and the "unity of time" that lies behind the conviction that when various specialists – including scholars in the humanities and artists – reflect on time, they are thinking about the same thing (Fraser 1981, xxi). His work and the work of the ISST illustrate that responding critically to the question "What is time?" is not a matter of scholars from a variety of disciplines finding some common ground on which to agree "*this* is what time is," but rather is about sharing a commitment to the unity of time that allows for fruitful engagement with its diversity.

Fraser proposes that time is not one thing foundational to reality but that reality is characterized by six temporal levels (described below); that these levels are related in a nested hierarchy, which is to say that there are "higher" and "lower" levels and that the higher levels encompass but cannot be reduced to the lower levels; that temporal reality is evolving and that the engine of this evolution is conflict, that is, higher levels come about as resolutions of the conflicts generated by the evolution of each level; and that evolution is in the direction of increasing complexity. This schema can be labeled both a hierarchical theory of time and a theory of time's conflicts (Fraser 1999, 21; also Fraser 1987, 280). This theory will be explicated further in the rest of this essay, and it will be supported practically by demonstrating how it helps us deal critically and fruitfully with time-related issues like entropy and mind-brain dualism. Fraser was convinced this work requires a reversal of our usual course as we

---

already existing disciplines; the latter focuses on the nature of the work that has to be done to get to this new place of understanding.

go deeper into what is temporal rather than rise above time to the timeless in search of stability.

1   Reversing Course

Fraser's image for his work (and his images always are useful to consider) is coming "out of Plato's cave," that is, reversing the course of thinking in Plato's allegory of the cave (*Republic* 514a–521a). Plato portrays our epistemological condition as caught up in a world of appearances – the shadows on and echoes off the back wall of the cave that we assume are real because they are all we know. Philosophy offers us liberation from this ignorance through ascent outside the cave, from appearances to truth, from darkness to light, from the constantly changing to the eternal forms – from the temporal to the timeless. In terms of an image that occurs earlier in the *Republic* (509d–511e), we need to cross the line from sensation to abstraction, first through number and then through dialectic, if we are to know truth. Truth cannot change; truth is timeless.

This is a metaphysical bias that continues to inform Western thinking in the physical sciences and mathematics as well as in philosophy. But the evolution of our (scientific) thinking calls on us to get beyond this stark dichotomy: with "the advance of evolutionary biology, psychology, and social science increasingly more doubt has been cast on the validity of any theory of knowledge which sees the world as divided into the temporal and the timeless" (Fraser 2007, 15; see Fraser 1981, xlvii). For Fraser, "the divided line is upside down" (Fraser 2007, 293) and we need to reconsider our journey from and into the cave and the turnings we make on this path. Our task is not to abstract from time into the timeless forms Plato provides but to reflect critically on (and *in*) "the immense wealth of whatever is temporal" (Fraser 2007, 13).

This is difficult: we are born into a worldly condition of constant change in which we need to find stable and secure values – truth, goodness, beauty – in order to develop as humans in the world. Fraser asks us to realize that our drive towards permanent values is rooted not in the nature of reality but in the strength of our need. "People seek truths and hold them for dear life not because truths are timeless but because all verities are temporal and vulnerable" (Fraser 1999, 229; see Fraser 2007, 293). This also is the case with goodness wherein "moral judgments ... are subject to relentless revisions" and with beauty "because a whiff of wind can collapse those configurations of conditions by which the feelings we associate with the presence of the beautiful have been generated" (Fraser 1999, 292). Our values are not about conserving the ideal but are revolutionary as they enable us to work through inevitable change (Fraser 1999, 2–3). Fraser compares them to "mental hungers" which like bodily hunger "can never be permanently satisfied" (Fraser 1999, 229). Hunger, mental as well as physical, can be satisfied by the appropriate food; but living means becoming hungry again. The only permanent response to hunger is death. Values are a response to the hunger for stability; they seek to establish islands of predictability in the middle of seas of uncertainty and conflict where they function as promises (Arendt 1958, 243–247). And their origin is us in the interplay between our needs and our experience of the world we live in. "The ideas of the true, the good, and the beautiful have not been copied but created. For that very reason they are immensely precious. There is no heavenly warehouse from which they may be replenished if lost through mismanagement" (Fraser 2007, 32).

Our mental hunger for truth turns knowledge into a process rather than a state, "a process of continuous, active recategorization of what is already known with what is being learned" (Fraser 1999, 25). Inverting the divided line and reversing course out of Plato's cave means recognizing there is no final reality to serve as a ground and guide for human knowledge. Fraser embraces evolution rather than trying to overcome it. If "time is constitutive of reality," then *reality*, understood by Fraser as "examined appearances," itself evolves. Fraser gives his essays titles like "The Many Kinds of Truth" and titles sections "Reality and Its Moving Boundaries" (Fraser 1999, 21) and "The Evolution of Causation and of Time" (Fraser 2010, 19). But Fraser is not a relativist: "reality is neither 'in the mind' nor is it 'out there.' Rather it is a relationship between the knower and the known. It is a family of examined appearances. It is a set of working assumptions that is continuously tested for its usefulness for making predictions about the future and for explaining the past" (Fraser 2007, 37). The dichotomy between objectivism and relativism is unreal (Bernstein 1983).

Philosophically, this is a phenomenological insight into time and experience, and in phenomenological terms, Fraser describes "an ontological plurality of times" (Schweidler 2013, 217; see Merleau-Ponty 2014, 432–457). Fraser himself appeals here to the work of the German theoretical biologist Jakob von Uexküll and to Uexküll's conceptualization of an *Umwelt* (Fraser 2007, 39–49; Fraser 1982, 19–23). Reality is not some neutral external "thing" that we sample in experience and try to reflect accurately in our concepts. Reality is experienced rather in the interplay between an organism's sense receptors (*Merkwelt*) and the responses to its environment of which it is capable (*Wirkwelt*). In other words, reality is experienced in a species-dependent manner, and the evolution of species is the evolution of reality. Human noetic activity has expanded the range of the bodily senses through instruments and conceptually, so Fraser makes use of "an extended umwelt principle." Note the implication of the extended umwelt principle: we are part of reality; we do not stand outside or above it even conceptually; there is no way to escape from reality including from time. Fraser marks the difference between "Received views" which "tend to regard time as a background to reality or equate it with the human experience of passage or define it through distinctness from the timeless," and his "hierarchical theory of time" which "regards time as constitutive of reality" (Fraser 2007, 48; 1999, 38).

The rest of this essay will describe what time looks like from "inside," according to Fraser, with no desire to escape with Plato into the timeless. It will demonstrate the usefulness of this model of thinking about and in time for addressing issues, like entropy and indeterminacy, that challenge our confidence in our ability to describe reality in meaningful ways; and it will suggest that thinking in time can give direction to our thinking in general and about thinking.

2   Fraser's Hierarchical Theory of Time / Theory of Time's Conflicts

Fraser labels his theory a "hierarchical theory of time" and a "theory of time's conflicts," and both are accurate descriptors of his work. On the basis of his reflection on what the sciences, humanities, and arts tell us about time, he proposes "that time had its genesis at the birth of the universe, has been evolving along a scale of qualitative changes appropriate to the complexity of the distinct integrative levels of natural processes and remains evolutionarily open-ended" – there is a hierarchy of levels of time, and these levels are nested. "Earlier temporalities are not replaced but subsumed by later ones" (Fraser 2007, 48; 1999, 38) in an evolutionary process driven by conflict and towards increasing complexity.

The claim that time *evolves* may seem odd insofar as evolution involves change and change is the marker of time and occurs in time. But the claim here is parallel to the concept of space in an expanding universe: the universe does not expand into previously empty space, but space itself expands. In like manner, time does not change in time so much as time itself changes in a process of evolution (Fraser 2007, 48). There is nothing "outside" or above the reality of expanding time or space.

Evolution has a direction: it is recognized in the increasing complexity of reality, where "complexity" means "the number of distinct states an autonomous system may assume" (Fraser 2007, 25; see Fraser 1999, 233–242 and Fraser

1982, 154–156). On this definition of complexity, the human brain is the most complex thing we know of. "As humans, we are at immense distances from the physical boundaries of the universe: from the limits of temperature, size, mass, speed, length, and periods of time. But, by virtue of having human brains we are at – we constitute as far as we know – the upper limit of complexity in nature" (Fraser 2007, 2–3). Here, too, we recognize ourselves as natural existents, as being in time; our thinking does not remove us from the temporal but is part of development in time.

Complexity can be defined in numbers, in measurements stated algorithmically. Doing so reveals both distinct levels of complexity in nature (see Fraser 1982, 28–29) and the evolution of increasing levels of organization in response to increasing complexity. Fraser describes six levels in the evolution of time and how each is organized or characterized by a principle of causation allowing us to understand it.

1. The *atemporal* is the world of space without time, an umwelt of absolute chaos in which everything happens at once – the world of an imaginary traveler riding a proton.

2. The *prototemporal* level is the world of elementary particles in which it is impossible "to say precisely *when* is *then*" in which "events do not hang together," as in certain pathological mental conditions. Causation can only be expressed in statistical or probabilistic terms (e.g., radioactive decay: the half-life of cobalt 60 can reliably be set at 5.3 years; but doing so does not allow us to predict when the next particle of the mass will decay or which will be the particle). Most importantly, this is not a matter of our ignorance but is the character of nature at this level. (See our examination of Fraser on indeterminacy, below.)

3. Large mass brings *eotemporality*, the deterministic world of classical Newtonian physics and of General Relativity Theory. Not everything happens at once; but time lacks a directional quality, inasmuch as the mathematical formulae describing this level treat time as moving backward as well as forward.

4. Living organisms bring *biotemporality* organized around the short-term intentionality associated with meeting the organic needs (like nourishment) of living beings.

5. The human mind introduces *nootemporality*. We are conscious of temporal flow as we move from birth to death. Our umwelt of language and culture is the world of long-term intentionality in which we can remember the past and plan for a distant future. Symbols allow us to set long-term goals.

6.   The language, culture and symbolic interaction that make possible reflec-
      tion on the past and setting goals for the future lead to the next level of
      development in *sociotemporality* organized around social intentionality
      or historical causation.

To sum this up in Fraser's words:

> The stable integrative levels of nature are as follows. First, the absolute
> chaos of pure becoming; this is the world of vacuum with its seething sea
> of creation and annihilation. Above that level we find the world of ele-
> mentary object-waves, then the world of massive matter, then the inte-
> grative level of life and finally, the world of the functions and collectively
> created structures of the mind.
>
> FRASER 2007, 59

Crucial to keep in mind is the way in which these levels are "nested" within
each other. "Each level subsumes the level or levels beneath it, it is restrained
by the principles of the lower level or levels, but each possesses certain new
and unique degrees of freedom" (Fraser 1981, xxx).

It is the nested character of the integrative levels that makes it possible to
use Fraser's theory as a way to think through the problems that resist resolution in the timelessness/temporal framework of Plato and so much of Western
thinking, as will be partially demonstrated below.

Fraser describes the "aesthetic adventure" of the "voyage among temporal
levels" that characterizes the "search for new knowledge" (Fraser 2007, 32) by
referring to the stories of two women, Lot's wife and Galatea: the former, turned
to a pillar of salt for the very human act of turning back to see the destruction
wrought by God on Sodom and Gomorrah, is an image of collapse from the
nootemporal all the way down to the prototemporal incoherence of dust; the
latter, carved and then loved by Pygmalion and brought to life by Aphrodite,
emerges from the prototemporal to the noetic (Fraser 2007, 32; also 175, 204).
Evolution does not involve the disappearance of earlier temporal levels but
their enveloping in the levels that emerge from them; creativity in responding
to this reality is a descent/ascent through temporal levels. "The unique gift of
man is to be able to repeat this journey in his mind" (Fraser 2007, 31).

The engine of evolution is conflict. For example, "[a]lthough the search for
truth is driven by the desire for permanence and stability, its historical function
has been the creation of conflicts and, through them, social, cultural, and personal change" (Fraser 1999, 73; also 165–166). This is where it becomes clearer
why Fraser describes his hierarchical theory of time as derived from a theory
of time's conflicts. There are characteristic conflicts at each hierarchical level:

at the *sociotemporal* level there is the tension between a society's goals and those of the individuals who comprise it; noetic conflict reflects struggles of individuals to maintain a sense of identity in the midst of identity-disrupting forces; life forces pit generation against decay; at the level of inanimate matter, the conflicts involve entropy (a topic to which we shall return); and at the most basic level, differing forms of ordering and disordering struggle against each other in seeking to overcome chaos. The difficulties resolving the conflicts at any particular level ultimately lead to the emergence of the next higher level and to moving beyond the level-specific conflicts. And what emerges is radically new, that is, the evolutionary process from level to level is discontinuous and disruptive.

Conflict is constitutive of reality and cannot be eliminated or otherwise overcome without negating what is real. Raji Steineck, in an essay in tribute to Fraser's work, states "the most crucial insight I have received from reading J. T. Fraser's work" is "[c]onflict sustains human life, and all that gives value to it. Without conflict there would be no life, and no way to make sense of it." Further, conflict is "not only a motor force for change, but also an essential factor for the stability and sustenance of our life and core elements of culture that are available" (Steineck 2013, 248–249). This is not a call for passive acceptance of the fact that conflict happens or praise of conflict for its own disruptive sake and a call to continual disruption, but a recognition that "some conflicts are *necessary and fundamental* in the sense that they are *constitutive* for certain modes of being" (Steineck 2013, 253). The critical response is to distinguish among conflicts as much as possible between those which are "accidental" and "can be defused, resolved, or avoided if stability is worth maintaining" and those "constitutive conflicts" that "need to be controlled instead of defused" (Steineck 2013, 259, 261). Conflict drives evolution, and evolution has a direction in the emerging of increasing complexity. But there is no end-point or goal to evolution like the symmetrical omega-state of perfected complexity envisioned by Teilhard de Chardin (Teilhard de Chardin 1964) that would mark the end of conflict. Our response to conflicts should not be to try to eliminate them, because "[t]o end *constitutive* conflicts means to lapse into less complex modes of beings, it means to lose what makes human life special and valuable" (Steineck 2013, 261). Instead we must learn to discern among conflicts those that are necessary as constitutive of reality and those that we can evaluate and defuse, resolve or avoid, based on our evaluation of what is most beneficial for the world as we perceive it.

Fraser acknowledges the influence of the evolutionary biologist Joseph Needham in his description of evolution and emergence (Fraser 2007, 273). Reviewing some of Needham's work will help clarify what Fraser means by a hierarchical theory and provide useful insights into the topics considered in the next section – even though Fraser and Needham disagree about the point just made: for Needham, there is an identifiable end point to evolution, although that point is better described in terms of political economy than biology. According to Needham, biological evolution results in differing levels of biological development in which new levels arise from conflicts in lower levels and result in the emergence of new life forms. This process is rooted in the materialist dialectic of history for Needham: he was a Marxist and one of the "Red Biologists" who were members of the Theoretical Biology Club at Cambridge in the mid-twentieth century; he believed in the eventual revolutionary victory of the proletariat as the goal and ending-point of conflict (Needham 1943, 272).

What Fraser and Needham share is more significant for understanding time than is this difference. Needham describes the levels of evolutionary emergence as "envelopes," a concept he gets from mathematical set theory (Needham 1968, 112). Both Needham and Fraser understand the relationship between the levels to be both hierarchical *and nested*. Higher levels supervene on lower ones *and include them* like larger envelopes may contain the contents of smaller ones. New and higher levels emerge through evolutionary conflicts, but they do not cancel out or eliminate the lower levels. Needham gives a clear statement of the importance of this insight:

> Meaning can only be brought into the natural world when we understand how the successive "envelopes" or "integrative levels" are connected together, not "reducing" the coarser to the finer, the higher to the lower, nor resorting to unscientific quasi-philosophical concepts.
>
> NEEDHAM 1968, VIII

Reductionism and mysticism/obscurantism are both banned: we can neither reduce higher levels to the lower levels we would use as foundations, nor can we turn to "higher" explanations that ignore the continued presence of "lower" realities in what is higher. This insight, shared by both thinkers, is the key to how Fraser's thinking might liberate our own.

3   Thinking in Time

Fraser helps us both recognize and work through the complexities involved in thinking about time and in general. The hierarchical and nested relationship of the stable integrative levels of nature and time can enable us to be critical without becoming reductionists. "The major fields of human knowledge ... display a division of concerns that corresponds to the integrative levels of nature" (Fraser 1999, 27) and there are ways of thinking appropriate to each level. The social sciences are the tools for studying social and noetic reality, biology responds to the biosphere, general relativity to astronomical reality, quantum theory to particle-waves, special relativity to light in ceaseless motion – sociotemporality/nootemporality, biotemporality, eotemporality, prototemporality, and the atemporal. Math and method apply across this spectrum, but only in level-appropriate ways. "It is the primitiveness of the roots of mathematics which guarantees its awesome universality, power, and beauty. It is the same primitiveness that makes mathematical tools increasingly useless for dealing with biological, noetic, and historical causations." This is because "As we rise along the integrative levels of nature from matter to life, to man, to society, the world becomes increasingly unpredictable – not because of our ignorance but intrinsically so" (Fraser 2007, 30). To return to Uexküll and the extended umwelt principle, there is not a single thing called "reality" against which we can measure our concepts; neither is reality something we construct. Reality – like nature and time – is arranged in hierarchical levels that, while integrated, remain irreducible to each other.

The "cash value" of the hierarchical theory of time for our thinking lies here. For example, Fraser was not troubled by Heisenberg's principle of *Unbestimmtheit*, a word Fraser translates as "indeterminacy" rather than "uncertainty" (Fraser 2007, 62–64). The fact that we are unable to determine with certainty the position and velocity of a subatomic particle at the same instant is only threatening if we want to build the house of our knowledge on eternal principles that are certain, indubitable and unchanging. We are far less troubled by the indeterminacy/uncertainty principle when we recognize that the reality where indeterminacy applies is prototemporal, and that the prototemporal level of reality is organized by statistical causation and not determinism. We are looking for the wrong truth if we expect to find deterministic causal laws at work in the prototemporal as we do in the eotemporal; the mathematics appropriate to each level differs. We are "trying to foist a higher order causation upon a lower integrative level which is not sufficiently complex to support it" and we should not be surprised to experience other relationships in nature that cannot be reduced to "simple mathematical forms:" these are "ontological indeterminacies – that is, aspects of nature – and not epistemic that is, tokens of human ignorance" (Fraser 2007, 63–64), as we just noted above.

A similar illustration of a temporal-level issue and of the usefulness of the hierarchical theory of time and the theory of time's conflicts in responding to it is entropy and the (apparent) problems it raises (Fraser 1999, 243–245).

The second law of thermodynamics entails that organized systems inexorably move to complete randomness (disorganization) or entropy. The British astronomer Arthur Stanley Eddington claimed entropy is responsible for our feeling that time flows in one direction, so that we can speak of time's 'arrow.' But Fraser responds that "the relationship between the second law of thermodynamics and the experiential passage of time is spurious" (Fraser 1987, 274). Entropy is a useful concept, but in popular literature on time and physics it suffers from a "mystic haze of implications" reflecting "a knee-jerk metaphysics that identifies reality exclusively with whatever may be explored through physics" and thereby has done "more harm than good to the study of time" (Fraser 1987, 275, 274). We fail to notice that we are again making a level-related mistake, in this case between the eotemporal and biotemporal and the principles of causation at work in each.

Applied to the biosphere of more or less highly organized living things, entropy bodes ill – life is doomed. But growth and decay (increase and decrease of organization) are equally characteristic of existence: "ordering and disordering, defining each other, not only emerged simultaneously from the primordial chaos but have remained two aspects of the cosmic process. Together they are responsible for the evolutionary character of the universe" (Fraser 1982, 104). Entropy increase obtains in eotemporal reality but does not describe biotemporal life. This is not due to some kind of immaterial force at work in the biosphere, and the problem of reductionism is not simply a product of materialism. Rather, the "material" reality of the biotemporal world, while inclusive of the eotemporal, is characterized by other conflicts like growth and decay, life and death. Living systems – as in our case – are self-organizing, which is a matter of entropy *decrease*. "Until the day we die, our bodies oppose the cosmic 'running down.'" Moreover, "entropy decrease – the growing of the body, the healing of wounds – proceeds along time's experienced arrow" (Fraser 2010, 28); this is the same temporal arrow described by Eddington as originating in our sense of entropy increase. Finally, as the ones who constitute the noetic level, we are positioned "at the complexity boundary of the universe" and "are subject to natural selection that favors increased efficiency in opposing the cosmic entropy increase" (Fraser 2010, 34–35).

Here, too. Fraser reflects the influence of Needham. Needham makes a distinction between the concept of *order* that obtains in the physicists' world and that of *organization*, used by biologists, and concludes that "thermodynamic order and biological organisation are two quite different things" (Needham 1943, 227). Scientists in different disciplines are socialized to different world-views.

For the astronomers and the physicists the world is, in popular words, continually "running down" to a state of dead inertness when heat has been uniformly distributed through it. For the biologists and sociologists, a part of the world, at any rate (and for us a very important part) is undergoing a progressive development in which an upward trend is seen, lower states of organisation being succeeded by higher states.

NEEDHAM 1943, 207

Or as Fraser puts it, "I dare speculate that if Eddington had been a biologist, he would have asserted with conviction and authority that, for the source of our sense of passing time, we must look to the capacity of living systems to selforganize" (Fraser 2010, 28).

One more conundrum in modern thinking that Fraser and Needham help us work through is (in philosophers' language) the mind-brain identity problem bequeathed to us by Descartes: how does something immaterial (the mind) relate to something material (the body)? Needham's language of envelopes helps us realize how this dualism is misleading. "There is a sense in which minds include and envelop bodies, for the boundaries of thought are far wider than those of what the special senses can record" (Needham 1943, 185). Rather than mind and body as two separable things, if something is alive, it is alive "all the way down," so to speak: "an atom, or an electron, if it belongs to the spatial hierarchy of a living organism, will be just as much 'alive' as a cell" because "the whole requires its components (at all levels) in order to be 'alive,'" and "the parts require the whole in order to make their particular contribution to it by virtue of which it is 'alive'" (Needham 1968, 117). According to Fraser, the mind is not a thing at all. "*Mind* is a noun that suggests an object in space. But do not ask what mind *is*, ask what it *does*. It would be better to speak only about minding, using a gerund, instead of speaking about the mind" (Fraser 1999, 17–18). The complexity of the brain requires description in terms of what it does over what it is. "One must have a brain to be minding something, but it is no more necessary to have a body part called 'mind' for remembering, thinking, attending, or intending than it is necessary to have a body part called 'quilt' for quilting or one called 'wink' for winking" (Fraser 1999, 18).

4 Conclusion

None of the above is meant to suggest these problems in understanding have been resolved. But the hierarchical theory of time and theory of time's conflicts provide us with ways to think through them more critically and effectively than is possible if we start from a Platonic metaphysics and understanding of truth. For Plato, truth is grounded in reality and reality ultimately is universal, unchanging, eternal and stable. Plato offers us the images of the cave and the divided line as a kind of epistemological therapy to aid and encourage us to find this truth. This is important to us in part because we hope to find rest in the truth. But "[a]s long as the Platonic heritage of the time-timeless dichotomy is retained and, accordingly, we keep seeking the foundations of the universe in the rigid beauty of mathematics alone, all attempts for an integrated understanding of time are bound to fail" (Fraser 1981, xlvii). The same can be said of attempts to understand existence in time.

What we learn from Fraser is that "[al]though searching for truth is driven by the desire for permanence and stability, its historical function has been the creation of conflicts and, through them, social, cultural, and personal change" (Fraser 1999, 73; see 19). The levels of nature, of time, of reality are stable, and it is possible to speak of them as "higher" and "lower." But the role of stability revealed to us in the hierarchy is to foster and respond to change: the new emerges in discontinuous processes driven by conflicts, conflicts that are themselves created by our desire for permanent stability. Coming out of Plato's cave and inverting Plato allow us to realize that "human knowledge does not, for it cannot reveal pre-existent, eternal verities. Instead it raises the incomprehensible to the level of the obvious, then it shows us that the new obvious is incomprehensible. Under 'knowledge' I include all of its many forms: the arts, the letters, and the sciences" (2007, 62). Responding to existence at the complexity boundary of reality means recognizing that "the highest integrative levels, in life, in mind, and in society ... are the most temporal" and that "The Socratic ideal of the liberation of the human soul ... resides in the ceaseless striving to get away from the timeless and reach the temporal" (Fraser 1981, xlvii). Rather than reducing complexity to a lower temporal level in search of a foundation or base or abstracting to a higher level in searching for a synthesis, time itself calls us to work through temporal conflicts in their complexity.

References

Arendt, Hannah. 1958. *The Human Condition*. Chicago and London: University of Chicago Press.

Augustine. 2006. *The Confessions*. Translated by F. J. Sheed; edited by Michael P. Foley. Indianapolis and Cambridge: Hackett Publishing Company.

Bernstein, Richard J. 1983. *Beyond Objectivism and Relativism: Science, Hermeneutics, Praxis*. Philadelphia: University of Pennsylvania Press.

Fraser, J. T. 1981. *The Voices of Time: A Cooperative Survey of Man's Views of Time as Expressed by the Sciences and by the Humanities*. Second Edition. Amherst: University of Massachusetts Press.

Fraser, J. T. 1982. *The Genesis and Evolution of Time: A Critique of Interpretation in Physics*. Amherst: University of Massachusetts Press.

Fraser, J. T. 1987. *Time: The Familiar Stranger*. Amherst: University of Massachusetts Press.

Fraser, J. T. 1999. *Time, Conflict, and Human Values*. Urbana and Chicago: University of Illinois Press.

Fraser, J. T. 2007. *Time and Time Again: Reports from a Boundary of the Universe*. Leiden and Boston: Brill.

Fraser, J. T. 2010. "Founder's Address: Constraining Chaos." In *Time: Limits and Constraints*. Edited by Jo Alyson Parker, Paul A. Harris and Christian Steineck, 19–36. Leiden and Boston: Brill.

Merleau-Ponty, Maurice. 2014. *Phenomenology of Perception*. translated by Donald A. Landes. London/New York: Routledge.

Needham, Joseph. 1943. *Time: The Refreshing River (Essays and Addresses, 1932–1942)*. London: George Allen and Unwin.

Needham, Joseph. 1968. *Order and Life*. Cambridge, MA and London: M.I.T. Press.

Parker, Jo Alyson. 2013. "A Brief History of the International Society for the Study of Time," *KronoScope: Journal for the Study of Time* 13, no. 2: 269–293.

Schweidler, Walter. 2013. "The Time of Freedom," *KronoScope: Journal for the Study of Time* 13, no. 2: 217–227.

Steineck, Raji C. 2013. "Appreciating Conflict: Lessons from J. T. Fraser's Theory of Time," *KronoScope: Journal for the Study of Time* 13, no. 2: 248–267.

Steineck, Raji C. and Claudia Clausius. 2013. "Obituary [for J. T. Fraser]." In *Origins and Futures: Time Inflected and Reflected*, edited by Raji C. Steineck and Claudia Clausius, ix–x. Leiden and Boston: Brill.

Teilhard de Chardin, Pierre. 1964. *The Future of Man*. Translated by Norman Denny. New York and Evanston: Harper & Row.

CHAPTER 3

From the Biotemporal to the Ecotemporal in Atilio Caballero's *La última playa*

Lucia Cash Beare

Abstract

Climate change and the environmental crisis more generally confront us with the need to think survival beyond the single organism. This project proposes that storytelling, emplotment, and the connective impulse of the human mind, have the potential to awaken our awareness of our dual existence as individuated beings and, simultaneously, as assemblages embedded in the ecologies of which we are a part. I turn to the novel *La última playa* (*The Last Beach*), by Atilio Caballero, to explore this potentiality and argue that, while the life of the single living organism is guided by the linearity of J. T. Fraser's biotemporal, the individual being aware of its embeddedness is open to the "ecotemporal." Ecotemporality is not limited to the singular body; it is the experience of the shared inhabitation of a moment in time by living organisms and non-living matter who are, for better or worse, bound by the same material fate.

Keywords

Anthropocene – Bateson, Gregory – biotemporality – Caballero, Atilio – climate change – Cuban literature – deterritorialization – ecological present – ecotemporality – emplotment – island

Living organisms die; there is, at least not currently, any way around that fact. Yet the awareness of human-caused ecological crisis is demanding that we think survival beyond the single organism. Storytelling and the possibilities opened by narrative emplotment have a crucial role to play in the recognition that local ecologies, not to say the planet as a whole, are not only our home but also constitute, in fact, a shared body and memory. I argue that the novel *La última playa* (*The Last Beach*), published in 1999 by Cuban author Atilio

Caballero,[1] figures a form of temporality that I will call the "ecotemporal," in dialogue with J. T. Fraser's hierarchical theory of time. This temporal experience, as we will see, is concurrent with a dual understanding of the self that foregrounds our individuated existence and, simultaneously, acknowledges that we are part of an interconnected and interdependent web of living and non-living matter.

For those who are not directly impacted by the manifestations of climate change, the human mind is the locus of the scalar imagination needed to seriously engage it, as it allows us to think beyond our embodied experience. The ecological crisis demands that we think relationships in time beyond those which we can perceive with our individual senses, that we engage in narrative thinking to explore the interconnections and interdependencies of living and non-living matter and explore a time scale beyond that of our lifetimes. To engage in narrative thinking is to take the information available – provided by the senses, prior knowledge, and intuition – and use language to create a story, illuminating relationships of causality and correlation that might not be immediately available to us. This is what Paul Ricoeur has termed, in the literary context, "emplotment." These narratives have the potential to help us face what Amitav Ghosh calls "the uncanniness" of the ecological crisis. As he explains, the changes brought about by the ecological crisis "are not merely strange in the sense of being unknown or alien; their uncanniness lies precisely in the fact that in these encounters we recognize something we had turned away from: that is to say, the presence and proximity of non-human interlocutors."[2] Non-human interlocutors can be living beings like animals and trees, but also, as we find in *La última playa*, rocks, soils, and waters. Caballero delves into this uncanniness by figuring what I will call the "ecological present," revealing, in the process, the human in its ecological embeddedness.

The novel begins at the end, when the body of an old man, swollen by the sun and the sea, floats onto the shore of Cayo Arenas, a small island or key off the coast of Cuba. It is only upon finishing the novel that we realize that the body belonged to Andy Simons, the protagonist about whom we know very few biographical details. Through the voices of other, unnamed characters – a

---

1 Atilio Caballero, *La última playa* (Madrid: Hypermedia Ediciones, 2016). All English translations are my own. Atilio Jorge Caballero Menéndez is a novelist, poet, playwright, screenwriter and founder of the alternative cultural project, "Paideia." He has won several awards for his work, including the UNEAC (Unión de Escritores y Artistas de Cuba) award for *La última playa*.

2 Amitav Ghosh, *The Great Derangement: Climate Change and the Unthinkable* (London: University of Chicago Press, 2016), 40.

sailor, a geologist – we find out the little that was known about him. He was an orphan who somehow found himself aboard a cargo ship as a baby, where he grew up until he was nine years old. Details about his time in Europe during World War II and stints aboard cargo ships traveling up and down the coast of the United States emerge in a disjointed manner throughout the text, giving the reader a fragmented sense of the main character's life history. Instead of providing a linear biographical account of Simon's life, the novel centers on his inhabitation and embodiment of the *Arenas* key in the present of the text.

I say "embodiment" because, as I will show, Caballero progressively blurs the boundary between Simons and the rest of the ecology of the island. In the narrative, Simons emerges as an organism-in-its-environment rather than a completely discrete body whose fate is bounded by an assemblage of flesh and bones. Already in the 1970s, Gregory Bateson argued that the understanding of the human as an isolated unit of survival is one of our most pervasive and dangerous epistemological pathologies. "The flexible environment," he stated, "must also be included along with the flexible organism because [...] the organism which destroys its environment destroys itself. The unit of survival is a flexible organism-in-its-environment."[3] This call for ecological thinking that recognizes the interdependency of living and non-living matter becomes even more pressing in the context of our current awareness of the human impact on climate change and the toxification of Planet Earth, a call that Caballero's novel reinforces.

The setting of *La última playa*, the Arenas key, is particularly productive to this manner of trans-matter ecological thinking. The island, Josefina Ludmer argues, is a territorial regime of signification that can erase differences by means of an opposition to an outside.[4] It is clear that within a geographical or social island, differences based on species, ethnicity, socio-economic group, etcetera, do in fact exist and matter. My purpose here is not to ignore internal diversity, as well as variation, in how islands and their inhabitants relate to the world around them to repeat paradigmatic notions of insularity. Rather, I turn to the space of the island because it can make evident to human perception the fact that the survival of any living being is bound to the wellbeing of the ecology it inhabits.[5] Islands can be productive spaces to think ecologically because, despite the existence of difference, the island can symbolically and materially

---

3   Gregory Bateson, *Steps to an Ecology of Mind: Collected Essays in Anthropology, Psychiatry, Evolution, and Epistemology* (New Jersey: Jason Aronson Inc, 1987), 320.

4   Josefina Ludmer, *Aquí América Latina/Una especulación* (Buenos Aires: Eterna Cadencia, 2010), 133.

5   Cuauhtémoc Pérez Medrano refers to the recognition that impermanence traverses living and non-living matter the "internalization of insular symbology." See Cuauhtémoc Pérez

blur those differences in favor of an opposition to an outside, which, in the case of an actual island such as *Arenas*, takes the form of an endless ocean.[6] Faced with that ultimate outside, everything in the inside becomes part of the same material world. Therefore, this territoriality can figure Bateson's unit of survival, the "organism-in-its-environment" more explicitly. What is more, the materially defined space of the island serves as a place that is felt as "home," intensifying affective ties that might otherwise be dependent on the imagined boundaries of a neighborhood, town, or country.

In the narrative present of *La última playa*, the time in which the novel's main plot take place, Simons is in his seventies. For decades, he has lived on the island of Arenas and has become obsessed by its pending disappearance into the ocean. Interested in insects since he was a child, Simons pays close attention to the ground. Because of this, he is very knowledgeable about the geology of the island. The unnamed geologist who visits the island at some unclear point in the narrative's past states that, for Simons,

> everything had an immediate meaning. Every discovery had to work towards what, in that moment, he did or projected. And assuming he was crazy, it was an enviable craziness. A man for whom a simple rock had so much value, must feel surrounded by treasures everywhere [...].[7]

Due to his careful and watchful inhabitation of the island's ecology, Simons notices that Arenas is mainly formed out of basalt that is being slowly eaten away by the ocean. Thus, the catalyst of Caballero's narrative is an embodied awareness of the linear temporality of decay, which affects not only living bodies but also non-living matter that disintegrates, transforming into something else. Simons perceives this natural impulse towards decay in the slow but irreversible material disappearance of the island. This process becomes a unifying material fact that manifests as trees falling into the ocean when their roots are exposed because of the progressive disappearance of the island's substratum, as the plant-world taking over unused walking paths on the island, and as the aging of Simons' physical body, of which he is keenly aware. Therefore, this apocalyptic fate exceeds any one body but no body can escape it. Paying attention within an ecology as small as the contained island of *Arenas* makes

---

Medrano, *Ficción herética: Disimulaciones insulares en la Cuba contemporánea* (Potsdam: Universitätverlag Potsdam, 2019), 152–153.

6   As Ludmer puts it, islands can define a common positionality determined by an organic substratum upon which the islands' unifying territorial regime of signification rests. Ludmer, *Aquí América Latina*, 133.

7   Caballero, *La última playa*, 78–79.

Simons very aware of this tendency towards death, which only the amorphous mass of the ocean, the outside, seems to escape.

The human awareness of decay is central to Caballero's construction of the island's ecology and is key to something that we might, building on the work of J. T. Fraser, call the "ecotemporal." Fraser's theory of a hierarchy of nested temporalities is particularly suited to ecological thought of the kind being carried out here in that it rejects the outright separation between the time of the natural sciences and human time.[8] Instead, different temporal relationships coexist as matter interacts at varying scales. Fraser identifies six levels of temporality, defined by the size and mass of matter and the directionality of time. Protons and other particles with zero rest mass are "atemporal." Causality at this scale has no meaning. Waves and particles with nonzero rest mass exist at the level of the prototemporal, where causality is probabilistic. The eotemporal is the direction-less flow of galaxies and the physical domain, where causality is deterministic.[9] Because I am particularly interested in human sensory experience, this analysis foregrounds variations in human temporal experience and dialogues specifically with J. T. Fraser's last three levels of temporal complexity in examining Caballero's novel: the biotemporal, the nootemporal, and the sociotemporal.[10] The biotemporal is the temporality of the biological functions of living organisms. At this level, time has a unique direction towards death. The nootemporal is the domain of the human mind, which allows for jumps and discontinuities between pasts, presents, and futures. It is characterized by "a sharp distinction between future and past as well as by long-term expectations and memories in respect to the mental present."[11] Finally, the sociotemporal is the temporality of human culture and history and is the manifestation of collective coordination.[12]

---

8   J. T. Fraser, "Human Temporality in a Nowless Universe," *Time and Society* 1, no. 2 (1992): 159–173, 163.

9   Paul Harris, "Time and Emergence in the Evolutionary Epic, Naturalistic Theology, and J.T. Fraser's Hierarchical Theory of Time," *KronoScope* 12, no. 2 (2012): 147–158.

10   Another approach to thinking ecologically could focus on the interplay between prototemporality and biotemporality at the molecular scale, where forms are not clearly defined but are instead in a constant process of exchange with everything else that is, in a continuous process of "becoming." However, while this level of material existence might occasionally be available to the senses, it often necessitates the mental power of abstraction. In this project, I choose to foreground the possibilities for an embodied sense of ecological embeddedness.

11   Fraser, "Human Temporality in a Nowless Universe," 163–164.

12   J. T. Fraser, *Time, the Familiar Stranger* (Amherst: University of Massachusetts Press, 1987), 190–196.

Clearly, Fraser's theory of time already accounts for variation in human time. However, the global ecological crisis has made evident that the variety of human temporal reality is even more complex. Fraser's "sociotemporal" can be reassessed in light of our awareness of human-caused climate change in the context of the Anthropocene, a term used to identify the era in which humanity, in its collective existence, has a geological impact on the earth. This would be the line of thought of "deep history." However, we can push our understanding of the "sociotemporal" even further when thinking human bodies as organisms-in-their-environment. Doing so demands that we think biotemporality beyond individual bodies and sociotemporality beyond species boundaries, expanding the realm of the "social," bridging the divide that Fraser identifies between the directed time of life and the non-directed time of non-living matter, and recognizing the embeddedness of organic bodies and their linear temporality in an ecological reality that encompasses and exceeds them. However, this extended bio- and sociotemporality, or, as I will call it, ecotemporality, is rarely available to human, embodied experience. It requires the flexibility of the nootemporality of the human mind to create narrative emplotment for its material manifestations to become available to human senses.

Emplotment, according to Ricoeur, entails connecting within the world of the text events whose interrelation might not be apparent to human senses in the world outside of the text.[13] In *Time in Literature*, Hans Meyerhoff makes a similar argument with regards to the formation of memory and the contextualization of experience within the human mind when he states that

> the inner world of experience and memory exhibits a structure which is causally determined by significant associations rather than by objective causal connections in the outside world. To render this peculiar structure, therefore, requires a symbolism or imagery in which the different modalities of time – past, present, and future – are not serially, progressively, and uniformly ordered but are always inextricably and dynamically associated and mixed up with each other.[14]

Storytelling can mirror the process of forming memories by extrapolating significant associations from the flow of events that make the present. When this is done in such a way that it preserves the rhizomatic and dynamic connections

---

13 Paul Ricoeur, "Mimesis and Representation" in *A Ricoeur Reader: Reflection and Imagination*, ed. Mario J. Valdés, (Toronto: University of Toronto Press, 1991), 143.
14 Hans Meyerhoff, *Time in Literature*, (Berkeley: University of California Press, 1955), 23.

between events, emplotment can highlight the extent to which we are all organisms-in-our environment. In other words, not only does it construct the sequence of events of the story but also has the potential to make visible the imbrication of organic, mental, and social presents that collectively compose the "ecological present."

In *La última playa*, it is not the inevitable end, the material disappearance of the island, that pushes the narrative forward and preoccupies Simons most. Rather, it is the disappearance of its existence in memory, in the collective imaginary. This is why he is so keen on extending the history of the island at the level of the sociotemporal by recreating out of cardboard the fronts of the houses that used to be there when the island was a more popular tourist destination. The comfort that he derives from these mnemonic artifacts stems from the fact that they take him back to a time when the burden of affirming the island's existence did not fall on his shoulders. The island was part of a wider socio-temporal frame that included humans better suited than he was to be carriers of memory, humans who might have family and friends who would remember not only the fact that they lived but also the places they passed through. In contrast, Simons has very few interactions with the human social world outside of the island. Although a priest visits him, his only other interactions occur during his occasional travels to mainland Cuba to get the few provisions he needs, for which he barters natural treasures he finds on the island.

Aware that his limited human social world would result in his and the island's erasure, as tourism to Arenas starts to decrease, Simons decides that he needs to ensure the islands survival in memory by building a bridge that will connect Arenas to the mainland. "To exist, [the island] has to radiate";[15] it has to connect to a continent. Simons' implied ontological argument is that existence, understood in its material and immaterial manifestations, can exceed the life of a singular organism or singular environment when its relationship to a wider social and environmental ecology is made materially and/or discursively explicit. Thus, working against the isolation of Arenas – and, therefore, his own – becomes Simons' life project.

The narrative time of *La última playa* alternates between the protagonist's time of action, narrated in the present tense by Simons and an omniscient narrator, and the atemporal testimonies of disembodied voices that provide the little biographical information we know about the main character. Together with Simons' reflections and remembrances, these testimonies figure the

---

15   Caballero, *La última playa*, 20.

sociotemporal frame of the story, locating Simons' final years at some point in the early nineties, after the fall of the Soviet Union.[16] The first time he started to build the bridge, his investors left after the beginning of World War II. The second time, they left after the success of the Cuban Revolution. The final time, closer to the narrative's present, occurred after the fall of the Soviet Union and the decrease in Russia's support of Cuban infrastructure. All of these are hinted at, never explicitly stated because the historical events in themselves are never a part of Simon's mental present. When he sees a submarine appear off the coast of the island, his thoughts do not turn to the Cold War, the repercussions of the Cuban Revolution and its alliance to the Soviet Union, or his economic stability and ability to continue to sporadically work onboard cargo ships in the United States. Instead, he thinks of the bridge and the possibilities that the mysterious presence of the submarine might afford. Through Simon's subjectivity, defined by the awareness that he is materially and symbolically one with the island he inhabits, the sociotemporal takes on novel meanings. The historical moment in which Simon lives affects not only his emotional and mental states but also his actions in the island because they either forward or inhibit his ability to build the bridge. Thus, his relationship to a wider social and material world is mediated by his intimate relationship with the ecology that he inhabits. In other words, the degree to which he cares about events in the world is directly related to the extent to which they affect the place he calls "home."

The immediate social present of this home is characterized by inter-species coordination that includes non-living matter. This social present is created, for example, through musical conversations between Simons and the material world he inhabits. It is while playing the flute at sundown that Simons feels the most content. In that moment, sight and hearing come together in a liminal space, a diffuse territory between the inside and outside of Simon's person. His eyes see, his being feels, and he produces sounds through the flute that respond to the living and non-living matter with which he is engaging. The world offers him possibilities with which he can choose to interact in this way:

---

16   For a more detailed engagement with the novel's construction of the socio-temporal frame in the context of a postmodern literary effort to respond to the legacies of the Cuban revolution, particularly as it relates to its construction of historical time and revolutionary subjectivity, see Nanne Timmer, "La crisis de representación en tres novelas cubanas: La nada cotidiana de Zoé Valdés, El pájaro, pincel y tinta china de Ena Lucía Portela y La última playa de Atilio Caballero." *Revista Iberoamericana* 73, no. 218 (2007): 259–274.

Once the motive that serves as inspiration – a reflection, a rock, an aban-
doned object, the atavistic rhythm of someone that works, a leaf moved
by the wind or a fish that sways in the water – , Simons imagines that
the melody, as it comes out of his flute, travels in the air until it bounces
on the exterior surface of its interlocutor, creating in the space between
them both an arc of resonances, an exchange of sensations, an exclusive
conversation.[17]

Simons inhabits a world of resonances where he is the center and at the same
time he is not. Through trans-species and trans-matter interactions like the
one described above, he experiences the ecotemporal, in which he is not only
the contained persona that is Simons, but also a body-event that is made and
unmade in the constant flow of matter, contacts, and dissociations. "To look is
to eat," he states in one of his reflections.[18] Following his train of thought, we
can argue that to look at something – to touch it, smell it, hear it, taste it – is,
to a certain extent, to incorporate it into the self. Just as food feeds our body,
other sensory experiences feed our holistic self, adding to our notion of what
the world is and what we are within it. It means that we add to what Felix
Guattari calls our "mental ecology."

According to Guattari's concept of "ecosophy," mental, social, and environmental ecologies ("the three ecologies") should be thought of together if one
is to approach existence from an eco-philosophical perspective.[19] There are
clear points of contact between Guattari's ecosophical approach and Fraser's
theory of time. Both maintain that subjectivity, sociality, and the environment outside of the self interact, clash, and coexist to make the present. Fraser
argues the following:

> If the coherence of the social present is lost, it is still possible to have per-
> sons around but it is not possible to have a functioning group. Similarly, if
> the mental present breaks down it is still possible to have a living human
> body but it is not possible to have a person with identity. And again, if
> the organic present breaks down one will still have matter, but not living
> matter. The organic, mental and social presents are synchronous because
> any social group must consist of thinking humans who, *eo ipso*, must also
> be alive.[20]

---

17   Caballero, *La última playa*, 9.
18   Ibid., 115.
19   Félix Guattari, *The Three Ecologies* (London: Athlone Press, 2000), 28.
20   Ibid., 168.

While the mental present varies from person to person depending on mood and attention, the organic present is species-specific.[21] The organic present comprises the synchronous actions designed to maintain the conditions of life. It strives, in other words, to maintain itself. "The striving directed to the maintenance of the organic present," Fraser states, "is recognized as purposeful behavior."[22] Yet *La última playa* confronts us with a character for whom maintenance of the organic present extends beyond species-specific considerations. Simons inhabits a present that inscribes the "nowless" non-living matter into an ecotemporal flow towards death and transformation. While Fraser's theory creates a distance between "the nowless world of the physical universe"[23] and the organic present, I argue that the notion of an "ecological present" blurs that boundary, focusing instead on the shared inhabitation of a moment in time and space by animate and inanimate, organic and inorganic, living and non-living matter and organisms. Following the logic laid out above, if the ecological present breaks down, one can have living organic beings and non-living matter, but not a functional ecosystem working to support the conditions for survival of the whole.

In the ecological present, the synchronization aimed at maintaining the organic present extends beyond the lifespan of any singular species or organic life, a point that *La última playa* reinforces. This concept is in dialogue with Norbert Elias definition of the present as "the timing of a living human group sufficiently developed to relate a continuous series of events, whether natural, social or personal, to the change to which it is itself subjected."[24] In Elias' definition, inhabiting the present means to recognize patterns of change that occur to our body, while at the same time recognizing that they are affecting everything else simultaneously. In the same vein, ecological time is beyond biological systems and includes non-living matter, such as water, air, and soil, that make a material present. However, unlike Elias' conceptualization of the present, the ecological present does not require a human group around which time organizes itself.

In Simon's world, the ecological present is defined not only by its beauty, but also by its pending doom. Because of this, he spends much of his time planning the bridge that would connect the island to the mainland and engaging in activities that could delay the island's ultimate destruction. "No espero

---

21   Fraser, *Human Temporality*, 167.
22   Ibid., 166.
23   Ibid., 165.
24   Norbert Elias, *Time: An Essay* (Oxford: Blackwell Publishers, 1992), 78.

[la muerte]: resisto"[25] (I don't wait [for death]: I resist), he states. One of these activities is tying up trees that are falling into the ocean due to the disappearance of the island's material substratum. Soon after he is introduced to the reader, he is in anguish because, after a day of hard labor, he wakes to find that the tree he had tied is again falling into the ocean. His awareness of ecotemporality manifests as the mix of grief and desperation caused by being in the presence of the dying of his island home and, therefore, of himself.

The affect of deterritorialization explored by Caballero emerges out of the experience of the death of a territory we consider ours, that is a part of us, that is "home." In this context, Gilles Deleuze and Félix Guattari's theorization of molarity and molecularity is very productive. The molar is the scale at which humans experience the world. It is the scale at which materiality is organized into distinct and bounded forms. In contrast, the molecular is the scale at which everything is always in exchange with everything else and forms are not clearly defined but are instead in a constant process of "becoming."[26] In this theory, "deterritorialization" has a positive connotation in that it allows us to move beyond calcified notions of how things are and how we can move in the world. The movement of deterritorialization occurs when established vectors of determination, established affective relationships between things, suddenly find themselves untenable as new ones manifest, breaking down previous forms of material and relational organization.[27] In other words, deterritorialization can open us up to new connections and new ontological experiences of the self. While I do not deny the value of deterritorialization as discussed by Deleuze and Guattari, I argue that the experience of deterritorialization proposed by *La última playa* challenges us to engage the possibility of negative manifestations at the scale of the molar. Granted, the breakdown of the material ecological present might bring about new forms of relationship. Trees that used to be firmly anchored in the island, providing a habitat for animals and insects, might become parts of fish as their disintegrating particles float into the ocean and feed them. Yet, when a place feels like "home," as Arenas does to Simons, we cannot be indifferent to its disappearance. When the home disappears, so does a little of the self, and that disappearance causes suffering. For Simons, this suffering is exacerbated by the fact that, because he is the island's only inhabitant and has no significant human relations that will hold

---

25   Caballero, *La última playa*, 108.

26   Gilles Deleuze and Félix Guattari, *A Thousand Plateaus: Capitalism and Schizophrenia*, trans. Brian Massumi (Minneapolis: University of Minnesota Press, 1987).

27   Deleuze and Guattari, *A Thousand Plateaus*, 10.

the island and himself in memory, without a material marker their combined existence will be erased.

Simon's most meaningful relationship is with Arenas. A close second occurs in a turn towards the fantastic, when Caballero suggests that a manta ray that appears in the shallow waters of the island turned into a young woman for whom Simons develops strong, although tragic, quasi-romantic feelings. Simons' brief relationship with this mysterious inter-species woman momentarily eases the urgency he felt when the reality of the approaching end of his life combined with his sensory experience of the disintegrating ecology. In addition to the sensory conversations shared with the rest of the ecology of Arenas through his flute-playing, this relationship allowed Simons to prolong the time of devastation, even if only at the level of the nootemporal.

To have "awareness of time," Simons reflects at the end of the novel, is to be aware that it has passed.[28] When he glimpses the remote possibility of a romance with the young manta-ray-woman, he quickly dismisses it because, given that he does not have much time left, it is not worth acting upon. He no longer has time to live life because everything he starts in the present would remain unfinished. When Simons was fully immersed in the ecological present, he did not feel time pass. Change occurred, yet his mind inhabited the same present as everything else around him. Yet, when the specter of disappearance emerged, when paying attention to the present made him jump to the future at the level of the nootemporal, the apocalyptic fate of Arenas, which included him, infused the present. In other words, when, at the level of the nootemporal, he created a narrative, he was able to see the ecological present more clearly and frame events such as trees falling into the ocean within the same generalized impulse towards disappearance, thus inhabiting the ecotemporal. The interrelations between material changes suddenly became available to the senses. Combined with his careful perception of the material conditions of the ecological present, narrative thinking allowed him to see the present of Arenas in its catastrophic complexity.

After this realization, which occurred before the present of the novel's events, Simons only has time for building the bridge. In the survival of the island's memory, something of him, of his life's work, would remain. He would leave a mark of his passing through this earth. The mark he sought, however, was not one tied to his individual subjectivity but to the whole ecology of Arenas. If there is a mark of the island's existence, then there is a mark of his, even if no other human knows it. Thus, *La última playa* demonstrates how Simons' and the island's fates are intertwined not only in Simons' mental present but also

---

28   Caballero, *La última playa*, 108.

in the organic present of the text's world. Their interdependence manifests not only at the level of the nootemporal in Simons' mind but also at the level of the ecotemporal. In other words, Simons and the island are not only symbolically but also materially entangled.

When he felt the end of his life approaching, and he discovered that there already was, in fact, a bridge underwater, the inevitable end of his and the island's lives no longer cause him desperation. Simons' last act is to die as the island will, dissolved by the sea. Released from the burden of leaving a trace of the island's existence, he walks into the ocean and gives in to the water, mirroring Arena's trees and sediments. In this moment, Simons' being is completely subsumed into the ecology of Arenas and its surroundings while, at the same time, his individual being is actualized in the assertion of his choice to die. In other words, he disintegrates into the flow of matter of which the island is also a part, while also having a last performance of individual volition. He is calm in the knowledge that he is an individuated self, inhabiting a body ruled by the rhythms of the biotemporal, but is also more. He is a part of the ecology of Arenas, and, together, they materialize the ecotemporal. As an "organism-in-its-environment," Simons exists in complete interrelation with his living and non-living surroundings, as we all do. However, it is important to note that this interrelation is marked by hyphens in Bateson's conceptualization of ecological existence. It is not that the organism as a contained being does not exist. Rather, it is that its individual existence cannot be thought of outside of the environment of which it is a part and on which it depends. Thus, these hyphens signify the coexistence of individuality and the fact of interrelation and interdependence with the material world that we emerge from, navigate, and disintegrate into. They signify, in other words, the coexistence of biotemporality and ecotemporality within a singular body, as Caballero explores in *La última playa*.

According to Fraser, expectations and memory "extend the horizon of purposeful behavior beyond the temporal boundaries of biological effectiveness."[29] There is, in this statement, the seed of the notion that the nootemporal, the temporality of the mind and, more powerfully, of the human ability to create narrative, could act as a bridge to extend the biotemporal beyond the individual body and the organic present beyond the species. Simons' actions in the organic present incorporate the island as part of the unit of survival whose memory, if not material existence, he strives to protect. Experiences of deterritorialization of the kind explored here can occur because of natural processes such as those depicted in the novel or because of human action. Caballero

---

29   Fraser, *Human Temporality*, 167.

helps us to think how our responses to that experience rest on the extent to which we feel the world we inhabit shares in the same ecological present we do or, in other words, whether we are aware of our simultaneous bio- and ecotemporalities.

One of the existential problems faced by humans today is the fact that we all know that we are not here to stay. That awareness allows us to defer dealing with the repercussions of the ecological crisis to future generations. We tie our permanence in time to human memory, to genealogy, to human ingenuity in the form of, for example, buildings that will outlive us. We care about what we feel is our home. Yet, because most of us do not have an affective, embodied feeling of the whole world as a home, the world becomes what Marc Augé calls a "non-place," a space, like airports and malls, that we usually only pass through and with which we form no emotional attachment.[30] In the context of conversations about the "Anthropocene," I argue that the key to ecological awareness at a world scale is, paradoxically, a return to the intimately local. As *La última playa* illustrates, it is in places that we consider "home" that we can experience the ecological present because we notice its changes. Therefore, it is these places that hold the potential for an embodied understanding of our embeddedness in the world and the interdependence of our fates.

References

Augé, Marc. *Non-Places: Introduction to an Anthropology of Supermodernity*. London: Verso, 1995.

Bateson, Gregory. *Steps to an Ecology of Mind: Collected Essays in Anthropology, Psychiatry, Evolution, and Epistemology*. New Jersey: Jason Aronson Inc., 1987.

Caballero, Atilio. *La última playa*. Madrid: Hypermedia Ediciones, 2016.

Deleuze, Gilles, and Félix Guattari. *A Thousand Plateaus: Capitalism and Schizophrenia*. Minneapolis: University of Minnesota Press, 1987.

Elias, Norbert. *Time: An Essay*. Oxford: Blackwell Publishers, 1992.

Fraser, J. T. "Human Temporality in a Nowless Universe." *Time and Society* 1, no. 2 (1992): 159–173.

Fraser, J. T. *Time, the Familiar Stranger*. Amherst: University of Massachussets Press, 1987.

Ghosh, Amitav. *The Great Derangement: Climate Change and the Unthinkable*. London: University of Chicago Press, 2016.

---

30   Marc Augé, *Non-Places: Introduction to an Anthropology of Supermodernity*, trans. John Howe (London, Verson, 1995).

Guattari, Félix. *The Three Ecologies*. London: Athlone Press, 2000.

Harris, Paul. "Time and Emergence in the Evolutionary Epic, Naturalistic Theology, and J. T. Fraser's Hierarchical Theory of Time." *KronoScope* 12, no. 2 (2012): 147–158.

Ludmer, Josefina. *Aquí América Latina/Una especulación*. Buenos Aires: Eterna Cadencia, 2010.

Meyerhoff, Hans. *Time in Literature*. Berkeley: University of California Press, 1955.

Pérez Medrano, Cuauhtémoc. *Ficción herética: Disimulaciones insulares en la Cuba contemporánea*. Potsdam: Universitätverlag Potsdam, 2019.

Ricoeur, Paul. "Mimesis and Representation." In *A Ricoeur Reader: Reflection and Imagination*, edited by Mario J. Valdés. Toronto: University of Toronto Press, 1991.

Timmer, Nanne. "La crisis de representación en tres novelas cubanas: La nada cotidiana de Zoé Valdés, El pájaro, pincel y tinta china de Ena Lucía Portela y La última playa de Atilio Caballero." *Revista Iberoamericana* 73, no. 218 (2007): 259–274.

CHAPTER 4

Founder's Lecture: Is Time Out of Joint? Or at a New Threshold? Reflections on the Temporality of Climate Change

David Wood

Abstract

J. T. Fraser articulates five different organizational levels of time: proto-temporality (disconnected fragments of time); eotemporality (physics, the fourth dimension); biotemporality (self-organization, life, direction); and nootemporality (human mind, including language). He later added a sixth – sociotemporality. What impact would impending catastrophic climate change have on this schematization? We argue first that, while change is central to time, change in the very shape of change marks a new threshold in Fraser's sense. We work through what it means to be a passive spectator to radical transformation, how our human experience of time is intrinsically tied up with language, representation, and money (can we afford to prevent the end of the world?), and the impact of a shrinking future horizon on our identity, on the Enlightenment project, and on any hope of progress. Finally, inhabiting time historically is subject to many strange loops, including the breakdown of the inductive assurances that the past traditionally supplied. Extending Fraser's scheme and (following Keller's adumbration of a kairological time), we endorse the possibility and indeed necessity of a new threshold, a new temporal dispensation.

Keywords

catastrophic climate change – extinction – future generation – Fraser, J. T. – kairological time – new threshold – species loss – temporal phronesis – virtual temporality

© KONINKLIJKE BRILL NV, LEIDEN, 2021 | DOI:10.1163/9789004470170_006

1   Introduction: After Fraser – A New Threshold?[1]

In June 2019, the United Kingdom set 2050 as its target for net zero carbon emissions. Just in time, one might say. For a headline the same day reads: "New Report Suggests 'High Likelihood of Human Civilization Coming to an End' starting in 2050." This was written by a former fossil fuel executive and backed by the former chief of Australia's military.[2]

As Heraclitus made clear, change is intrinsic to time. And this last item is clearly worrying. But *could* things change in such a way as to fundamentally put in question how we think about time? I spend this paper asking whether this rethinking is indeed a major import of the projected climate catastrophe.

Our understanding of time continues to change. A major step forward can be found in J. T. Fraser's distinguishing a number of different organizational levels of time.[3] He distinguished atemporality (chaos); proto-temporality (disconnected fragments of time); eotemporality (physics, the fourth dimension); biotemporality (self-organization, life, direction); nootemporality (human mind, including language). To this first five he later added sociotemporality. "Each of these organizational levels" he writes in shorthand, "matter, life, the human mind, and society – has its own temporality."[4]

I will be suggesting a number of ways in which climate change, especially its possibly catastrophic form, provokes reflection on what shapes our human understandings of time, and I will conclude by asking whether we might not now be at a new threshold, calling for an extension of Fraser's scheme.

2   Could the Real Prospect of Catastrophic Climate Change Shake Our Understanding of Time?

Headlines like those above are now common. And there are countless urgent responses we may have to them. Political, scientific, religious, sociological, historical, practical, ethical. Each deserves and has received a stream of books

---

1   This paper is an extended version of the Founders Lecture presented at the International Society for the Study of Time conference, Loyola Marymount University, Los Angeles, California, July 2019.

2   Nafeez Ahmed, "New Report Suggests 'High Likelihood of Human Civilization Coming to an End' starting in 2050," *Motherboard*, June 3, 2019, https://www.vice.com/en_us/article/597kpd/new-report-suggests-high-likelihood-of-human-civilization-coming-to-an-end-in-2050.

3   J. T. Fraser (author of *Time: The Familiar Stranger* [*TFS*], and many other works) was the founder of the ISST, and this lecture was delivered in his honor.

4   J. T. Fraser, *Time: The Familiar Stranger* (Amherst: University of Massachusetts, 1987), 6.

and papers. I have contributed to this stream. But like many of you, I suspect, I long ago contracted what for some is an ailment, a tendency to a certain chronomania, seeing everything through the lens of time. I think of phenomena as made up of tissues of temporalities, multi-stranded, folded in on each other. Catastrophic climate change is no exception.

It is commonly observed, for example, that much of what will happen is "already in the pipeline" – if we wait for proof it will be too late to change course – that we are (or are not) simply witnessing natural climatic cycles and that these changes are irreversible. I will not set these ideas aside. Rather I want to ask a second order question: whether such a change, or our contemplation of it, might not challenge our very understanding of time itself. As an incomplete analogy, consider the idea of the birth of the universe, the idea that time might have "begun" at some point. That is not just a puzzling thought, but it can provoke reflection of what a time that could have a beginning could possibly mean. Apocalyptic time, or what are sometimes called the End Times, is not puzzling in quite the same way. We assume that even if humans are all wiped out, the sun will still rise on that scorched earth the day after, and clocks will tick on. And if we can think this without too much trouble, doesn't that suggest that our temporal categories would survive the meltdown – even if we did not? The phrase catastrophic climate change would seem to confirm that intuition. What could be more quintessentially temporal than "change"? And catastrophes too, are surely lodged securely within our ordinary understanding of time, like surprise, reversal, or disruption. Indeed, these occurrences all seem to presuppose a certain underlying continuous process in order to stand out against it. So is there really anything special in a temporal respect about catastrophic climate change?

I take my cue from Samuel Scheffler's claim that belief in the continuation of the world after one's death is a necessary condition for the intelligibility of much of what we do while alive.[5] It may make sense for me to plant a plum tree whose fruit I will not live to eat. But it would not make sense if I believed there would be neither humans nor squirrels nor fruit flies around to eat them. Who or what would I be planting it for?

At this point it is tempting to claim that without humans there would be no future.

Here I suggest we can draw on the early Heidegger to expand Scheffler's intuition more broadly. Heidegger argues that our temporal categories are neither "subjective" nor "objective" in the traditional sense but rather reflect and are grounded in our ecstatic being in the world. Our practical engagement with

---

5   See Samuel Scheffler, *Death and the Afterlife* (Oxford: OUP, 2013).

the world provides the schematic basic for our temporal categories having any meaning and (with Fraser) I take that to be inseparable from a naturalistic, evolutionary understanding of time. I concur with him that there are nested dimensions or regimes of time, which makes our treatment of it unavoidably tied to context, levels, and so on.

Such accounts of time are reducible neither to conceptual nor empirical claims. The intrinsic temporality of human existence (and many animals') is simply what gives time language any significance, however abstract it may get.

However, even if we accept this situation, it surely does not follow that we cannot apply that language beyond our own mortal span, or beyond the span of our own species. We seem to be able to say: After human extinction there will be a renewed flourishing of some of the species whose survival we threatened. We do not need to be around *then* for it to make sense now. So have we answered our original question in the negative? Doesn't our temporal language happily survive the prospect of civilizational collapse and extinction? To be clear I am asking – before such collapse has occurred – whether the prospect or possibility of such a collapse, when seriously reflected on, would bring about any sort of collapse with respect to our temporal conceptuality.

I offer now a series of linked meditations probing this possibility.

3   Change Is Changing

Heraclitus told us we can never step into the same river twice, and he explained why: New waters are ever flowing in; change is a constant. But there is something oddly stable and reassuring about this image. We are still dealing with a riverbank through which new water is flowing. There are, as it were, boundaries to change. If so, this raises the question of whether there are not different kinds or scales of change, and it seems obvious that there are and that it matters. There is a difference between reform and revolution, between a glass getting fuller and it overflowing, between growth and maturation and death, between water getting colder and turning to ice. One technical name for such changes is catastrophe (from the Greek, meaning a sudden turn). There are changes within an envelope and changes that breach the envelope altogether. Obviously, everything depends on how we identify an envelope. A comedian who "goes too far" for some will be seen as working within the "raunchy" envelope by others.

Our ordinary lives are, by and large, adjusted to change. The most common nonlinear events – births, marriages. and deaths – are normalized through ritual, drawn back into the social envelope. Even disasters like wars and hurricanes are things from which most people more or less recover. They are deep wounds after which healing happens. I do not want to make light of historical trauma. Palestinians call their 1948 expulsion from their homeland *Nakba* – disaster or catastrophe. But even there, a restitution, a healing process, remains a possibility. Those who prophecy End Times expect salvation to follow. Everything may seem to be lost, but then there is a recovery.

Worst-case scenario climate change would be different. We are already witnessing rapid irreversible species loss, and we can envisage the breakdown of many of the natural and social support services that are the background conditions for anything like what we call civilization. Radical Gaia enthusiasts will extend the narrative of restitution and healing to a post-human world in which cockroaches inherit the earth and evolution gears up again for a new adventure. But for anything we humans care about, this outcome would be a catastrophe, not just normal accommodation to change. Our world would suffer an earthquake. The carpet would have been pulled out from under us. If the shape of change changes, we will not be able to step in the same river again, because new waters will have stopped flowing in, or the whole river will have washed away in the flood.

4 Transience: Watching the World Go By

Existentially, psychologically, the idea that the character of change itself changes brings to prominence the idea of transience. The significance of this is not absolutely specific to catastrophic climate change, but is shared by other dire historical moments. Every moment is essentially transient – that is the mystery of time. But when we realize that "we cannot go on like this," we can start to see everyday life as if it were an old movie. We realize, viscerally, that this way of living is unsustainable and will not last. It is like watching horses and carts go by after the introduction of automobiles. The difference here is that it is hard not to see this transience as tragic illusion, living on borrowed time, like Wile E. Coyote still going strong past the edge of the cliff. Don't look down. This sense of transience is not the joyful appreciation of a flowing present, nor is it Nietzsche's *ressentiment* at the very passage of time, to which he opposed willing the eternal return. It is transience as a specific tragic spectacle, tragic because sometimes it seems one can only watch and fiddle while Rome burns. With more time, I would try to connect this with points made by Sonia Front about apocalyptic films in her paper "In the Forest of Realities" (in this volume). What if we come to experience our own reality as a movie? Finally, this sense of transience can easily bleed into a sense of the loss of collective agency. We are watching a spectacle we feel powerless to affect.

5   Nootemporality and Language

Fraser's remarkable section on language in *Time, the Familiar Stranger* occurs under the heading of nootemporality. He writes that *"Language* [*is*] *the architect of noetic time,"*[6] and he adopts an explicitly evolutionary perspective, starting with animal languages and then touching on the variety of human languages and yet I did not find any reference to tense or mood. Language makes narrative possible, which gives sense to experience. But in terms of the architecture of noetic time, surely we need to give tense and mood a special treatment. Here, I must say, I share something of Hegel's view of a cow's limited temporal horizons. Mooing and chewing grass offer no entry to the complexity of human language.

But first a word about nomenclature. Climate change is at one level just a phrase. It has competed with global warming, and now climate catastrophe, for public prominence in attempts to tune its resonances. Now it has been subjected to explicit censorship:

> A recent missive from Bianca Moebius-Clune, director of soil health in the US, lists terms that should be avoided by staff and those that should replace them. "Climate change" is in the "avoid" category, to be replaced by "weather extremes."[7]

The reality to which climate change refers is also subject to broad-brush disagreement. Is it just part of a natural cycle, or a new development driven by humans? Then there are obviously questions about when climate change will happen, or when it started, how it will evolve, how it will "end" and so on. And then there are questions about what we can or should do about it, why we did not start earlier, when it will be too late, what too late means, and so on. It is here that tense and mood come into play as the main players in language's role as the architect of noetic time.

---

6   *TFS* 172; my emphasis.
7   Oliver Milman, "Federal Agencies Discourage Saying 'Climate Change,'" *The Guardian*, August 8, 2017. Reported in https://www.hcn.org/author_search?getAuthor=Oliver%20 Milman/The%20Guardian&sort_on=PublicationDate&sort_order=descending.

Consider the following claim: We must act now! We were told thirteen years ago that, were we not to act promptly on climate change, we would face a situation in which attempts to prevent it would turn out to be too little too late. If we do not act, we will look back on what might have been with deep regret.

The astonishing temporal and modal somersaults in these sentences do not begin with climate change. Any rich life testifies to such shapes of hope, regret, memory, et cetera. But climate change is a peculiarly vital and intense public space for them to be deployed. We look ahead to looking back on what we hoped might happen. We weave future perfects, past conditionals, continuous presents, subjunctives, questions, and imperatives into skeins of meaning that would otherwise be impossible. Temporal language is a virtual dimension of unparalleled sophistication. Anxiety about catastrophic climate change (and indeed brave action plans) may have their free-floating inchoate forms, but formulations like this articulate our situation remarkable well. When we speak of "living in denial," it is as if the challenge of such complex formulations defeats us. We can think baroque but then sink back into rough and ready backwoods slogans. It's too late. What can we do?

It is worth thinking through here how this base-level capacity for complex temporal linguistic formulation intersects with political, scientific, and economic analyses – with predictions, prescriptions, promises, warnings. And how these various discourses – some calm and descriptive, others programmatic, yet others dire and prophetic – get translated into action and change. Between climate science and politics, there are many rifts, including one of translation.

6   Time, Measurement, Calculation, and Big Numbers

In 2012, Bill McKibben wrote a landmark essay followed by a national "Do the Math" tour.[8] I want to take up the broader question raised here about time. While Aristotle connected time to the measurement of motion "with respect to before and after," I cannot set aside how Augustine, McTaggart, Bergson, Husserl, and Heidegger linked time essentially to experience, to human existence and to language, especially tense and mood. On this view, measurement would not be central to time and indeed might threaten our sense of it. Measuring time would be like investigating the swimming habits of fish by

---

8   See Bill McKibben, "Global Warming's Terrifying New Math," *Rolling Stone*, July 19, 2012, https://www.rollingstone.com/politics/politics-news/global-warmings-terrifying-new -math-188550/.

taking them out of the water to check more closely. More broadly there is a
strong legacy of philosophers for whom calculation is something to be very
suspicious of, especially when it seeks to substitute itself for *thinking*.

And when it comes to catastrophic climate change we are bombarded
with big numbers, calculations about how dire our situation is, many of them
directly connected to time. Here are some random examples:

1. [As we began] A new report suggests "High Likelihood of Human Civil-
    ization Coming to an End" starting in 2050.
2. UK Chancellor Philip Hammond says meeting the UK's 2050 net zero
    goal will cost $1 trillion.
3. James Anderson, "We Have Five Years to Save Ourselves from Climate
    Change." "The chance that there will be any permanent ice left in the
    Arctic after 2022 is essentially zero," Anderson said, "with 75 to 80 percent
    of permanent ice having melted already in the last 35 years. [...] The level
    of carbon now in the atmosphere hasn't been seen in 12 million years
    [...] pushing the climate back to its state in the Eocene Epoch, more than
    33 million years ago. By 2050, markets for low-carbon technologies could
    be worth at least $500bn."[9]
4. To keep within the 1.5 degrees temperature rise [*Carbon Tracker
    Initiative*] we cannot burn more than 565 gigatons of fossil fuels. The
    known oil reserves (2012) when burnt would release five times that num-
    ber: 2,795 Gigatons. They cannot be used without destroying the planet
    (McKibben).
5. We are amidst the largest period of species extinction in the last 60 mil-
    lion years. We are now losing species at 1,000 to 10,000 times the nor-
    mal rate, with multiple extinctions daily. Multiple species will disappear
    before we learn they exist.
6. Insect populations have decreased by more than 75% in Germany over
    the last 28 years.

These numbers, figures, percentages are all guestimates, based on assumptions that may or may not turn out to be justified. Many people are allergic to
the ability of numbers to capture the truth. And many of these numbers are
frankly hard to comprehend. We can relate to a bird dying because it cannot
find enough insects to eat. But the reality of a species dying out is harder to
grasp. If the CO_2 levels in the atmosphere are at a 12 million year high, how do

---

9 Jeff McMahon, "We Have Five Years to Save Ourselves from Climate Change, Harvard Scientist
   Says," *Forbes*, January 15, 2018, https://www.forbes.com/sites/jeffmcmahon/2018/01/15/
   carbon-pollution-has-shoved-the-climate-backward-at-least-12-million-years-harvard
   -scientist-says/#43aa8b1c963e.

we process this fact? There is an understandable tendency to switch off when these numbers are bandied about. I can perhaps reach for the net zero target of 2050 by thinking that my youngest grandchild will be 33 years old. But 2100? The weakness or poverty of our imagination with regard to the future is visible in our well documented inability to assess high long-term risks. Proverbially we are aware of the contradictions here. "A stitch in time saves nine" suggests thinking ahead. "I'll cross that bridge when I come to it" suggests waiting.

The future can bring wonderfully unpredictable things – for example, falling in love through a chance meeting. A life fully planned out seems like a mistake. And as I have said, these numbers above are unlikely to be precise. Some see it as a deficiency in cost-benefit or utilitarian arguments that possible benefits (happiness) are weighed against each other when they typically cannot be commensurably quantified in the first place. But what such proposals do is to present broad-brush options in graphic terms. The point of taking seriously these dire long-term predictions is that one morning or afternoon they will bite. And here and now we can anticipate looking back and asking – "How did we let this happen?" Or our children asking: "Daddy what did you do in the Great War?" (UK WWI recruitment poster).

7   Time Is Money

In this section I ask the question of whether preventing the end of civilization might be too expensive, or not "cost-effective." But more thoughtfully, I ask whether time might have been monetized (and what that could mean) in a way that was not just a metaphor.

This may not seem like a serious question to pursue in connection with climate change, but it bears exploration. For while it is normally a casual slogan, it seems to me it contains a powerful truth hard to integrate with Fraser's schema. Its everyday meaning could be connected with rents, interest rates, mortgages, the price of labor, the costs of delayed completions, and so on. It arises out of the fact that people sell their labor and that people and institutions lend and invest money. From what one can tell, many political and economic decisions about action to address or anticipate catastrophic climate change are driven by considerations of cost, profit, returns on investment and so on – and obviously, then, the question of who pays, who benefits.

We know that many things happen or do not happen for reasons of cost. I want that car, but I cannot afford it. That hospital in rural Tennessee has closed because it's not economically viable. My question put provocatively is this: Could it be too expensive to prevent the end of civilization? The image

I have in my head would be of time as a silver money stream flowing into the future that, with our binoculars, we can see drying up. Could time monetized in this way then be considered a causal agent?

It might be said that thinking of time as monetized in this way is a misunderstanding. Money is just a way of representing actual and possible exchange relations and nothing in itself. It belongs to the symbolic universe. A dollar bill is worthless unless others are willing to accept it in exchange for something else. Think of those scenes of people carting boxes of paper currency in wheelbarrows to the grocery store in countries suffering hyperinflation. It is said that the cost of the UK going net zero on carbon by 2050 could reach a trillion dollars. "Can we afford this?" the Chancellor of the Exchequer asked.

The Economics of Climate Change: The Stern Review, a 700-page report released for the UK Government in 2006, argued that the investment of billions of dollars at that time would save multiples of that sum in the decades to come.[10] Sir Nicholas Stern, who was chief economist of the World Bank wrote:

> If we don't act, the overall costs and risks of climate change will be equivalent to losing at least 5% of global GDP each year, now and forever. If a wider range of risks and impacts is taken into account, the estimates of damage could rise to 20% of GDP or more. In contrast, the costs of action – reducing greenhouse gas emissions to avoid the worst impacts of climate change – can be limited to around 1% of global GDP each year.[11]

It would seem obvious that faced with such dire economic consequences, let alone a climate disaster, making these large investments today rather than tomorrow would be a no brainer.

To be up front, it is worth adding here that Stern took a position that assumed a close to zero discount rate – making the ethical assumption that future values should count no less than current ones. The real interest rate – the actual cost of borrowing money – was at that time about 6%. And the question of who is to make these investments was clearly heavily political. With gas prices fairly low and stable, US car companies in the last decade have made their greatest profits from large SUVs. Only now are they contemplating the huge investments needed to go electric, anticipating future legislation, changes in public sentiment, and perhaps the reliability of dependence on oil supplies from unstable parts of the world.

---

10   *The Economics of Climate Change: The Stern Review* (Cambridge: Cambridge University Press, 2007).

11   Ibid., xv–xx.

The Stern report was over 14 years ago. Seven years later Stern wrote:

> I got it wrong on climate change – it's far, far worse. Looking back, I under-estimate the risks. The planet and the atmosphere seem to be absorbing less carbon than we expected, and emissions are rising pretty strongly. Some of the effects are coming through more quickly than we thought then.[12]

It is perhaps worth noting that we are increasingly in a position to be able to recall earlier dire scientific warnings and predictions about the past that have turned out to underestimate the future, which is now our present. I just recently read that permafrost in Canada is thawing 70 years earlier than cli-mate change models had predicted. Quite different from the history of false religious prophecies of doom.

The picture I am painting is this: actual major decisions on carbon emis-sions by corporations, and international agencies clearly impact the scope and speed of climate change. These decisions are often driven by investment models working with discount rates that maximize short-term profit or politi-cal expediency, flying in the face of long-term rational behavior. If so, it is not exactly "money" that is hastening disaster, but a depreciating monetizing of the future. Obviously, things are never simple. My own decision about whether to convert to solar power is an investment decision based in part on the likely return on a significant capital investment and my anticipation that this cost will go down substantially in a year or two, coupled with my wanting to do the right thing. One interesting counterbalancing trend is worth commenting on. Insurance companies are increasingly worried about future liabilities with respect to storm and flood damage, for example. This concern has seen the rise of re-insurance companies that will insure the regular ones. Lloyds of London is active here, I understand. And it is becoming obvious that the risks they are taking on could wipe them out. One suspects that much of the future will be uninsurable, that governments will be forced to insure the re-insurers, just to maintain the markets, and that when it comes to it, they will wash their hands of impossible liabilities, blaming their irresponsible predecessors.

I do not know whether I have exactly shown that time is money in relation to climate change. I have tried to show that the geo-historic process of climate change is the product of a range of causes in which temporal considerations

---

12   Heather Stewart and Larry Elliott, "Nicholas Stern: 'I Got It Wrong on Climate Change – It's Far, Far Worse,'" *The Guardian*, January 26, 2013, https://www.theguardian.com/environment/2013/jan/27/nicholas-stern-climate-change-davos.

are central, and that many of these are critically bound up with investment narratives concerning money and profit.

Jim Yong Kim, the former president of the World Bank, said there would be no solution to climate change without private sector involvement and urged companies to seize the opportunity to make profits: "There is a lot of money to be made in building the technologies and bending the arc of climate change."[13]

The question we leave open here is whether it is not money as such that is driving climate change, but capitalism, and the commitment to inexorable growth – and whether we could not devise other forms of money, or radically different economies of exchange, that would encourage sustainable incentives.

8   Should We Have Children?

By far the most significant way of combatting climate change is not to have children. But if we do, can we justify what we are letting them in for? For us humans, the future is an unavoidable consideration whether we choose to ignore it or reflect deeply on it – no more so than when it concerns our own mortality. We know that the time of our lives is not all there is to time. I assume that the sun will rise the day after my funeral and that my watch will keep on ticking after my heart stops beating. It is not just that our lives are finite. That's true of rain-showers, the shelf life of cheese, the playtime of a song. The difference is that we inescapably *are* these finite lives, that our being alive is a sheer contingency, that the end point is uncertain, that the shape and span of our time on earth matters to us, and that we cannot seriously make sense of our ceasing to exist. We could say – well, all that is true, but these existential concerns are not really about time as such, just our way of living or experiencing it. I will not rehearse here my reasons for rejecting this claim. Suffice it to say that without beings that live temporally, what we call "time" would either be an expression devoid of meaning, or so conceptually poverty-stricken as to be of little interest. In the same vein, when John Michon asks whether Fraser is a naturalist and realist about time or whether his different levels are just our human ways of making sense of the phenomena, I would try to have it both ways.[14] In the case of time, there is no It without Us.

---

13   Speech at World Economic Forum in Davos (2014).

14   John A. Michon, "J. T. Fraser's Levels of Temporality as Cognitive Representations," in *Time, Science and Society in China and the West: The Study of Time V*, ed. J. T. Fraser, N. Lawrence, and F. C. Haber, 51–66 (Amherst: University of Massachusetts Press, 1986).

If then our interest in the future is inseparable from our mortality, it does not end there. I mentioned Scheffler's *Death and the Afterlife*. As I said earlier, he argues not just that time continues after our deaths but that we have to believe it for much of what we do to make sense. We not only plan for the future but also act in ways that we hope and believe will have long-term consequences. The broader phenomenological argument would be that meaning typically requires temporal horizons, that the future is key to this, and that it extends beyond my lifetime. Our current deeds project future meaning.

In this context, there is something distinctive to be said about our relation to our children, and, by further extension, to future generations, with respect to climate change. Having a child is and is not like planting a plum tree whose fruit I may never taste. This latter is unselfish, a gift to the world after me. Rearing a child is both a responsibility and the occasion for a certain identification that takes one beyond oneself. There are many stories of parents, including animal parents, risking or losing their lives to protect their offspring. It can be assumed that this act has an evolutionary explanation. But that has little bearing on how, reflectively, we respond to its urgency. The point here, of course, is that while I myself, living on high ground in Tennessee, may well shuffle off this mortal coil before disaster hits, my love and commitment to my children and theirs is not negotiable. I assume, of course, that they will meet challenges and hopefully overcome them. I would like them to be able to go hiking, bird-watching, swimming in the sea, and traveling abroad, and to eat plentiful, wholesome food. I can up to a point accept some change here, albeit with a certain sadness and regret. But I cannot contemplate with equanimity the implications for health and welfare of catastrophic climate change. It's not that human beings have not faced such troubles before and survived. Some British folk said that the solidarity and camaraderie of WW II made it the happiest time of their lives. Disasters have their silver linings. And it is not as though we all have to imagine what the future might bring. For the people in New Orleans it was very real. And for the climate refugees – since 2018 an official UN category – who are camped along unwelcoming borders in Europe, Central/South America, and the Middle East, the impossible future has already arrived.

Our current privileged prosperity insulates us from the reality of existing climate change suffering. I am grateful for it. "There but for the grace of God go I." But I doubt this insulation can last. Catastrophic climate change will put at risk even the children of the well-heeled; gated communities and extra sunscreen won't protect them. And if, as is true for most of us, even our own death and suffering can be less important than those of our children, catastrophic climate change poses a direct and immediate threat to our ongoing identity, extended as it is to our children. They extend our personal identities beyond our own lives and give us a direct interest in ensuring that we provide the conditions for their flourishing.

If children put little pressure on our ability to identify with the needs of future others – resting, say with Adam Smith, on natural sympathy for our offspring – there are strong ethical and political arguments that would give to future generations more generally equal consideration, on a par with people in distant lands with whom we have no contact or connection. Such equal consideration is not entirely unproblematic. We do not know what their preferences will be. And, as Derek Parfit has argued, it is fundamentally unclear who these people will be, or indeed whether they will be, which makes looking to their interests tricky.[15] I have to say, I find these arguments unconvincing. Acquiescing in the annihilation of future generations would indeed resolve the problem of how to address their needs, but is hardly a recipe for action. And whoever inherits the earth should be able to expect some minimum opportunities for a good life.

The generally established rule for how to move forward in a way that does justice to our successors was formulated in the 1987 Brundtland Report, which appropriately avoids too much precision: "Sustainable development can be defined as development that meets the needs of the present without compromising the ability of future generations."[16] This formula from over 30 years ago addresses the question of climate change only indirectly. If the threat of catastrophic climate change can be laid at the feet of a development ideology – unregulated global free markets blind to the limits of natural capital – then what is needed is a kind of development limited by its own conditions of possibility – sustainable development.

For our purposes, thinking about the shape of time, this principle breaks with the idea of infinite growth, in which the future would just be more and bigger, in favor of a kind of steady state. The linear conveyor belt of historical progress as growth will take us over the cliff. Sustainable development leaves a world our kids could hopefully live with.

I have tried to show through thinking about our children that even when we think selfishly about our own lives, catastrophic climate change is a pressing matter. And ethical arguments that take into account the interests of others quickly include future generations of humans, indeed of other species. It may

---

15   Derek Parfit, *Reasons and Persons* (Oxford: OUP, 1984).
16   The Brundtland Commission, *Our Common Future* (Oxford: Oxford University Press, 1987).

be that, leaving aside ethical arguments, what is at stake is a deeper picture or story of evolutionary world history as a creative streaming living flux that we find ourselves individually and as a species to be part of, and firmly want to continue.

Having a child, especially in the developed world, is a highly extravagant carbon luxury, dwarfing any other lifestyle choice one might make. On the other hand, we have responsibilities to our progeny that we seem not to be discharging. I suggest here that those obligations are not just abstract but build on the natural sympathy we have for our children and theirs, through which our own identity is extended. We cannot think about the future without thinking about future generations.

Leaving aside ethical arguments, what is at stake is how we relate to the deeper picture or story of evolutionary world-history as a creative streaming living flux of which we are each a part.

9   Living in Denial

Psychologists have discussed people's need to avoid cognitive dissonance, specifically with respect to catastrophic climate change. One of the ways in which this works is for people to question the science of climate change, as I mentioned. Are we not dealing with mere predictions? Why should we believe them when meteorologists cannot even be sure about whether it will rain here tomorrow? It is tempting to respond that *You don't need a weatherman to know which way the wind blows.* But that would be too glib.

There *are* logical puzzles about statements concerning the future. Aristotle's famous sea-battle argument suggests that it must be true now that it will or will not occur tomorrow. But that tells us nothing about what will happen, even if it is predetermined. Sceptics will argue that science generally is not to be trusted, that apocalyptic claims rarely turn out to be true, that the history of the planet is a singular idiographic sequence not amenable to law-like explanation, that climate change scientists have their funding dependent on their scary predictions, that climate change activists are fear-mongering, and so on.

There is little doubt that prediction is an essentially contestable practice, even as it is unavoidable. Stock market predictions can be self-fulfilling and subject to criminal investigation.[17] And where predictions turn into

---

17   See Elon Musk, Tesla, and the SEC (2018).

promises, such as candidates for electoral office are prone to make, things get murkier still. The future can be gamed in this way precisely because in a strict sense it is unknown until it happens. And yet we absolutely need to make projections, anticipations, predictions, preparations, however contingent they may be. Whenever we walk down the street, one leg is already leaning into the future. This very sentence cannot be written or understood without some unrolling anticipation of how it will be completed. Businesses make investments, and individuals take out 30-year mortgages, with an eye on the future. It is the necessity to imagine, bet on, rely on the future that makes prediction such a vulnerable practice, open to genuine miscalculation as well as disingenuous manipulation. All this is to say that climate change skepticism feeds off genuine logical, epistemological, and metaphysical problems about prediction. But it is itself equally vulnerable to fallacious reasoning.

One cannot rule out utterly unforeseen events that would reverse or mitigate the direction or outcome of climate change: meteor strikes, a planetary plague that would wipe out most of us, a carbon-capture technological breakthrough, the return of Jesus. This is especially true, in principle, with respect to nonlinear processes, spawning tipping points, accelerations, feedback loops, and so on. But if these events and processes make inductive generalizations more tricky, there looks to be another level of induction that makes things worse. For as much as we might hold out hope for a nonlinear salvation, the inductive evidence so far is that the feedback loops are accelerating disaster. The melting of the Greenland ice sheet is a clear case. Arctic ice is not far behind. Climate models are consistently being outpaced by changes on the ground.

The politics and economics of climate change are clear enough. Oil reserves lose their value if they cannot be burned. (See McKibben.) And strategists explicitly seek to generate "uncertainty" in public debate, as once happened with tobacco and cancer, trading on a standard of certainty that could never be reached.

The obvious response to this notional uncertainty, when the climate science consensus overwhelmingly points in the same direction, is at minimum to adopt the precautionary principle. As with Pascal's Wager, the downside of believing in God and being wrong is so much less than not believing and being wrong.

None of this is news, I appreciate. What I am trying to make clear is that while we may think about time and our knowledge of the future in abstract terms, logical distinctions and insights are, in public discourse always already materialized, compromised by political and other interests, often with dire consequences.

10   Climate Change in the Future? No. The Future Is Happening Now

A few years ago, we were worried about the future, about the possibility of climate change happening some time down the road and what we might do about it. Even then it was said that much of the expected rise in CO_2 levels was already in the pipeline, which complicates references to the future. Time and history include lag effects, uneven development, multiple strands of time. I have argued elsewhere for a temporal phronesis fluent in recognizing and working with these shapes of time.[18] Now when people speak of climate change, we often have to admit that it is already happening. Droughts, hurricanes, heat-waves are not (just) signs of what is to come; it is here already. Many of us find ourselves in a time pretzel in which we remember anticipating, warning against and screaming about, what is happening at this very moment.

11   Enlightenment Progress – Is the Dream Over?

The prospect of catastrophic climate change is arguably an awakening of the most sobering sort for what one might call the Enlightenment project, one that imagines human history as one of progress in education, health, freedom, human rights, and so on, peppered perhaps with setbacks. There have been other signs that all was not well. In the 1930s, Husserl wrote "The dream is over."[19] Since then, we have seen the Holocaust, two world wars, Vietnam, Iraq, and now the rise of unapologetic rightwing populism – and what William Connolly has called the "capital-evangelical resonance machine."[20] This machine has been critiqued as cloaking an imperialist, even racist, agenda, justifying Western expansion and political and economic domination. But catastrophic climate change attacks the unlimited growth agenda on which even the innocent version of Enlightenment rests. *Decroissance* – degrowth –

---

18   See my *Reoccupy the Earth: Notes Toward an Other Beginning* (New York: Fordham University Press, 2019), ch. 3.

19   "Philosophy as science, as serious, rigorous, indeed apodictically rigorous science – the dream is over." In Edmund Husserl, *The Crisis of European Sciences and Transcendental Phenomenology*, trans. David Carr (Evanston, IL: Northwestern University Press, 1970).

20   See William Connolly, The *Fragility of Things* (Durham, NC: Duke University Press, 2013), 27. Quoted by Catherine Keller in *Political Theology of the Earth* (New York: Columbia University Press, 2018).

is the new slogan.[21] And while there are those – like me – who *imagine* a convergence between stepping off the growth pedal and political and economic reforms that would bring about social justice, it is hard to discount the consequences of increased scarcity, desertification, rising sea levels, crop failures, antibiotic resistance, and violent storms, which directly threaten human health, safety, and welfare. But if "enlightenment," as some think, was a secularization of religious eschatology, what of the plain religious versions?

I once directed a five-year-long interdisciplinary workshop on Ecology and Spirituality, and in the course of our research we met with and interviewed some local evangelical Christians.[22] To our surprise, many were skeptical about climate change. They mistrusted science, and they trusted that they were being looked after by a benign God who would not let this happen. But there were others who embraced catastrophic climate change as proof that we have entered the Last Days, as predicted in Daniel and in the Book of Revelation. There will be wars, plagues, and conflagrations, and Satan's empire will be defeated, inaugurating a new age of redemption. These beliefs have long been widespread in Abrahamic religions, and they make sense of troubled times. The pure in heart will be saved (Rapture), and good will triumph. For some it makes sense to accelerate the end to bring down the house of evil as soon as possible. It is easy, and perhaps right and proper, to treat these beliefs with a heavy dose of skepticism. But what they reflect is perhaps a demand for intelligibility with respect to history and our fate, which we should acknowledge. Or else it's "a tale told by an idiot, full of sound and fury, signifying nothing" (*Macbeth* 5.5.26–28). Is the apocalyptic eschatology of the Book of Revelation not just an extra strong narrative filling a space we ourselves fill in other ways? I'm suggesting that catastrophic climate change makes these deepest questions about the meaning and direction of history, of human time, absolutely unavoidable. And if we think the End Times crowd is crazy (and they are), what do we say instead? Can we really reconcile ourselves to the idea that Gaia will be fine, that life will regenerate even if we disappear? That humanity might just be, have been, a finite project, with design flaws?

---

21   See the *Degrowth* movement, reflected in Greta Thunberg's address to the UN – "We are in the beginning of a mass extinction, and all you can talk about is money and fairy tales of eternal economic growth." September 24, 2019.

22   With Beth Conklin, Professor of Anthropology, Vanderbilt University.

12   The Shadow of the Past

With regard to temporal phronesis, a familiarity with the strangenesses of time, I have developed a taxonomy of the structures of time that we would do well to bear in mind.[23] A list of these time pretzels: The "always already," the evidence lag that makes it too late to act if we wait until we know we must, the pluri-dimensionality of time, the role of the past in projecting futures, ones that we can now recall, uneven development, tipping points, the virtual temporalities made possible by compounding tense and mood, and various other strange loops. In thinking about climate change, we find it hard not to focus on the future. But as so often happens we find that the past casts a long shadow. Consider five ways in which this is so.

First, so much of our atmospheric carbon crisis is the product of our releasing into the air carbon sequestrated by plants and compressed into coal, oil, and gas from solar-photosynthesis over past millennia. What has been saved over the very long term is, on a geological time-scale, being spent almost instantly. This is a phenomenon in which, as they say, time is out of joint. Time-scales collide. The same is true of pumping water out of ancient aquifers like there's no tomorrow. A visual illustration is provided by the minutes it takes to fell a thousand-year old tree with a chain saw. By contrast, renewable solar, wind, and tidal energy draw anew on a daily supply. This offers a quite different attunement to energy input.

Second, when it comes to thinking about mitigating the effects of climate change and distributing the cost of doing so, developing countries argue, not unreasonably, that nations that industrialized early benefited unequally from being able to pollute the environment in a much less restricted way. And their current capital assets, political power, and technological prowess reflect that. So it is only fair that such nations shoulder a disproportionate burden going forward in tackling the problem. Carbon restitution we might call it.

Third, and perhaps less obvious, when we think of the impediments to the kinds of social, political, and economic changes we need to tackle climate change – our habits, our instincts, our desires, our mindset – these are formed biologically and historically, under conditions very different from those prevailing today. We may, for example, be well prepared to fight normal identifiable enemies, such as other tribes, but not so well prepared when, as Pogo said, the enemy is us, or our aggregated regular practices.

---

23   See my *Reoccupy Earth*, 229 fn. 11.

Fourth, the past is already programming the future in terms of lag times. The climate responds to rising CO_2 levels with a delay of decades. Dangerous increases are already locked in.

Lastly, in a slightly different vein, we have to reckon with the fact that there is a history of apocalyptic prophecy and a lot of sociological theory to explain why it reappears at times of stress. There are also historical precedents for the kind of general mobilization arguably needed to tackle this problem. Such as war. And there are geological antecedents for rapid and catastrophic climate change – such as that which wiped out the dinosaurs 65.5 million years ago, when either a volcanic eruption or a 6-mile wide asteroid hit the Earth. The sky was blacked out for years, possibly decades, killing most plant and microbe species, and all the animals that fed off them. Whether any of these sources of perspective help in dealing with an *unprecedented* situation is another matter.

13   The Politics of Climate Change

The emergence of rightwing populism in both the US and Europe is troubling on many scores. Its embrace of climate change denial is especially alarming. But the most plausible explanations for this phenomenon, at least in my view, shed light on another temporal corollary of climate change. The first references the huge influence of fossil fuel companies on the political landscape, financing climate-change-denial thinktanks and bankrolling politicians. This is an entirely rational practice on their part precisely for the reasons that Bill McKibben has made clear. Trillions of dollars of oil could not be burned, vast assets would be wiped out, and share prices would tumble if climate change were acknowledged to be true. The window of obfuscation and deception is, however, closing. Meanwhile it's burn, burn, burn.

My second observation has to do with what we might call the wrench being thrown into the wheel of politics. In the old days, politicians wanted to win, of course, but, if they lost, there was always the next time, and there would be regular switches of power, what Pareto called the circulation of elites. Everyone would get a turn. But this presupposes a long future stretching out ahead in which things would roughly balance out. Recent cynical political maneuvers seem incredibly short-sighted. Surely, they wouldn't want to give their successors such a precedent to use against them when they come to power. But this thought presupposes that stretched-out future we mentioned. If you believed that time was short, that the chickens might never come home to roost, that the best strategy for survival is to rob the bank now, hope you don't get caught, and buy all the protection you can get against the coming disaster, this "short-term" strategy – which we might call MitchMcConnellism – would be entirely ratio-nal. Such a strategy could well be coupled with pretending not to believe in climate change. At the same time, our geo-political commitments, sustaining undemocratic oil-rich Middle East regimes, a source of systemic corruption, all rest on resisting the end of oil as long as possible. All this in the context of the dire need for climate change leadership from the major powers, especially the USA. I am saying that it is the clear but unacknowledged long-term danger of climate change that is generating the short-term hyper-corruption of our politics and attempts to break the cycle of electoral succession. Again, time is *so* out of joint.

14   Conclusion: A New Threshold?

Are we then, or now, to use Fraser's word, at a new threshold? His original five levels got provisionally extended to include a new level – sociotemporality. Many of the observations I have made about climate change involve temporal structures not exclusive to it. But is a new emergent temporality perhaps now being born? There is recent official acceptance of the word *Anthropocene* to mark the human entry onto the geological stage. But that does not address or capture the character of the new regime. I have explored a number of the shapes of time's materializations – language, money, generational flesh – and at the same time borne witness to how the future is being gamed for profit and to how progressive ideals about future possibilities are waning. The word I am looking for would bundle a shared sense that business as usual is unsustain-able and will eventually lead to catastrophe, that our political and economic systems are not adapted to deal with this situation, that catastrophic climate change is transparently the consequence of our innocent daily routines within a globally consumerist society, and that many feel individually impotent to change things. Al Gore once nicely called our situation an addiction.[24] Plato's allegory of the cave walks again, as his shadows re-emerge as temporal fore-shadowings. We could call it a dark awakening. It is a time out of joint, one of potentially terminal uneven development. We can predict, foresee, and imagine what we feel powerless to change. We can even understand the evolu-tionary lag that makes this so. To extend Fraser's schematism, I'm tempted to a term like "ecogeohistorical." Whatever word we settle on, it would be riven by the tension or contradiction that we call unsustainability – a predictable

---

24   See Al Gore, *An Inconvenient Truth: A Global Warning*, dir. David Guggenheim (Paramount, 2006).

collision course perhaps between bio- and socio-temporality. But even that that leaves out a lot.

Suppose then we were to really stick our neck out and conjure a further stage. Let me suggest an unlikely source – Catherine Keller's extraordinary book *Political Theology of the Earth*.[25]

> Once upon a time ... we had time. Whatever the story of our individual mortalities, there extended out from all of us, from us all together, the space of a shared time, the time of a shared space. The sharing was rent with contradiction: we reached no consensus on the layout of the future. [...] But there would be time enough for the space of a more marvelous togetherness: New Heaven and Earth, utopic horizon, seventh generation, endless rhythm, r/evolutionary leap, fitful progress, sci-fi tomorrow. Or so the stories go. We had time. And now we seem to have lost it. Time, our time, the time of human civilization, appears to be running out.

These remarks deploy well the temporal gymnastics I have attributed to language. Keller is looking back nostalgically to how we used to look forward. But is our time running out? Is there still hope?

Keller's solution is to rework St. Paul's *kairological* time, cousin of Benjamin's *Jeztzeit*, a Now time of infinite possibility which proposes a break with *chronos*. Her distinction, as a process theologian, is to understand this possibility in terms of transformations in our social relations, that is, in secular and material terms. There is no Messiah at the gate, but we can imagine, remain open to and encourage a new social dispensation.[26] Or as Gore once put it, there are tipping points in public sentiment. Greta Thunberg and her ilk might be a harbinger of an utterly unexpected turn.

References

Ahmed. Nafeez. "New Report Suggests 'High Likelihood of Human Civilization Coming to an End' starting in 2050." *Motherboard*, June 3, 2019, https://www.vice.com/en_us/article/597kpd/new-report-suggests-high-likelihood-of-human-civilization-coming-to-an-end-in-2050.

---

25   Keller, *Political Theology of the Earth*, 1.
26   I toy with such possibilities in *Deep Time: Dark Times* (New York: Fordham University Press, 2018).

The Brundtland Commission, *Our Common Future*. Oxford: Oxford University Press, 1987.

Connolly, William. The *Fragility of Things*. Durham, NC: Duke University Press, 2013.

Fraser, J. T. *Time: The Familiar Stranger*. Amherst: University of Massachusetts, 1987.

Fraser, J. T., N. Lawrence, and F. C. Haber. *Time, Science and Society in China and the West: The Study of Time V*. Amherst: University of Massachusetts Press, 1986.

Gore, Al. *An Inconvenient Truth: A Global Warning*. Directed by David Guggenheim, Paramount, 2006.

Husserl, Edmund. *Crisis of European Sciences and Transcendental Phenomenology*. Translated by David Carr. Evanston, IL: Northwestern University Press, 1970.

Keller, Catherine. *Political Theology of the Earth*. New York: Columbia University Press, 2018.

McKibben, Bill. "Global Warming's Terrifying New Math." *Rolling Stone*, July 19, 2012, https://www.rollingstone.com/politics/politics-news/global-warmings-terrifying -new-math-188550/.

McMahon, Jeff. "We Have Five Years to Save Ourselves from Climate Change, Harvard Scientist Says." *Forbes*, January 15, 2018, https://www.forbes.com/sites/jeffmc mahon/2018/01/15/carbon-pollution-has-shoved-the-climate-backward-at-least -12-million-years-harvard-scientist-says/#43aa8b1c963e.

Michon, John A. "J. T. Fraser's Levels of Temporality as Cognitive Representations." In *Time, Science and Society in China and the West: The Study of Time V*, edited by J. T. Fraser, N. Lawrence, and F. C. Haber, 51–66. Amherst: University of Massachusetts Press, 1986.

Milman, Oliver. "Federal Agencies Discourage Saying 'Climate Change.'" *The Guardian*, August 8, 2017. https://www.hcn.org/author_search?getAuthor=Oliver%20Milman/ The%20Guardian&sort_on=PublicationDate&sort_order=descending.

Parfit, Derek. *Reasons and Persons*. Oxford: Oxford University Press, 1984.

Scheffler, Samuel. *Death and the Afterlife*. Oxford: Oxford University Press, 2013.

Stern, Nicholas. *The Economics of Climate Change: The Stern Review*. Cambridge: Cambridge University Press, 2007.

Stewart, Heather and Larry Elliott. "Nicholas Stern: 'I Got It Wrong on Climate Change – It's Far, Far Worse.'" *The Guardian*, January 26, 2013. https://www.the guardian.com/environment/2013/jan/27/nicholas-stern-climate-change-davos.

Wood, David. *Deep Time: Dark Times*. New York: Fordham University Press, 2018.

Wood, David. *Reoccupy the Earth: Notes Toward an Other Beginning*. New York: Fordham University Press, 2019.

CHAPTER 5

Slow Time: The Suspension of a Tension

Paul A. Harris

Abstract

This essay reflects on "Time and Variance" through the lens of slow time. Slow time is defined in both psychological and physical terms, as a contemplative mode of thought and geologic and cosmic timescales, respectively. It argues that slow thought facilitates "temporal phronesis," an ability to conceptualize and accommodate disparate timescales, and analyzes gardens and art installations that explore time in variance in different ways and facilitate temporal phronesis. The essay also argues that J. T. Fraser's hierarchical theory of time is complicated by considering "Gaiatemporality," constituted by the conflicts created by the intrusion of humanity into earth processes.

Keywords

Anthropocene – Gaiatemporality – slow time – labyrinth – Charles Jencks – J. T. Fraser – gardens – temporal phronesis – walking

This essay unfolds speculative arguments by assembling and analyzing rather heterogeneous materials in an interdisciplinary nexus of landscape and installation art, and philosophy. It begins and ends by discussing two very different gardens, The Garden of Slow Time at Loyola Marymount University and The Garden of Cosmic Speculation in Scotland. In between, it presents art installations produced for the "Time in Variance" conference, which lead into reflections on Anthropocene temporality. These materials serve as springboards for reflecting on "Time in Variance" from various perspectives, all the while engaging specific tenets of J. T. Fraser's hierarchical theory of time. Stretched to its furthest limits, "Time in Variance" ultimately invokes the question of how to comprehend, or think together, radically differing temporalities. Here, I will take up three examples of diametrically opposed temporal regimes: the planetary and the personal, the eternal and the ephemeral, and the cosmic and the contemplative. I approach these themes through the lens of "Slow Time," which carries both physical and psychological meanings. In a material sense, "slow time" refers to inhuman metaphysical (eternal) and physical (cosmic, geologic) timescales that the mind struggles to apprehend. In a cognitive sense, "slow time" refers to a mindful state of awareness, an intuitive, contemplative cerebral disposition that entertains and explores possibilities, imagines variegated scenarios and solutions, remains open to multiple potentialities.

The challenge of thinking about time in radical variance with itself is more than a philosophical problem to be resolved. Trying to wrap one's mind around temporal variance puts a pause on my habits of thought and pushes me to reconsider: to ask anew how we think, what thinking means, how we might approach tasks seemingly steeped in paradox. Grappling with time in variance demands what David Wood calls a "temporal phronesis," an ability to think time "in all its multistranded complexity" (Wood 2019, 17). The basic gambit explored in the following is that the physical and psychological dimensions of "slow time," taken in tandem, open a way to temporal phronesis. Simply put, in order to reconcile our habitual human timescales with inhuman ["slow"] timescales, we need to practice a "slow" mode of thinking.[1] This assertion is reiterated in the homonymic resonances of the essay's subtitle: accommodating radically variant timescales is accomplished through "the suspension of a [conceptual] tension," and this form of thinking is achieved by "the suspension of [conscious] attention."

There are many ways that one might approach slow thinking and a suspension of conscious attention. This essay grounds slow thinking in the specific practice of walking meditations, a fitting choice given the essay's focus on gardens. Precisely how and why perambulation provides a means to meditation is eloquently articulated by Rebecca Solnit in her book *Wanderlust: A History of Walking*:

> The rhythm of walking generates a kind of rhythm of thinking, and the passage through a landscape echoes or stimulates the passage through a series of thoughts. This creates an odd consonance between internal and external passage, one that suggests that the mind is also a landscape of sorts and that walking is one way to traverse it. A new thought often seems like a feature of the landscape that was there all along, as though thinking were traveling rather than making.
>
> SOLNIT 2001, 11

While it is difficult to convey these kinds of thought-experiments and -experiences in an essay, I invite the reader to read this essay as if following

---

1   See Walker (2017) for a treatment of slow philosophy as a reading practice.

an itinerary, undertaking a series of exploratory visits to sites to discern what
insights and ideas arise along the way.

1   The Garden of Slow Time, Loyola Marymount University

The Garden of Slow Time aspires to inspire temporal phronesis by inviting visitors to experience and ruminate over variant timescales and temporalities. The
garden (figure 5.1) was consecrated in 2016 as part of a "Slow LMU" initiative
I co-directed with my colleague Brad Stone.[2]

I designed the garden to conjure and negotiate a series of tensions between
opposites, including archaic and contemporary, playful and prayerful, local
and global. Sited on a bluff overlooking the Los Angeles Basin, it provides a
peaceful haven in stark contrast to the bustling metropolis below.[3] The garden evinces a "Slow Time" philosophy informed by the discourse of mindfulness. The garden's central feature is a meditative labyrinth, and labyrinth
walking has become a popular form of contemplative practice and spirituality.

FIGURE 5.1   The Garden of Slow Time
PHOTOGRAPH BY AUTHOR

---

2   The Slow LMU initiative and relevant documents may be accessed at https://digitalcom
    mons.lmu.edu/bellarmineforum2016_resources/14/.
3   The Garden of Slow Time was installed by LMU Facilities Management personnel led by
    Mario Arroyo, who are also responsible for the beautiful plantings, landscaping, and stability
    of the site.

Simultaneously, the garden's dominant material elements elicit associations with Neolithic sites. The labyrinth design is among humanity's most archaic shared symbols – its seven-circuit pattern, known as the "classical" or "Cretan" labyrinth, has been found on ancient sites around the world. Nine standing stones and boulders set around the labyrinth evoke a stone circle.

These rocks include the Solstice Stone, a tall megalith sited to casts its shadow at sunset on the Winter Solstice through the center of the labyrinth; the Cosmos Stone, which features concentric ring markings that evoke the solar system; and the Janus Stone, a tall spire with two faces looking to the past and future.

Mindfulness, defined by one's focusing awareness on the what one is feeling and thinking in the moment, is synonymous with a "suspension of attention," a turning away from the constant influx of things competing for our notice in the "attention economy" of hypercapitalism.[4] The Garden of Slow Time explicitly expresses this turn in the playful epigraphy on benches around the labyrinth, which transpose cliches of capitalist time efficiency into phrases invoking

FIGURE 5.2   The Solstice Stone (foreground), Garden of Slow Time
PHOTOGRAPH BY AUTHOR

---

4   As Tim Wu observes, "Our computers are ostensibly productivity-enhancing machines, but they also are loaded with platforms whose business model is to consume as much of your time as possible with ads and noise and distraction" (Wu 2016, 11).

FIGURE 5.3   Bench epigraphy, The Garden of Slow Time
PHOTOGRAPH BY AUTHOR

mindfulness: the urgent call for incessant productivity of "no time like the present" becomes an injunction to root oneself in the now ("KNOW TIME LIKE THE PRESENT"); the demand for peak performance and productivity sounded in "spend your time wisely" becomes a suggestion to stop time in its tracks ("SUSPEND YOUR TIME HERE WISELY").

Walking a labyrinth facilitates an embodied experience of a specific form of slow time. As Solnit observes, walking and thinking are activities that are useless within the capitalist time economy:

> Thinking is generally thought of as doing nothing in a production-oriented culture, and doing nothing is hard to do. It's best done by disguising it as doing something, and the something closest to doing nothing is walking.
>
> SOLNIT 2001, 5

The spatial itinerary of the labyrinth, which begins and ends at one point and involves a circuitous journey, may be interpreted as a diagram of temporal suspension (Harris 2014). The symmetrical repetitions and reversals of the sinuous route may be rendered verbally in the lines "In the end it turns out/The way out is the way in/In reverse, inside out" (Harris 2018, 12). The topology of the labyrinth walk evokes a phenomenological suspension of time, an aporia or pause in which the directional distinction between past and future is lost.

The slow time of the labyrinth walk may be conceptualized as a form of "eotemporality," one of the scales in J. T. Fraser's hierarchical theory. Fraser named eotemporality "for Eos, goddess of dawn" (2007, 17), befitting of an archaic form of time, born in the early universe, embodied in the older parts of the brain, and associated with the mind's "deeper" psychological levels.

In conceptual terms, the eotemporal is continuous and reversible; it could be represented as a simple line along which time does not have a preferred direction. In terms of the physical history of the universe, eotemporality corresponds to the scale of massive matter and thus emerges with the slowing and freezing of matter into galaxies. In relation to human psychology, eotemporality corresponds to the oceanic, the unconscious, and hence, Fraser observes, "such a temporality often infuses our dreams" (2007, 17).

Fraser's account of the eotemporal provides a concise conceptual description of the psychological dimension and experience of the "suspended attention" of "slow time." Suspension invokes a hanging between, an anticipation of something yet to come mixed with a temporary withholding, as well as a carry-over or resonance between a past thing (a thought, a musical note) and another. In Fraser's evocative phrasing, eotemporality is experienced as "the two-wayness of time. This is the feeling of that curious fore- and afterknowledge which resembles the listening to a composition already well known" (2007, 189). Eotemporality comprises a continuous line without direction, which from the human viewpoint means moving in a temporality in which there is a double or uncanny sense of anticipation and memory folding into one another. This temporal ambiguity finds analogous expression or metaphorical mirrorings in states of disorientation, the loss of distinctions and/ or sense of direction, all properties associated with labyrinths in general, and the Cretan labyrinth in particular. These characteristics capture a reflective or ruminative mode of thinking in which multiple and opposite scales or components can co-exist, or in which past knowledge and accumulated images mix with imagined possibilities or new insights.

When we walk the labyrinth, we not only revert to or experience an archaic form of temporality; we also reprise an ancient human practice. As stone labyrinths at Neolithic sites attest, humans have walked this exact itinerary for millennia. The ritual function of labyrinth walking is a matter of speculation and differs across cultures; it has been associated with the womb and birth (labyrinth as Earth uterus), weddings, and the tomb and funerals (Kern 2000). Walking the labyrinth has thus been a tool for reflecting on and confronting the confounding boundaries of existence and thresholds of life. The Garden of Slow Time recontextualizes the labyrinth walk in a site whose surroundings encompass sharply divergent timescales, from the urban sprawl of Los Angeles to the Santa Monica and San Gabriel Mountains. Directly below the bluff lies "Silicon Beach," home to prominent technology companies and start-ups, the epitome of the attention economy of hyper-capitalism. Yet, in stark contrast, the landscape also bears tangible traces of indigenous human settlement: the LMU Tongva Memorial, around a bend on the bluff from The Garden of Slow

Time, is dedicated to the Gabrielino/Tongva tribe, who inhabited the region from at least as far back as 1000 AD, and elements in the Ballona Discovery Park along the creek below also document the tribe's presence.

The Garden spurs "temporal phronesis" in the striking juxtaposition of urban development with the natural beauty of the ocean (Santa Monica Bay) and mountains (Santa Monica and San Gabriel ranges). And just as modern Los Angeles – and the iconic Hollywood Sign visible in the hills across the basin – stands for constant novelty, so too does its geologic context. The Pacific Plate on which the mountains rest is side-sliding north along the San Andreas fault (only 20 miles to the east) at a rate of two inches a year, stunningly fast in geologic terms. Bordering the fault, the San Gabriels feature steep slopes and fractured rocks, signs of the strain of ongoing, unusually fast uplift. Both the Santa Monica and San Gabriel mountains are part of the Western Transverse Ranges Block, whose name references their unusual east-west orientation. Between 35 and 15 million years ago, the Farallon Plate subducted under the North American Plate, and the Pacific Plate began side-sliding along the San Andreas fault, dragging rocks west of the fault northwest for several hundred miles. These developments caused the mountain ranges around Los Angeles to migrate from near where San Diego is now and rotate clockwise (Meldahl 2015).

I like to think of the labyrinth as a "Slow Time Machine." Stepping into it, and taking deliberate steps along its path, is a simple means of accessing or inducing a contemplative state in which one opens to different percepts, affects, and concepts. Walking the labyrinth relaxes the mind into a sort of lucid dreaming or wakeful sleep-walking. In his account of the eotemporal, Fraser observes that

> it cannot be by chance that when people turn strongly inward by way of meditation, or by concentrating on abstract tasks, they appear to be sleeping. The dominant umwelt are their minds are pre-noetic and their bodily behaviour reflects this fact.
>
> FRASER 2007, 190

Using the labyrinth as a tool for contemplative practice, I have designed various meditations for students to consider as they walk it that invite reflection on relations among personal, geologic, and cosmic timescales.[5] These prompts follow the spirit of Stoic thought-exercises such as the Meditations of Marcus Aurelius, which in turn influenced Ignatius of Loyola's conception of

---

5   For more about the labyrinth, pedagogy, and contemplative practice, see Harris (2018).

his Spiritual Exercises, a regimen of meditation, prayer and reflection (Hadot 1995). As a tribute to my experience with the Spiritual Exercises at LMU, I call the meditations for students "slow time exercises." In these contexts, The Garden of Slow Time provides a setting where ancient philosophical and religious traditions coalesce with a contemporary contemplative practice that integrates modern scientific accounts of time and cosmic evolution.[6]

2   Conference Installation: ... l'eternel et l'ephemere

The installation entitled Suspension of a Tension was created for the Time in Variance conference in collaboration with LMU colleagues Garland Kirkpatrick (Art and Art History) and Timothy Snyder (President).

The work enacts a literal, physical suspension of tension in the form of a fabric banner hung down the three-story stairwell in Hannon Library (the conference setting). The attempt to conceptualize conflicting temporal orders is expressed in the text of the suspended banner, which cites Georges Perec's sublime sentence, "Je cherche en meme temps l'eternel et l'ephemere" [I seek at the same time the eternal and the ephemeral]. Perec composed this sentence for *Les Revenentes* (1972), a novel-length monovocalism (the only vowel used is e). The sentence fuses existentially opposed concepts (eternal/ephemeral) by expressing them in a uniform medium or textural consistency, a sentence (and text) marked by all vowels being the same letter. The sentence reappears as an epigraph to the final chapter of *Life A User's Manual*, Perec's masterwork set at the instant of the protagonist's death, death being the moment when the ephemeral meets with eternity. The banner's typographic syntax responds to the phonetics of Perec's binary terms, creating a site-specific calligram referencing a book spine.

Hanging the banner bearing Perec's phrase in the staircase made a walking meditation available to conference participants. The suspended banner extended a standing invitation to conference participants to contemplate the variance between the ephemeral and the eternal while ascending or descending the three flights of stairs. The installation included *esceleer*, a musical soundscape composed by Timothy Law Snyder (accessed through QR codes posted on didactics).[7] In Snyder's words,

---

6 For a reframing of the Spiritual Exercises in the context of contemporary cosmology, see Savary (2010).

7 Esceleer may be accessed at https://soundcloud.com/timothylawsnyder/esceleer.

FIGURE 5.4 Suspension of a Tension
          PHOTOGRAPH BY AUTHOR

the soundscape was constructed using sTVC – Space-Time-Variant Composition – which manipulates the perception of physical spaces from which a given piece seems to be sourced, as experienced by the listener. Echoes and reverb can be manipulated to give a sense that a piece's space is changing in real time, which brings about time in variance. *esceleer* seeks to exemplify that, while also juxtaposing sonic perceptions of the ephemeral and the infinite.[8]

Ascending or descending a staircase provides a metaphorically and physically appropriate context for negotiating a passage between the temporal and the timeless. This work was designed as an homage to J. T. Fraser. In writing about the experience of timelessness, Fraser cites Freud's remarks about a sensation of "eternity" that he described as an "oceanic" feeling of limitlessness and boundlessness (Fraser 2007, 183). As indicated already, Fraser characterized the experience of timelessness as a regression of the nootemporal (consciousness) to the eotemporal. Fraser framed his hierarchical theory as a reversal of Platonic philosophy and its ascent from time to the timeless.[9] Yet in the last years of his life, Fraser advocated for scholarship and conferences dedicated to the theme of "Time and Eternity."[10] His deep concern with mortality and the question of time and eternity is apparent in a letter simply addressed to "Paul and Anita," which he gave us at our wedding in 2005. The excerpt he chose to read us aloud is from Rollin J. Wells' popular ninteteenth-century poem "Growing Old":

A LITTLE MORE TIRED at the close of day,
A little more anxious to have our way,
A little less ready to scold and blame,
A little more care for a brother's name;
And so we are nearing the journey's end,
Where time and eternity meet and blend.

---

8   Quoted from the description of the work in the Time in Variance conference program.
9   See Fraser, "Out of Plato's Cave," and Steve Ostovich's essay of the same name in the present volume.
10   Fraser expressed his interest in work on "Time and Eternity" in informal correspondence and conversations, as well as meetings of the ISST Council.

3   Conference Installation: "Time in Variance"

This titular conference installation was placed in the Hannon Library foyer as
a welcome to participants. It spells out and embodies the conference theme
in divergent materials that evoke an Anthropocene entanglement between
human and geologic timescales.

FIGURE 5.5   Time in Variance
         PHOTOGRAPH BY AUTHOR

The Styrofoam and stones express the encroachment of expanded polystyrene foam (EPS) products into the geologic rock record. While the black stones and white foam present contrasting colors and clashing materials, both are formed by processes involving heat and pressure. However, these processes unfold on variant timescales: while igneous rock emerges within the ongoing rock cycle, foamy EPS converts fossil fuels into plastics that cannot be recycled. The collected stones, basalt shards from Scotland and California and igneous rocks from the Yuha desert express geologic variance in age and context. Their recontextualized placement and arrangement into letters in a library exemplifies human transformation of the lithosphere. As Clive Hamilton argues, while modern philosophy deems thinking to be our species' defining feature, the Anthropocene implies that it is "the *transformation* of the Earth, rather than the contemplation of it, that accords with humankind's true calling" (Hamilton 2017, 119).

4 Conference Installation: *The Gravity of the Situation*

The title of this installation evokes an Anthropocene sense of urgency, coupled with geologic volatility and fragility in a period of climate change, mass extinctions, and environmental degradation.

FIGURE 5.6 The Gravity of the Situation
PHOTOGRAPH BY AUTHOR

The precariously balanced triangular boulder evokes a shaftless arrow of time, as if time itself were hanging in the balance because these cataclysmic processes are disrupting human and inhuman cycles, rhythms, and temporalities. Echoing Perec's ephemeral/eternal polar pairing, the free-standing stone sculpture stages a contrast between ephemerality, embodied in the precarious balance of stones, and the deep time of rocks. Balancing stones demands a meditative suspension of attention, which opens an aporia where the ephemerality of precarious cairns commingles with the aeonic calm of absorbing rocks. Stone stacks, like mortal human life, persist in a temporality that could be thought of as a suspended sentence: because of the literal gravity of their situation, stacks are doomed to fall at an uncertain time, just as we know we are sentenced to die but usually do not know when or how. By extension, the Anthropocene is infused with a dread, an uncertainty as to how and when geologic history will subsume human history, leaving us as another extinction to be exhumed in the rock record.

5   Postscript: Post-Conference Installation: *Leveled*

This kind of Anthropocene dread played out dramatically in a post-conference postscript.

Just a week after the conference concluded, Southern California was shaken by an earthquake of magnitude 7.1, whose epicenter in Ridgecrest was 150 miles from L.A. Earthquakes, of course, epitomize the unpredictable, disruptive eruption of geologic time in human history. The quake toppled the balanced stone stack in the library foyer. (Interestingly, nothing else – no books, furniture, or people – fell.) Thinking about the fitting irony of "The Gravity of the Situation" being destroyed by geologic forces, I realized that the earthquake had created a new installation. I called this installation "Leveled" and characterized it as a collaboration, an iteration of an original installation by seismic vibration.

The new title evokes an Anthropocene sense of submission, coupled with geologic volatility and fragility in a period of climate change, mass extinctions, and environmental degradation. The balanced stack has been leveled by tectonic waves; sculptures of sufficiently delicately poised sensibilities are especially susceptible to them. The rocks fell as they were designed to in case of perturbances – away from the floor and visitors. The vertical spine stone toppled into a horizontal bridge, spanning what becomes a de facto waterway or moat, linking the shorestone island to a rugged basalt coastline. Yet this formal fit of fallen stones bespeaks a restful harmony belied by the telluric shearing

FIGURE 5.7 Leveled
            PHOTOGRAPH BY AUTHOR

that shook them down. Aeonic calm suffers instantaneous obliteration when geologic time erupts abruptly into human history. It left one humbled, taken down a notch from aspiring to commingle with rocks, and oddly grateful to have gotten to collaborate (through passive reverberation) with the earth on a public piece.

6   Gaiatemporality

The "Time in Variance" and "Gravity of the Situation" installations were designed to invoke the Anthropocene as a placeholder for "time in variance" in the twenty-first century. The Anthropocene serves as a metonym for the convergence of and entanglement among radically different timescales, as the geologic processes of deep time mesh with the economics of carbon capital and consumerism in fossil fuels and rare-earth minerals.[11] The gradual unfolding of history in the Holocene is ruptured by rapid changes in atmospheric and

---

11   The most-cited articulation of Anthropocene in this sense is Chakrabarty (2009).

oceanic chemistry and mass extinctions that confront us with prospects of a near-to-distant future far different from our present. This variance, a being-at-odds-with-itself or internally divergent and inconsistent, defies simple conceptual mapping. It might be roughly imagined as a multi-dimensional or perhaps fractal-dimensional iteration of "time out of joint." Michelle Bastian and Thom van Dooren aptly describe the Anthropocene as being "about foldings and pleatings. About simultaneous and contradictory temporalities, about the breakdown and (re)formulation of new multitemporal relations" (5).

The Anthropocene, as a historical rupture that confuses disparate temporal scales and relations, would seem to complicate or even dismantle the conceptual edifice of Fraser's hierarchical theory of temporal levels.[12] However, drawing on Fraser's method of defining temporal levels in terms of "constitutive conflicts," the Anthropocene could be understood as a conflict between the temporality of advanced, carbon-fueled capitalism that has driven "The Great Acceleration" on the one hand, and the planetary temporality of the geophysical earth system on the other. In what seemed a tentative postulation at the time, in a paper for the 1998 ISST conference I proposed amending Fraser's hierarchical theory to include a "Gaiatemporal" level. Citing Fraser's theorizing of a "Time-Compact Globe" in 1987, I argued that his temporal level of "Sociotemporality" should be augmented by an emergent "Globotemporality" proper to global capitalism, and proposed a new temporal level, "Gaiatemporality," "defined by the conflicts between and mutual imbrications of the human species' technological development and the planet's ecology" (Harris 2001, 37).[13] This formulation was informed by Michel Serres' prescient presaging of the Anthropocene in his 1995 book *The Natural Contract*. Serres described a turn in time when "Global history enters nature; global nature enters history," in which nature "becomes a global objective, Planet Earth, on which a new, total subject, humanity is toiling away" (5).

Clearly, "Gaiatemporality" does not align with Fraser's hierarchical theory in a simple fashion. It does not constitute a level that can be added to Fraser's tightly nested hierarchy, which is defined by an evolution of complexity as one traverses levels "up" from physical to living to cultural domains: "Nature comprises a number of integrative levels which form a hierarchically nested and evolutionarily open system along a scale of increasing complexity" (Fraser

---

12   The question of how to adapt or reconfigure Fraser's theory to Anthropocene temporality
       is also explored in David Wood and Lucia Cash-Beare's contributions to this volume.
13   On the notion of Gaia and the Anthropocene, see Isabelle Stengers's reflections on "The
       Intrusion of Gaia" (Stengers 2015).

1999, 26). This synchronic conceptual hierarchy could be translated into a diachronic process consistent with accounts of cosmic evolution such as Big History in which "emergence" plays a key theoretical role analogous to "conflict" in Fraser's theory. In both accounts, "the evolution of complexity" moves from nature to culture; the "highest" levels of complexity are human consciousness and collective learning. I have argued previously that Fraser chose to posit "a hierarchical theory of time" rather than a history of emergent levels of complexity because such histories arrange events along a timeline, thereby reducing "time" to a metric (Harris 2012). As Walter Benjamin writes in his "Theses on the Philosophy of History," "universal history has no theoretical armature. Its method is additive: it offers a mass of facts, in order to fill up a homogenous and empty time" (Benjamin 1969, 262). From Fraser's standpoint, a timeline would falsely imply or construct a "time" metric "in" which time evolves. In his 1978 essay "Temporal Levels: A Fundamental Synthesis," Fraser states:

> I [...] maintain that the evolutionary emergence of new stable structures of increased complexity is best understood as developmental steps along the evolution of time itself, rather than evolutionary steps in the framework of a pre-existing "absolute, true and mathematical time [which] of itself and from its nature flows equally without regard to anything else" (Newton, 1687).
>
> FRASER 2007, 160

There is a stark conceptual contrast between Fraser's hierarchical theory, which moves from physical matter and natural processes to human consciousness and culture, and the Anthropocene and Gaiatemporality, which are defined by the inseparability of nature and culture and the entanglement of geologic and human histories. Gaiatemporality thus does not "add" a level at the top of Fraser's hierarchy but represents a folding of time's evolution back on itself, as it were. This temporality intermixes the physical, geological, biological, and social, crumpling up linear time like a handkerchief (Serres 1995, 58). It marks an epoch that leaves time's arrow shaftless and hanging in the balance because it marks the dissolution of exponentially accelerating linear human history and its deposition in the geologic rock record. Gaiatemporality replaces the image of the present as a point on a timeline of evolving complexity with an image of a present nested within a set of concentric cycles rippling out across scales of past and future. Earth history shows that deep time's linear sequence unfolds in an oscillating rhythm of uplift and erosion, explosion and extinction, emergence and collapse, in which periods of stability are punctuated by upheavals and cascades of sudden changes. The Anthropocene, shaped by research into human traces in the rock record, is haunted by ghostly images of the traces humanity will have left in a distant future rock record (see Zalasiewicz 2008).

On the broadest conceptual level, what Fraser would term the "constitutive conflict" of Gaiatemporality could be understood as an imbrication or entanglement of linear and cyclical temporalities. If we wish to situate the concerns of an Anthropocene present in the context of the evolution of time, we would first need to reimagine Fraser's hierarchical theory of time in tandem with Big History's linear account of cosmic history. Both Fraser's theory and Big History provide a map of time's variance, its timescales and development, but their diagrams distribute these temporalities in a vertical conceptual schema and horizontal timeline, respectively. What cultural iconography might *complicate* – in its etymological sense of folding together, entangling, intertwining – these comprehensive conceptual comprehensions of time?

7   Charles Jencks, the Universe Cascade

I would suggest that The Universe Cascade, a construction by Charles Jencks in The Garden of Cosmic Speculation, expresses precisely such a complicated iteration of time imagined as a nested hierarchy of scales and a history of evolving complexity. Jencks wrote over thirty books spanning architectural history and theory and landscape design, and he is most widely known as a proponent of postmodern architecture.

In 1978, Jencks married Maggie Keswick, daughter of Sir John Keswick, knighted for work on Sino-Chinese trade relations. Keswick grew up partly in Shanghai and visited many imperial and private gardens, and eventually wrote a landmark book, *The Chinese Garden*, which also features an essay by Jencks. Jencks's landscape design and gardens integrate these traditions with a postmodern architectural sensibility and deep reading in the sciences (cosmology, evolutionary biology, fractal geometry). The Garden of Cosmic Speculation represents a synthesis of these interests and is among the most ambitious attempts to express a contemporary cosmology and attendant concept of time in a physical artwork.

Built on the Keswick family estate of Portrack House near Dumfries, Scotland, The Garden of Cosmic Speculation encompasses fields and woods bordered by the River Nith. Landscaping that eventuated in the garden began in 1989, when Keswick wanted a pond made for their daughter. As the project evolved, Jencks sought to develop a "new grammar of landscape design" (Jencks 2005, 32) to express his vision. Perhaps the most distinctive element in this grammar is what Jencks termed landforms, sculpted mounds with winding paths. Jencks

FIGURE 5.8 Charles Jencks with Liesegang Stones
PHOTOGRAPH BY AUTHOR

landforms may be found at many sites in Scotland, including Landform Ueda outside the National Gallery of Modern Art in Edinburgh, Mounds of Life at Jupiter Artland, and the Crawick Multiverse, a public landscape garden in the countryside near Dumfries.

The Garden of Cosmic Speculation features several separate gardens and installations that compose a complex whole. Many garden elements (or sub-gardens) are dedicated to scientific concepts, including The Symmetry Break Terrace and The Black Hole Terrace, The Quark Walk (presenting the Standard Model of particle physics), The Comet Bridge (speculations on comets as agents of cosmic evolution and a source of terrestrial life), The Fractal Bridge, and The DNA Garden of the Six Senses. There are also gardens expressing Scottish history (The Railway Garden, The Bloodline).

Clearly, Jencks's gardens are works of "temporal phronesis" (Wood 2019), inviting and inducing the visitor to think across multiple spatial and temporal scales. For Jencks, gardens serve as expressions of natural philosophy: "When you design a garden, it raises basic questions: what is nature, how do we fit into it, and how should we shape it where we can, both physically and visually?" Thus for Jencks a garden represents "a miniaturization, and celebration, of the

FIGURE 5.9   The Snail Mound, The Garden of Cosmic Speculation
PHOTOGRAPH BY AUTHOR

place we are in, the universe" (Jencks 2005, 17). For cosmic orientation, Jencks
turns to "the ideas of contemporary science," which he believes "provide the
basis for a cultural awakening and a new iconography," but that for this vision
to come to fruition, ideas "must be made more tangible through art if it is to be
assimilated" (20). Gardens, in his view, should not merely represent scientific
ideas – he holds garden design to the standard of a

> cosmogenic art that [...] layers ideas and patterns into a complex whole.
> The layers should make one slow down, think, and wonder about received
> notions. It should celebrate the beauty and organization of the universe,
> but above all resupply that sense of awe which modern life has done
> much to deny.
>
> JENCKS 2005, 248

There are two tenets of Jencks's philosophy of landscape design particularly
germane here. First, "A garden should not only present [a] worldview but also
heighten our relationship to it, through the senses" (2005, 5). Jencks playfully
expands the notion of the garden's appeal to the senses in different ways.

FIGURE 5.10   The Universe Cascade, The Garden of Cosmic Speculation
               PHOTOGRAPH BY AUTHOR

He created features called "Taking Leave of Your Senses" and "The Sense of
Fair Play" (a tennis court) and adds to the five physical senses the "sixth sense"
of women's intuition (162) and "the sense of humor" (35). The simple premise
of deepening understanding of abstract ideas by expressing them in tangible
forms that appeal to the senses presents an immense challenge: how can theories of time and matter be given material form? how can complex concepts

be expressed in physical elements and itineraries in a landscape? The second premise is Jencks's stipulation that, "Understanding demands a certain slowing of time – why else enter a garden?" (5). Jencks's garden design aesthetic encourages slowing down, suspending disbelief, and suspending attention through several strategies including meandering pathways (such as those winding up and down landforms), hidden surprises waiting to be discovered, enigmatic verbal and physical elements (ambigrammis, words etched in the ground that read the same upside down and backwards) and a formal grammar of "waves, twists, and folds" (2005, 5). Here, as before, walking and thinking intermingle in synergetic relations.

Situated on a steep hillside just below Portrack House, The Universe Cascade realizes Jencks's ambition to construct "a structure that would portray the story of the universe" (187).

Jencks's interest in this story as a "metanarrative" conveying the universe as a single unfolding event was influenced by Thomas Berry and Brian Swimme's *The Universe Story* (1992) and informed by interacting with scientists, including Lee Smolin, Brian Goodwin, Paul Davis, and Roger Penrose. Jencks's synthesis of different accounts and elements of cosmic evolution led him to an image of "the jumping universe," whose underlying conceptual principle is that "the cosmos unfolded as a series of symmetry breaks," marking fundamental changes in organization (189). The Universe Cascade presents the history of the universe in a series of 25 "jumps" marking seminal events in cosmic and terrestrial history. Each jump is depicted in a display composed of rock arrangements, industrial materials, and etched words or symbols. The rocks include large Chinese Taihu and Lingbi stones, and pebbles and small boulders Jencks collected in local rivers, including red jasper, greywacke, yellow and black rhyolite, and "Liesegang" stone.

The latter, also called Goethite, features red concentric rings made by iron pulsating in the sandstone. Jencks prizes these patterns as examples of self-organization, similar to that seen in the famous Beousov-Zhabotinski reaction in slime molds. On the one hand, geologic materials are fitting materials to embody an aeonic story: "What better than the artistic wonders of the cosmos – crystals, fossils, and rocks eroded into extraordinary clouds – to explain the cosmos?" (189). On the other hand, Jencks attests, "recounting the story of the cosmos with rocks and objects" presents a tremendous challenge, as "detail, narration, and nuance are severely limited by these media." The resulting compositions culminate in a creative process Jencks describes as "approaching the art of mosaic combined with Japanese rock arrangement, Constructivist collage, and word art (epigraphy)" (2005, 191).

FIGURE 5.11   Taihu Stone, The Universe Cascade
               PHOTOGRAPH BY AUTHOR

Structurally, The Universe Cascade combines the timeline of cosmic history with the nested levels in Fraser's hierarchical theory: one walks a path that marks events in the evolving cosmos, moving upwards through platforms/ levels of increasing complexity. In Jencks's marvelous conceptual diagram of cosmic history, the universe is rendered as a vertically oriented expanding light cone, held together by the four fundamental forces and observed by the eye of consciousness.

Time unfolds as "a spiraling evolution" (193) that integrates linear and cyclical movement. The diagram's vase-like form expresses the nested nature of time and evolving complexity, the dependence of each stage on what preceded it, in a way that a timeline does not.

Like any rendition of the story of the cosmos, the Universe Cascade reflects a series of selective choices about what to include and emphasize in the narrative. The Cascade's platforms include familiar fundamental "jumps" referenced in scientific and Big History narratives, such as the emergence of matter, formation of the Solar System, evolution of life and DNA, and human culture. Jencks's style of thinking and interpretation of cosmic history is evident in the more idiosyncratic platforms such as "Color, Music, Perfume," "Symbiosis," and "Feeling, Society, Gossip." His metanarrative sensibility is expressed in the "Eye of Consciousness" looking down from atop the structure, which signifies that "the mind's eye can grasp the universe, the imagination can see its narrative as a whole – up and down" (228).

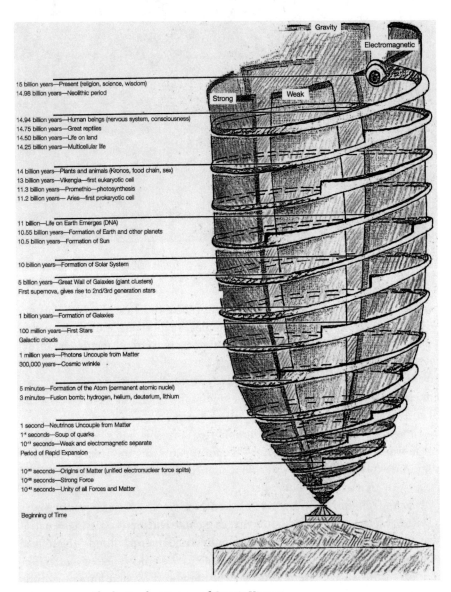

FIGURE 5.12   Charles Jencks, Diagram of Cosmic History
COPYRIGHT CHARLES JENCKS. USED BY PERMISSION OF LILY JENCKS

Consistent with Jencks's philosophy of landscape design, The Universe Cascade does more than "present a worldview" through scenarios that represent events or ideas. The structure also "heighten[s] our relationship to it, through the senses" because of what Jencks calls "delightful parallels" between fundamental tenets of cosmic evolution and the Cascade's architecture: "breaks in symmetry are like discrete levels; evolution in nature is like climbing a ladder of organization, with definite steps and platforms" (189). Its itinerary also intentionally impedes the viewer's progress to spur reflection: "Ascending the hill, one would be forced to go slow, both by the water channel cutting through the steps – forcing one to take care – and the difficulty of interpreting each stage of the jumping universe," reinforcing his view "that experiencing a garden should take time, that multiple meanings should slow down transition while they speed up appreciation" (189–190). Benches on platforms invite viewers to pause, rest the body and contemplate a "jump" in the cosmic story, as well as see it in relation to those already passed and those to come.

This immersive, embodied experience of traversing cosmic history resembles projects like *A Walk Through Time* (an exhibition of 95 illustrative panels accurately placed along a one-mile walk, where each foot represents one million years of evolution), which subsequently spawned a book (Liebes 1998) and a Deep Time Walk App. But Jencks's work sharply differs from this form of staging cosmic history as an educative itinerary. One fundamental contrast is that Jencks does not map time as a metric – one step does not correspond to a "length" of time. While *A Walk Through Time* traverses a quantitative, absolute time along which factual information is arranged and conveyed in pictures and prose, the 25 platforms of The Universe Cascade move through history unevenly: the first six platforms recount the first five minutes; the next four progress from 300,000 to 50 million to 100 million to two billion years. Rather than presenting written information about what was happening at a given date, The Universe Cascade shapes a story of seminal stages in cosmic evolution. By presenting the "jumps" in the story as enigmatic, emblematic assemblages on platforms at the ends of stairways, Jencks also shifts the sense of the narrative from following a sequence of equally weighted events (what happens next?) to arriving at and pausing to peruse a series of qualitatively unique scenarios (what is going on here?).

Jencks designed The Universe Cascade to depict a comprehensive scientific vision that "provide[s] the basis for a cultural awakening and a new iconography," answering his own call for the science to "made more tangible through art" so that it can "be assimilated" (20), internalized through the senses and therefore felt as well as understood. In ways that Jencks could of course not have anticipated, The Universe Cascade takes on new meanings and resonates

differently in a contemporary Anthropocene context. As Jencks's model of
The Universe Cascade clearly shows, the cosmic story unfolds along a meandering path where one retraces steps in order to progress, representing "a universe of forked paths, progress and regress, of two steps forward and one step
back" (191).

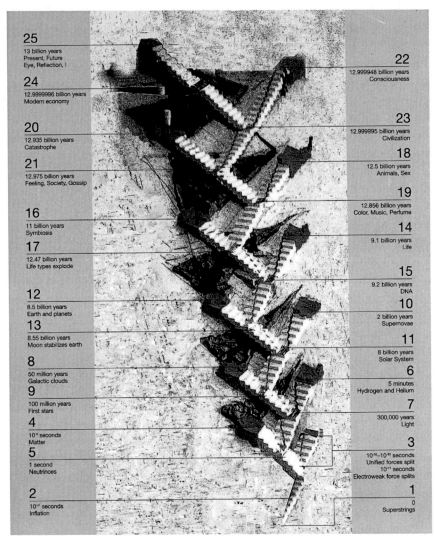

FIGURE 5.13   Charles Jencks, Diagram of The Universe Cascade
COPYRIGHT CHARLES JENCKS. USED BY PERMISSION OF LILY JENCKS

This non-linear, bifurcating itinerary provides an embodied experience of a complex temporality, a time in which the "forward progress" evocative of emergence, creation, and evolving complexity is complicated by the "regress back" connoting collapse, extinction, and entropic dissolution. In this sense, the Cascade conveys a concept of time consistent with Gaiatemporality's "foldings and pleatings," "simultaneous and contradictory temporalities," and "the breakdown and (re)formulation of new multitemporal relations" (van Dooren and Bastian, 5).

This "reading" of the Cascade – an interpretation that coalesced out of various impressions, associations, and drifting thoughts – came to me on my most recent visit in 2018. I remember being struck by how different this *sense* of the experience of walking the Cascade was from my first time doing so some years before, when I imaginatively projected the climb up the hillside onto the narrative of qualitative jumps and increasing complexity. Of course, the shifting meanings that a garden assumes over time are an integral part of their beauty and value, and why they nurture "temporal phronesis" in ways that other forms of expression do not.

8   Conclusion

The essay has proposed a slow time practice and philosophy in which a suspension of conscious attention enables a suspension of conceptual tensions among variant timescales. The efficacy of this approach in deepening a "temporal phronesis" is of course a matter of conjecture and subjective experience. But making time for mindfulness and being able to be mindful of radically disparate temporalities both seem more important than ever, for one's own and the planet's wellbeing. Slow time as delineated here represents a contemporary form of "self-care" grounded in "the care of the self" (Foucault 1988) in Stoic philosophy and thought exercises. The Anthropocene as a rupture, our period as a possible "sixth extinction," has given rise to an apocalyptic discourse of crisis. As Joseph Masco observes,

> [i]f you tune in to the mass-mediated frequency of crisis today, it quickly becomes overwhelming [...] the everyday reporting of crisis proliferates across subjects, spaces, and temporalities today and is an ever-amplifying media refrain.
>
> MASCO 2017, 65

Consequently, it is difficult not to experience the present period as a *"convergence of catastrophisms,"* in the pithy phrase of Richard D. G. Irvine (2020, 168). Irvine makes a plea to resist the pull such a sense of time exerts on us, a wrenching out of temporal continuity altogether:

> we need to adopt a time perspective that enables us to perceive actions in the present with the fully diachronic recognition that they are actions in geological history. From this perspective, to speak of a coming catastrophe is therefore not to cut ourselves off in time but to project ourselves into the future in order to recognise our responsibilities in relation to it.
>
> IRVINE 2020, 170

Mindfulness, suspending attention, attending to the present moment, is a first step in shaping and deepening such a perspective. And within this perspective, to hold in suspended tension, the multiple dimensions of time in variance.

References

Bastian, M., van Dooren, T. 2017. "The new immortals: Immortality and infinitude in the Anthropocene." *Journal of Environmental Philosophy* 14(1).

Benjamin, Walter. 1969. *Illuminations*. Edited and with an introduction by Hannah Arendt. Translated by Harry Zohn. Schocken Books.

Chakrabarty, Dipesh. 2014. "The Climate of History: Four Theses." *Critical Inquiry* 35 (Winter 2009): 197–222.

Foucault, Michel. 1988. *The care of the self*. Translated by R. Hurley. New York: Vintage.

Fraser, J. T. 2007. *Time and Time Again: Reports from a Boundary of the Universe*. Leiden: Brill.

Hadot, Pierre. 1995. *Philosophy as a Way of Life: Spiritual Exercises from Socrates to Foucault*. Translated by Arnold Davidson. Wiley-Blackwell.

Hamilton, Clive. *Defiant Earth: The Fate of Humans in the Anthropocene*. Polity Press, 2017.

Harris, Paul A. 2001. "Ten Soundbytes for the Next Millenium: Mutations in Time, Mind and Narrative." In J. T. Fraser and Marlene Soulsby, eds. *Time: Perspectives at the Millenium*. Westport, CT: Greenwood Publishers, 2001: 35–48.

Harris, Paul A. 2012. "Time and Emergence in the Evolutionary Epic, Naturalistic Theology, and J. T. Fraser's Hierarchical Theory of Time." *Kronoscope: Journal for the Interdisciplinary Study of Time* 12(2) (2012): 147–158.

Harris, Paul A. 2014. "Tracing the Cretan Labyrinth: Mythology, Archaeology, Topology, Phenomenology." *Kronoscope: Journal for the Interdisciplinary Study of Time* 14 (2014): 133–149.

Harris, Paul A. 2018. "In the Labyrinth of Slow Time: 'A Perturbation in the Deep Stream' (David Mitchell, co-author) and 'A Perambulation in the Deep Stream'." In *C21 Literature: Journal of 21st-century Writings* 6(3) (Fall 2018). doi: https://doi .org/10.16995/c21.61.

Irvine, Richard D. G. 2020. *An Anthropology of Deep Time: Geological Temporality and Social Life*. Cambridge University Press.

Jencks, Charles. 2005. *The Garden of Cosmic Speculation*. Frances Lincoln.

Kern, Hermann. 2000. *Through the Labyrinth: Designs and Meanings over 5,000 Years*. Prestel Publishing.

Liebes, Sidney, Elisabet Sahtouris, and Brian Swimme. 1998. *A Walk Through Time: From Stardust to Us – The Evolution of Life on Earth*. Wiley.

Masco, Joseph. 2017. "The crisis in crisis." *Current Anthropology* 58(S15) (2017): 65–76.

Meldahl, Keith Heyer. 2015. *Surf, Sand, and Stone: How Waves, Earthquakes, and Other Forces Shape the Spouthern California Coast*. University of California Press.

Savary, Louis M. 2010. *The New Spiritual Exercises: In the Spirit of Teilhard de Chardin*. Paulist Press.

Serres, Michel. 1995. *The Natural Contract*. Ann Arbor: University of Michigan Press.

Serres, Michel, and Bruno Latour. 1995. *Conversations on Science, Culture, and Time*. University of Michigan Press.

Solnit, Rebecca. 2001. *Wanderlust: A History of Walking*. Penguin.

Stengers, Isabelle. 2015. *In Catastrophic Times: Resisting the Coming Barbarism*. Open Humanities Press.

Walker, Michelle Boulous. 2017. *Slow Philosophy: Reading against the Institution*. Bloomsbury Academic, an imprint of Bloomsbury Publishing, Plc.

Wood, David. 2019. *Deep Time, Dark Times: On being Geologically Human*. Fordham University Press.

Wu, Tim. 2016. *The Attention Merchants: The Epic Scramble to get Inside our Heads*. Knopf.

Zalasiewicz, Jan. 2008. *The Earth After Us: What Legacy Will Humans Leave in the Rocks?* Oxford University Press.

PART 2

Variant Narratives

∵

CHAPTER 6

Temporal Otherness and the "Gifted Child" in Fiction

Adam Barrows

Abstract

This essay explores the representational trope of the "Disabled Time Child": the child with disabilities who is "gifted" with superhuman or extrasensory powers involving time and temporality. Often occurring in the genres of science fiction, fantasy, and horror, the disabled child who occupies a disjunctive temporality could be understood as a mechanism whereby creative literature is able to confront alternative, non-human forms of temporal being. Situating this representational figure within cultural histories of the child and of disability, I argue that the trope depends upon highly problematic associations between childhood and disability alike as forms of alien otherness. Examining the work of science fiction writer Clifford D. Simak as a case study, I explore the ways in which the trope of the "Disabled Time Child" recapitulates racial fantasies of the primitive in an attempt to gesture towards ecologically inflected futurities.

Keywords

childhood – disability – disjunctive temporality – futurity – science-fiction – Clifford D. Simak – temporal primitivism

The title of the seventeenth triennial conference of the International Society for the Study of Time, "Time in Variance," gestures towards temporalities and time-scales that are at odds with one another or that exist in disjunctive relationship to one another. This essay explores a narrative trope of fiction and film that combines a number of those disjunctive temporalities into one figure: the disabled child whose disability grants the child fantastical temporal powers, effectively rendering him or her temporally alien or non-human. The intellectually or physically disabled child who can see through time, who is haunted by ghosts, or who makes contact with aliens existing on another temporal plane is a recurring figure, especially in the genres of science fiction, fantasy, and horror. Situating this representational trope within cultural histories of childhood and of disability, I examine as a case study the work of Clifford D. Simak, a science fiction author of the 1950s and 1960s whose work made frequent and productive use of the trope.

Simak's fiction is significant in that it represents disability as a gateway to non-human or ecological modes of temporal belonging, thus vividly demonstrating the close associations between childhood, temporal giftedness, and non-human temporality that continue, in more contemporary representations, to be bound up with the figure of the "Disabled Time Child."

While the temporal experience of disability has been a persistently compelling subject of disability writing and activism since Alison Kafer's discussion of "crip time" in her book *Feminist, Queer, Crip* (2013), it has thus far had little impact within time studies more broadly.[1] This essay attempts to redress that gap while also making sense of a pervasive cultural trope in popular fiction and film. While that trope represents an attempt to think beyond narrow definitions of human temporality and its limits, it also depends upon some problematic associations between disability, childhood, and primitivism that time scholars might not be eager to embrace uncritically.

1   The Disabled Time Child

Children with intellectual or physical disabilities have long been represented in fictional narratives as temporal "others" with access to types of temporal experience outside of human frameworks.[2] Children "gifted" with the ability

---

1   Kafer, *Feminist, Queer, Crip*, 25–46. A recent collection of essays on time and literature from Cambridge, for example, provides an impressively comprehensive representation of approaches to time in literary criticism, but disability does not feature anywhere in the collection (Allen, *Time and Literature*).

2   The trope is certainly not limited to children, extending as it does to any and all disabled persons in literature who perceive time differently or whose relationship to the conventional temporal plot of a narrative is itself disabled. Michael Bérubé identifies this fictional scenario of a disabled temporality as an "intellectual disability chronotope" whereby the disablement creates a "productive and illuminating derangement of ordinary protocols of narrative temporality" (Bérubé, *The Secret Life of Stories*, 83). In the case of children, however, the disability chronotope manifests itself uniquely as a source of power or strength on the part of children to intervene in the narrative, in some cases heroically but just as often diabolically, as in the case of the horror television and film trope of the evil child or bad seed, whose very existence is a temporal throwback to archaic or pre-civilizational cultural forms with all their accompanying savage superpowers. See, for example, the fictional character of Ben Lovatt, a violently powerful homunculus child with temporal links to a distant pre-human past, whose

to communicate with ghosts from the past or entities from the future, or the ability to make first contact with alien beings that exist on alternate temporal planes are longstanding generic tropes.[3] As one familiar example, the mythopoesis of the *X-Men* comic book series depends upon the association of disabled children with a variety of superpowers involving other forms of temporal being. One of the first students enrolled in "Dr. Xavier's School for Gifted Youngsters" is Jean Grey, a young girl who is psychiatrically disturbed but under Xavier's tutelage learns to hone her powers of telepathy and telekinesis.[4] In a number of X-men plots, Jean uses her telepathic gifts to see into the future and in particular into the futures of some of her teammates.[5] In Philip K. Dick's 1964 novel *Martian Time-Slip*, to give another example, a young autistic boy named Manfred is coveted and manipulated by unscrupulous capitalist speculators on Mars, who recognize that Manfred's autism gives him the power to see into the future and thus reveal future untapped markets for land development. Manfred is ultimately sheltered by and integrated into an outcast society of indigenous Martians, who seem to share with the autistic Earth boy a power of temporal prognostication and manipulation. This is a common association: between the presumed temporal otherness of non-Western or indigenous peoples and the disabled Westerner (who is invariably either a child or child-like in mental capacities as traditionally understood). In George R. R. Martin's *A Song of Fire and Ice* novels (adapted into HBO's *Game of Thrones* television series), Bran Stark seems to be a "normal" child until he is physically crippled after being pushed from a window ledge. When he loses the use of his legs, he simultaneously gains powers of foresight that link him paranormally to the old

---

childhood and adolescence is chronicled by Doris Lessing in *The Fifth Child* and *Ben, in the World*. The Evil Child trope has been analyzed by Karen J. Renner, *Evil Children*.

3 I use "gifted" here not in the sense of children with exceptionally high IQs, as defined by narrow psychometric methods, but rather in terms of what educational psychologists have identified as the "special needs" child who is simultaneously "gifted" with particular talent, intelligence, task commitment, and/or creativity. The terms in educational policy and theory are: G/LD (Gifted Students with Learning Disabilities) or the "twice-exceptional student" (Beckmann and Minnaert). It should be noted, however, that there exist robust critiques against the models of normalcy in education and in child development upon which such terminology depends. See, for example, Karmiris, "De-centering the Myth."

4 Jean's psychological problems stem from her psychic link with a childhood friend whose pain she feels after her friend is struck in a hit and run accident (Claremont, *Bizarre Adventures*). Her early days with Dr. Xavier are chronicled by Roy Thomas, *X-Men*.

5 See, for example, her time travel into an alternate future involving Cyclops and her future daughter, Rachel (Lobdell, *Adventures of Cyclops and Phoenix*).

gods of Westeros and also to the minds of animal beings, including notably his adult intellectually disabled companion/servant, Hodor.[6]

The close associations between cultural ideas of the paranormal, indigeneity, childhood, and physical or mental disability are clearly on display in this old representational cliché that I am calling the *Disabled Time Child*. They are so deeply entrenched that it is often hard to distinguish the terms from one another or to trace a narrative of causal relation. While in the case of Bran Stark, disability seems to lead causally to the acquisition of his paranormal abilities and thus to his association with indigenous powers, in many other examples the terms seem to presuppose each other: to be a child is already to be disabled, to be disabled is already to be paranormal, to belong to an alienated indigenous population is already to be childlike, et cetera. This is the reason it is often fruitless to ask which came first, the paranormal powers or the disability, since in the case of Jean Grey and indeed all of the X-Men, it is clear that their powers *are* their disabilities and vice versa (Jean is only psychiatrically disturbed as a child because her untrained telepathic powers force her to share in the agony of a friend's death when that friend is killed in a hit and run incident).[7] This accounts for a character like Danny Torrance in Stephen King's *The Shining*, who is never explicitly identified as cognitively disabled but whose paranormal power to "shine" renders him socially abnormal and thus in a sense disabled.[8] To be in touch with the temporality of the ghosts of the past or aliens of the future is often a vivid metaphor in these texts for what it is like to be disabled.

In some ways, we can understand the figure of the Disabled Time Child as simply a translation into the genres of science-fiction, horror, and fantasy of a representational figure that has been frequently critiqued by disability activists and scholars: the so-called "supercrip."[9] People with disabilities are only acceptable and tolerable in normative discourses and institutions, according to this critique, if their disabilities can be prosthetically supplemented with some form of superpower or achievement. We praise and idealize Paralympic

---

6   Bran's manipulation of Hodor is an especially pernicious use of temporal superpowers, when it is revealed (in the HBO television adaptation) that Hodor's name and identity have been bestowed upon him by Bran when Bran time travels to the past, enters Hodor's mind, and instills him with a new name, a purpose in life, as well as a predetermined cause of death: all of which constitute the act of "holding the door" that is keeping Bran away from an army of the dead pursuing Bran and his companions in the present (Benioff and Weiss, "The Door").

7   Claremont, *Bizarre Adventures*.

8   Danny's mother, Wendy, fears that his shining will tip over into autism, a word that "frightened her; it sounded like dread and white silence" (King, *The Shining*, 219).

9   The term "supercrip" dates from the 1990s (also the era of the passage of the Americans with Disabilities Act). For an overview of the term, see Shapiro, *No Pity*, 16–18.

athletes, for example, or disabled people who become academic super-achievers because they compensate for their disabilities with these achievements, but we have a hard time with disabled people who do not subsequently become "gifted" as a compensation for their disabilities. The "cripple" must become a superhuman form of individual in order to be accepted or legitimized. This has particularly been the case with disabled children, who are routinely labeled with the euphemism "special." An entire branch of new age philosophy depends upon the idea that mentally or physically disabled children are remnants of a hyper-real fantasy world: this is the so-called "crystal" or "indigo" child, whose alleged otherworldly attributes often match many of the symptoms on the autism spectrum disorder.[10] This reading of disability as something that can be compensated for through the acquisition of extraordinary or other-worldly capabilities features not only in numerous comic book superhero backstories but also in public service campaigns that use the figure of the *supercrip* to celebrate the heroism of overcoming disability.

2 Disability and Non-human Time

Is the supercrip fantasy the only explanation for the popularity of this figure of the Disabled Time Child? Or is that figure doing additional cultural work in terms of translating non-human temporality into human terms? In his recent book *The Secret Life of Stories*, the literary scholar of disability Michael Bérubé argues that intellectual disability is often the mechanism whereby fictional narratives are able to gesture towards the limits of intelligible human time and temporality. Disability in narrative could be understood as a Bakhtinian "chronotope" that, in his words, "offers an outlet into realms of temporal experience that exceed human perception, bringing animal consciousness and/or geological time into play."[11] Bérubé acknowledges that the linking of disabled otherness to paranormal temporal indigeneity is "a bit embarrassing, politically," as in the case of Dick's *Martian Time-Slip* and what Bérubé jokingly refers to as the "Subaltern subchannel" that the indigenous Martians in that novel use to communicate telepathically with an autistic Earth child, mimicking as it does the trope of the "magic negro" so familiar from Hollywood cinema.[12] Yet, despite how closely bound up disability discourse has been with racist and eugenicist

---

10   For representative books in this new age tradition of thought, see Carroll and Tober, *Indigo Children*. See also Virtue, *Crystal Children*.
11   Bérubé, *The Secret Life of Stories*, 84–85.
12   Ibid., 100.

narratives throughout history, Bérubé suggests that it is still imperative to try to understand the ways in which the disability chronotope in narrative – and disabled narrative practice more generally – helps us to "imagin[e] other ways of being human that transcend the limitations of our own space and time."[13] Indeed, if we are to take seriously the challenge that disability criticism and activism poses (that we rethink sociocultural models of space and time that have enabled some bodies while disabling others), then it is incumbent upon us as time scholars to take the disability chronotope seriously.

If disability really does provide a means of conceptualizing or imagining temporalities that go beyond our limited and socially constructed ideas of what it means to be human, it is going to be difficult to extricate those conceptions or imaginations from the cultural history of childhood itself. Disability and the child are inextricably intertwined in so many narratives of disability. Indeed, many of the examples Bérubé uses to exemplify the disability chronotope in literature are characters who are either children or who are treated as children (by the narrator and/or other characters) because of their intellectual disabilities. Faulkner's Benjy Compson from *The Sound and the Fury* or Daniel Keyes' Charlie Gordon from *Flowers for Algernon* stand as "child-adults" alongside child characters like the autistic Manfred or Mark Haddon's protagonist in *The Curious Incident of the Dog in the Night-Time*.[14] Unpacking the theoretical relationship between disability and nonhuman time, then, requires a sustained engagement with cultural histories of childhood. Eighteenth-century conceptions of children as equivalent to the wild, the untamed, or the "savage" were hallmarks of both English Romantic poetry and post-Enlightenment educational reform theory. Children who are possessed of and by the powers of elemental non-human forces, before they are timed and tamed by adult civilization, have appeared throughout the literary history of the past two centuries or more, from Wordsworth's and Coleridge's *Lyrical Ballads* through the early twentieth-century weird tales of Algernon Blackwood (whose story "The Temptation of the Clay" features a child literally possessed by an entire forest) to the trope of the "evil child" so pervasive in the horror film genre.[15] Making sense of the disabled child as a gateway to different forms of temporal being necessitates an engagement with the extent to which children more generally

---

13   Ibid., 116.

14   Ibid., 73–83; 79 and 129; 128–138.

15   See for example, "The Idiot Boy," one of Wordsworth's contributions to the original *Lyrical Ballads* collection, and Rousseau for the founding text of Romantic educational theory. Rousseau's hypothetical protégée, Emile, is frequently compared throughout that text to a hypothetical "savage," with the latter often serving as an example of natural instincts, virtues, and capacities that little Emile should be allowed to cultivate.

have been cast since the Enlightenment as temporally outside of adult civilizational norms and standards.

3   Early Child Psychology and Temporal Primitivism

In the earliest days of child psychology as a codified field of scientific inquiry at the end of the nineteenth-century, researchers argued that the best analogy for understanding the behaviors and practices of children were the behaviors and practices of savage or primitive humans. Constant references to the savage in the construction of childhood as a field of scientific study linked child psychology to a host of nineteenth-century Western sciences dependent upon racist and colonialist schemas of global difference, such as anthropology and physiology.[16] James Sully, for example, in his foundational *Studies of Childhood* from 1896, compares figure drawings of children under the age of 10 to cave drawings of the human figure by prehistoric adults, pointing out the similarities in anatomical figuration.[17] What is most important about infancy, for Sully, is its "primitiveness."[18] The first years of a child, he writes,

> with their imperfect verbal expression, their crude fanciful ideas, their seizure by rage and terror, their absorption in the present moment, acquire a new and antiquarian interest. They mirror for us, in a diminished distorted reflexion no doubt, the probable condition of primitive man.[19]

The temporal dynamics of statements like these are revealing in that they suggest a kind of time traveling that is implicit in the study of children: children bring out the antiquarian in us in that they bring into the present an antiquated past; thus, the very being of the child embodies disjunctive temporality. For Sully, the child's individual development recapitulated the ascent of the child's ancestors from primitivity to Enlightenment, a developmentalist version of

---

16   Certainly, Sigmund Freud's characterization of infancy as marked by the primordial desires of an Id that is only tempered by ego formation through development and external conditioning plays a role in the evolving narrative of childhood's association with the primitive and with antiquity – although as the work of James Sully (which predated most of Freud's major texts by a decade) demonstrates, such ideas were part of the zeitgeist well before Freud was articulating his ideas on the Ego and the Id in his paper of the same name from 1923.

17   Sully, *Studies of Childhood*, 331–398.

18   Ibid., 4.

19   Ibid., 8–9.

what was then a widely accepted biological fallacy that ontogeny recapitulated phylogeny. Sully, whose concept of evolution seems more Lamarckian
than Darwinian, posits that each new generation of infants starts a little further along the path out of savagery:

> It gives a new meaning to human progress to suppose that the dawn of
> infant intelligence, instead of being a return to primitive darkness, con-
> tains from the first a faint light reflected on it from the lamp of racial
> intelligence which has preceded; that instead of a return to the race's
> starting point, the lowest form of the school of experience, it is a start in
> a higher form, the promotion being a reward conferred on the child for
> the exertions of his ancestors.[20]

Sully suggests here that the child serves to exemplify, in its very being, the
primitive struggles of its distant ancestors. While much of his work has been
supplanted by more sophisticated, although no less teleologically driven or
colonially inflected models of child development,[21] that idea of the child as
intrinsically linked to primitive forms of ancestral temporal being has persisted
in twentieth and twenty-first century imaginations.[22] I want to suggest that the
trope of the Disabled Time Child is at least in part the product of these ideas. It
vividly captures the racially inflected temporal politics that continue to linger
in child studies, in which the child is somehow not temporally coexistent or
coeval with adult human beings, but rather brings into the adult's present time
a glimpse, sometimes piquant and sometimes unsettling, of a distant ancestral past, with all its attendant strange and savage gods and powers. Rendering

---

20   Sully, 9. In a sense, Sully's antiquarian treatment of the temporally primitive child trans-
      lates into the nascent field of child psychology the kind of racist and colonialist temporal
      frameworks that Edward W. Said analyzed in *Orientalism* or that Johannes Fabian dis-
      cusses in *Time and the Other* in that, in the discourses of orientalism and of anthropology,
      cultural difference is often translated into the language of temporal disjunction. In Sully,
      it is the strange culture of the child that is made analogous to the anthropological or ori-
      ental other.
21   The persistence of such models has motivated some educational theorists to engage in a
      "decolonization" of educational theory. See, for example, Erica Burman, who summarizes
      efforts to delink human "growth" from the discourse around civilizational advancement
      in "so-called less developed" states (Burman, "Child as Method," 13).
22   These ideas lend, for example, persistent pathos and eeriness to the figure of the child
      ghost or haunted child in fiction and film, whose haunting evokes times and temporali-
      ties that have been suppressed or repressed in the historical record. A powerful example
      of the child ghost as reminder of historical repression is the figure of Beloved, who stands
      as a reminder in the present of generations of slavery, in Toni Morrison's novel of the
      same name.

this child figure disabled accomplishes one of two things for writers: one, it heightens and makes particularly traumatic or pathos-laden the temporal disjunction already latent in the so-called healthy child, since the disabled child signifies a broken link in teleological narratives of successful child and thus racial development; and/or two, it can serve as an easy figuration of exaggerated racial or ancestral progress into futurity in that the disabled child will, in true supercrip form, often become even more powerful or far-seeing than the temporally limited healthy humans. That kind of figuration is rife in science fiction and fantasy narratives, where the disabled child becomes the great hope of humankind in that he or she is often the only character in the story who can communicate with the alien and thus mediate or negotiate for the human race. In the latest *Predator* reboot, for example, Rory, a child with Aspergers Disorder can communicate with the Predator and is described in the film as being, because of his disability, "the next evolutionary step."[23] In this way the genetic or ancestral temporal regression that disability signifies in developmentalist notions of childhood is turned on its head, with the disability becoming a sign of future health and ability for the species.

4   Simak, Time, and Disability

All of these ideas of disability, time, childhood, and futurity are bound up in the work of the mid twentieth-century science fiction grand master, Clifford D. Simak, whose body of fiction (consisting of 30 novels and more than 100 short stories) often draws on the same ideas: disabled children or childlike adults, time travelling aliens, and wild or untouched natural landscapes. In some senses, Simak may seem like a strange choice for a case study in that his readership has declined since the 1960s (although his collective body of work has been in the process of being reprinted over the past few years). Yet Simak, whose popularity in the late 1950s and early 1960s coincided with a dramatic rise in the number of reported cases of autism in children, can be seen as one of the forefathers of this trope in its early literary articulation. That spike in autism cases, arguably corresponding to changing practices in psychiatric epidemiology,[24] increasingly prompted fiction writers to postulate autism

---

23   The character Casey, discussing Rory, states: "A lot of experts think being on the spectrum's not a disorder. Some think it might even be the next evolutionary step" (Black and Dekker, *Predator*, 77).

24   See Evans, "How Autism Became Autism."

as an alternate mode of temporal being.[25] In Simak's work, we see not only autism but cognitive disability more generally as pretexts for the "giftedness" of a variety of temporal powers: the ability to time travel, receive precognitive visions, or communicate with aliens on alternate temporal planes. More particularly, we find in Simak's fiction a clear articulation of the impulse that Bérubé associates with the disability chronotope: to find in human disability a gateway to articulating and envisioning non-human or ecologically inflected forms of temporal being. Simak's version of "pastoral science fiction," in which aliens and other science-fictional elements are incongruously located in rural settings and among simple country folk, deeply informs his ethos as a writer. The disabled have a natural association for Simak with the wild and untamed landscape of his native southwestern Wisconsin (still today referred to as the Driftless region for its having been largely untouched by the glaciations of the Pleistocene era). In that sense, Simak more acutely captures than any other writer of his generation the close associations between childhood, temporal giftedness, and non-human temporality that are bound up in the figure of the Disabled Time Child.

In many ways, the exemplary Simak text for exploring the trope is his 1965 novel, *All Flesh is Grass*. The novel is about a species of sentient flowers that invades the earth by establishing a landing point in the small town of Millville, Wisconsin (where Simak was actually born and raised). The flowers, which can see through the "time matrix" that encompasses multiple parallel earths, are able to enlist, forge telepathic links with, and speak through the mouths of, only certain individuals who are receptive to their impulses, the most receptive of which is a cognitively disabled young man named Tupper Tyler.[26] Tyler, characterized by the narrator in alarmingly offensive terms as a "slobbering, finger-counting village idiot" in fact possesses "some sensory perception that the common run of mankind did not have [...] a form of compensation, to

---

25   Certainly, the persistent tendency to represent people with autism as temporally other
       has some foundation in the demonstrated symptomology of so-called "Autism Spectrum
       Disorders" (ASD). As the autistic writer Naoki Higashida argues: for people with autism,
       "time is as difficult to grasp as picturing a country we've never been to" (*The Reason
       I Jump*, 63). Higashida mentions that "time intervals and the speed of time" are difficult
       for him to gauge and indeed, as Falter and Noreika note, "there are anecdotal and clinical
       reports suggesting that individuals with ASD lack an intuitive sense of time [...]. Instead,
       they have an abnormal experience of passage of time [...] and an abnormal perspective of
       themselves in the past, present, and future" ("Time Processing," 577). Certainly the mani-
       festation of these symptoms is in part what sparks the imagination of creative writers
       to make comparisons between "abnormal" autistic time and the time(s) of the alien or
       non-human.
26   Simak, *All Flesh is Grass*, 117.

make up in some measure for what he did not have."[27] Tupper's ability to speak with and for the alien flowers puts him in a sense within their time matrix and outside of the temporal boundaries and constraints of human subjective time in Millville. He is primitive in the sense that Sully characterized childhood as fundamentally primitive. Despite his biological age (he is in fact older than the narrator), Tupper is eternally childlike. He ran away from home as a child and we are first introduced to him through the eyes of his grieving mother, whose temporality appears to be frozen in the moment that her son disappeared, rendering him forever as a lost child who had "just stepped out the door."[28] Tupper's age "made no difference at all," the narrator explains, because "in his own mind Tupper had never outgrown childhood."[29] This alienation of Tupper from adult human rational temporality aligns him with the literally alien vegetation that is trying to get a stranglehold on earth. Tupper's disjunction from human temporality and engagement with, or enlistment on behalf of, a non-human botanical form of vegetal temporality demonstrates for Simak the potential of the figure of the Disabled Time Child in terms of conceptualizing alternative forms of time. Yet Simak's characteristically degrading character-ization of Tupper as a "village idiot" indicates the extent to which the disabled have to be radically othered and dehumanized in order for the trope to work.

The figure of the disabled child-man appears again in Simak's later novel *Mastodonia* in the character of Hiram, a "simpleton" who, like Tupper, is the only character able to communicate with a crash-landed alien being (nick-named Catface for its feline appearance) that has the power to open up cor-ridors to any point in past or future time. In this case, the alien temporal being serves not as invasive threat but rather as financial opportunity, since the narrator (by means of Hiram's child-like communion with Catface) exploits the time corridors in order to profit from high-ticket dinosaur hunting safari expeditions. In another example, from perhaps his most well-known novel, the Hugo-Award-winning *Way Station*, the price at stake in communing with alien temporality is the fate of the universe itself, when a galactic war is almost sparked by the human kidnapping of the body of an alien delegate. In *Way Station*, a deaf child-like young woman (usually referred to as "the girl" despite her age being given as twenty-two in the text) serves as the hyper-empathic communicator with an alien emissary, who has come seeking a representative of the human race able to carry a talisman that will prevent galactic war. Lucy, the "beautiful deaf-mute," who is also referred to as a "wild thing" and a "little

---

27   Ibid., 138.
28   Ibid., 18.
29   Ibid., 19.

fairy person," has a profound affinity with the natural world as well as with alien worlds beyond the stars.[30] Indeed, her "deaf-mute" status is made explicitly analogous to that of a greater alien world that struggles to be understood by provincial human beings (for example, one of the alien races in the text is described as a "sightless, deaf, and speechless race from the mystery stars of the far side of the galaxy."[31]).

In all of these texts, the intellectually disabled child-like character holds the key to communicate with the aliens and thus provide the human race with the existential choice to advance or regress. While the males, Tupper and Hiram, are clichéd village idiots and Lucy is a typical ingénue in distress, all of them embody pre-modern or ancestral primitive traits (typified by their status as throwbacks to an eternal childhood) that Simak, a committed pastoralist, would like to see projected into the future. Simak is clearly operating under the supercrip assumption that all disabled people must have compensatory superpowers that link them with animals, aliens, or other non-humans. This ecologically inflected characterization of the eternal child is essential to the disability chronotope in Simak's work.

5   Disability, Ecology, and the Driftless Region

Although Simak lived most of his life in the Minneapolis suburbs as an editor for the *Star Tribune*, almost all of his narratives are set in the rural southwestern Wisconsin region of his birth, which he describes in a letter to his editor at Doubleday as

> the most beautiful country you have ever seen – high sharp hills, deep ravines, forests, and the Wisconsin and Mississippi rivers. The land is old. It stands in the Wisconsin driftless area and was never touched by the Pleistocene glaciations. It stands much as it has stood since Mesozoic times.[32]

---

30   Simak, *Way Station*, 121, 45.
31   Ibid., 44.
32   Letter from Clifford D. Simak to Lawrence P. Ashmead, January 10, 1966. Clifford D. Simak papers, Minneapolis: University of Minnesota archives. Thanks to the Elmer L. Anderson Research Scholars Program for funding my trip to the archives.

For Simak, the Driftless region's beauty lies in its ecological disjunction from the rest of the Midwest. It is a temporal throwback, untouched not only by modernity and urbanization but also by the processes of glaciation that defined the rest of the contemporary landscape. Disability or childishness (for Simak, as for so many writers, they are one and the same thing) renders his "simple" characters at one with their temporally thrown-back topography. They are more part of the animal and vegetable worlds than they are human. Lucy and Hiram are first introduced to the readers in scenes of communion with non-human creatures (Lucy with a butterfly and Hiram with a dog), while Tupper literally becomes one with the vegetation. This too was an old idea in child studies. James Sully noted that "infants of civilized races," like "the lowest races of mankind" are in "close proximity to the animal world."[33] Simak rarely visited his hometown of Millville, Wisconsin, in part because he preferred to think of it as a lost and pristine Eden, an illusion that he acknowledged was harder to maintain from close up than from afar. While the Driftless region is indeed a special place (I own a house there), Simak seems less interested in it as a real place and more as a state of mind or a vision of a paradise lost that his now unfashionably optimistic and pastoral fictions are always trying to regain. Simak's vision of a human race bound in philosophical and democratic communion with advanced species predates Gene Rodenberry's by a decade or more, but unlike *Star Trek*, which often used time travel as a means to satirize the cultural and technological backwardness of the twentieth century as compared to an idealized future, Simak's fiction does not fetishize technological modernity. Rather, he locates the gateway to human advancement in ecologically rooted pre-modern wisdoms and rural values, all of which are located in the unique topography of unglaciated southwestern Wisconsin and all of which are best emblematized by his child-like intellectually disabled characters.[34] While never themselves the protagonists, they unconsciously guide the protagonists and thus the human race into a future communion with advanced alien species, while simultaneously anchoring that futurity to a homely, pastoral simplicity rooted in nature.

Simak's work provides just one example of the ways in which disability, childhood, and time can interrelate, but it is a telling example in that, characteristically for this genre, it yokes the temporal disjunction inherent in

---

33   Sully, *Studies of Childhood*, 5.
34   Father Flanagan, in *All Flesh is Grass*, best captures Simak's suspicion of modernity when he says that he "do[es] not hold with the modern cynicism that seems so fashionable. There is still, I think, much room in the word today for a dash of mysticism" (188).

disability and in the figure of the child to a fantasy of alien temporal alter-natives. Disability in narrative might then be a means for writers to gesture towards the limits of intelligible human time, as Michael Bérubé suggests, but that gesture all too often rests on some very old cultural constructions of time and childhood that nowadays we might not be so eager to embrace as scholars. "I make occasional pilgrimages [to Millville]" Simak continues in his letter to his editor, "but I never can stay for long [...]. There are too many memories and nothing to tie the memories to the present."[35] At a time when laudable efforts are being made by time scholars to think past and beyond the time of the human (to think of the time of the earth or of non-human temporalities), we need to be cognizant of the cultural (and personal) histories of the tropes we use to mediate between the human and the non-human, in order to ensure that we aren't simply returning to our own private Millvilles: our own unrecon-structed fantasies of the primitive.

References

Allen, Thomas M., ed. *Time and Literature*. Cambridge: Cambridge University Press, 2018.

Beckmann, Else and Alexander Minnaert. "Non-cognitive Characteristics of Gifted Students with Learning Disabilities: An In-depth Systematic Review." *Frontiers in Psychology* 9, no. 504 (2018): 1–20.

Benioff, David and D. B. Weiss. "The Door," *Game of Thrones*, season 6, episode 5. New York: HBO, 2016.

Bérubé, Michael. *The Secret Life of Stories: From Don Quixote to Harry Potter, How Understanding Intellectual Disability Transforms the Way We Read*. New York: New York University Press, 2016.

Black, Shane and Fred Dekker. *The Predator*. Film script. 2018. https://www.scriptslug .com/assets/uploads/scripts/the-predator-2018.pdf.

Blackwood, Algernon. *Pan's Garden: A Volume of Nature Stories*. Freeport: Books for Libraries, 1912.

Burman, Erica. "Child as Method: Implications for Decolonising Educational Research." *International Studies in Sociology of Education* 28, no. 1 (2019): 4–26.

Carroll, Lee and Jan Tober. *The Indigo Children: The New Kids Have Arrived*. Carlsbad: Hay House, 1999.

Claremont, Chris. *Bizarre Adventures* 27. New York: Marvel Comics, 1981.

Dick, Philip K. *Martian Time-Slip*. Boston: Mariner Books, 1964.

---

35   Letter from Clifford D. Simak to Lawrence P. Ashmead, January 10, 1966.

Evans, Bonnie. "How Autism Became Autism: The Radical Transformation of a Central Concept of Child Development in Britain." *History of the Human Sciences* 26, no. 3 (2013): 3–31.

Fabian, Johannes. *Time and the Other: How Anthropology Makes its Object*. New York: Columbia University Press, 1983.

Falter, Christine M. and Valdas Noreika. "Time Processing in Developmental Disorders: A Comparative View" in *Subjective Time: The Philosophy, Psychology, and Neuroscience of Temporality*, edited by V. Arstila and D. Lloyd, 557–597. Cambridge: MIT Press, 2014.

Higashida, Naoki. *The Reason I Jump*. Translated by K. A. Yoshida and David Mitchell. New York: Random House, 2016.

Kafer, Alison. *Feminist, Queer, Crip*. Bloomington: Indiana University Press, 2013.

Karmiris, Maria. "De-centering the Myth of Normalcy in Education: A Critique of Inclusionary Policies in Education through Disability Studies." *Critical Disability Discourses* 8 (2017): 99–119.

King, Stephen. *The Shining*. New York: Random House, 1977.

Lessing, Doris. *The Fifth Child*. London: Jonathan Cape, 1988.

Lessing, Doris. *Ben, in the World*. London: Harper Collins, 2000.

Lobdell, Scott. *The Adventures of Cyclops and Phoenix*. New York: Marvel Publishing, 1994.

Martin, George R. R. *A Game of Thrones*. New York: Bantam Books, 1996.

Morrison, Toni. *Beloved*. New York: Knopf, 1987.

Renner, Karen J. *Evil Children in the Popular Imagination*. Basingstoke: Palgrave, 2016.

Rousseau, Jean-Jacques. *Èmile: or, on Education*. New York: Penguin, 2007.

Said, Edward W. *Orientalism*. New York: Random House, 1979.

Shapiro, Joseph P. *No Pity: People with Disabilities Forging a New Civil Rights Movement*. New York: Three Rivers Press, 1994.

Simak, Clifford D. *Way Station*. New York: Open Road, 1963.

Simak, Clifford D. *All Flesh is Grass*. New York: Open Road, 1965.

Simak, Clifford D. *Mastodonia*. New York: Open Road, 1978.

Sully, James. *Studies of Childhood*. London: Longmans, 1896.

Thomas, Roy. *X-Men* 2.30. New York: Marvel Comics, 1968.

Virtue, Doreen. *The Crystal Children*. Carlsbad: Hay House, 2003.

Wordsworth, William and S. T. Coleridge. *Lyrical Ballads 1798*. Oxford: Oxford University Press, 1969.

CHAPTER 7

The Seductive Quality of Variable Time in *Elder Scrolls V: Skyrim*

Sue Scheibler

Abstract

This essay uses existing scholarship on time in video games to analyze the ways that the video game *Elder Scrolls V: Skyrim* uses time to offer a richly rewarding game experience for players. It approaches the game as a blend of progressive and emergent play and argues that these two approaches map onto Jesper Juul's categories of playtime and fictional time and Jenkin's argument about embedded narrative to explain the reason why the game, originally released in 2011, continues to be popular, praised for its gameplay and story. In addition, I read the use of time in the game through my own experience as an autistic player with ADD to comment on the specific pleasures offered to players who have a heightened sensitivity to temporal variations.

Keywords

autistic stimming – embedded narratives – emergent play – fictional time – gameworld time – ludic temporalities – lusory attitudes – neurodivergent time – open world games – playing time – RPG

1 An Introduction to *Skyrim*

At this moment in time, the video game *Elder Scrolls V: Skyrim* is running on all four of my game consoles. More specifically, approximately twenty or twenty-five separate and unique versions of the game are running, illustrating Christopher Hanson's observations:

> [G]ames may be in play even as they appear suspended temporally or indefinitely. Although a game may appear to be stopped to an outsider observer, it remains animated as far as the players are concerned. [...] [W]ithin the imaginations and minds of its players, a game is in play even

at times when the game's diegetic temporality is halted. Such a game is active even when it appears to be still. [...] This state of suspended animation and potentiality in games demonstrates the nuances of game temporalities.[1]

It is in the nuances of game temporalities that I, an autistic gamer, find so much pleasure.

Like many autistic people, I often find that there are too many temporalities buzzing around me. Everything becomes part of a temporal cacophony that can, if left unchecked, lead to an autistic meltdown or shutdown. When this happens I, like many other autistic people, turn to my special interests – in this case, gaming in general and the video game *Skyrim* in particular. *Skyrim*, in its multiple games, suspended yet in play across my many consoles, is my great stim. Stimming is the way that autistic people bring calm, comfort, and balance to their lives. Through repeated actions, an autistic person is able to slow down the sensory onslaught that can be so overwhelming. The game's nuanced temporalities help me manage the complicated and complex temporalities that exist in the "real world."

Video game scholar Grant Tavinor puts it well when, in his essay "Art and Aesthetics," he observes that the

> video game *The Elder Scrolls V: Skyrim* can be considered a beautiful representational artifact. The naturalism and rich detail of its environments, the evocative nature of its music, and the exploratory role of the player, makes playing this game a frequently rewarding experience.[2]

In this essay, I expand on Tavinor's observation by arguing that the "rewarding experience" provided by the game is due, in great part, to the way it invites the player to explore, navigate, manipulate, and master time. In his book *Game Time*, Christopher Hanson points out the following:

> [I]n a video game, the player attains agency over that which cannot be controlled in the real world: time. Video games enable the players to experience and manipulate time in ways that transcend other media. The temporalities of video games are numerous: players preserve, pause, slow down, rewind, replay, reactivate, and reanimate time as part of the play

---

1 Christopher Hanson, *Game Time: Understanding Temporality in Video Games*, 22–23.
2 Grant Tavinor, "Art and Aesthetics," 59.

mechanics. [...] When we play a game, we are already subject to multiple temporalities.[3]

Hanson goes on to say that "games offer new modes of temporal control that fundamentally alter our experiences and understanding of time in contemporary culture."[4] *The Elder Scrolls V: Skyrim*, generally referred to as *Skyrim*, immerses the player in a massively open world. It exists to be experienced as a topologically and narratively diverse world with a deep mythology and complex temporalities that enrich the player's investment in and pleasure from the game.

Developed by Bethesda Game Studios and released in 2011, the game has won hundreds of awards, topping most "best games of all times" lists and selling, at last count, over 30 million copies. The game continues to remain popular, due, in great part, to the fact that it employs a massive open world structure. Open world games are often likened to sandboxes that allow children the freedom to create nearly anything they want within the confines of the sandbox. In the same way, an open world game provides players the freedom of movement and progression to such an extent that an open world game can never be said to be fully completed. As a single player role playing game (RPG) *Skyrim* privileges the encounter between the player as character in the game, the denizens of the world, including the various AI characters (NPCs, or non-player characters), and the player as gamer, engaging a game-world through specific game mechanics. In this way, the player as character, embedded in the gameworld, becomes familiar with the numerous timescales within the game through exploring the world, completing tasks, and talking to non-player characters. In addition, as the player completes the game objectives, they deepen their character identity while expanding their knowledge of the game mechanics, including the various ways the game enables the player to manipulate and control time. Finally, because the player chooses a gender and race for their avatar at the beginning of the game, they are embedded in Skyrim's racial, economic, geographic, religious, and cultural histories, most of which inform the numerous quests that structure the game.

Skyrim's gameplay consists of two hundred and seventy-three quests. Some of these quests comprise a variety of main quest lines while others are classified as side or miscellaneous ones. While many of the quests are completed once and only once in any particular game, others can be undertaken an infinite number of times. At the same time, specific player actions will foreclose

3   Hanson, *Game Time*, 2–3.
4   Hanson, *Game Time*, 3.

some quests. In order to experience all of the possible quest lines, a player would need to play several versions of the game. The player would need to save each game version so that it could exist as a separate, loadable game. As the player plays each version, they would be replaying and repeating some aspects of the game as played while encountering new aspects of the game as yet-to-be-played.

In *Skyrim*, players have a choice as to which, if any, quests to complete and in which order. In fact, a player may opt not to take up any quests or to finish any of the quest lines, replacing the experience of completing the game with the experience of inhabiting a world. Any given game session may involve the completion of one or more game objectives or it may involve simply wandering around, chatting with people or engaging in non-quest related activities such as chopping wood, reading books, enjoying a cup of mead, listening to a bard sing, harvesting crops, learning alchemy, or even building and furnishing a home. While the game does include a series of "Main Quests," the game does not end when these are completed as there may be other quests in play. The player may opt to continue exploring the gameworld, or they may decide to re-play a portion or portions of the game. It is possible that, at any given time, the player is involved in all of the possible questlines available in the game simultaneously, moving back and forth between them as they choose.

While the player does need to complete each portion of a quest sequentially, they need not do these within a specific time frame. For example, the main quest line is composed of seventeen required quests that must be completed in sequence and three optional quests. Completion of the main quest line generally takes fifteen to thirty hours of gameplay, which may unfold in a fairly linear fashion or may unfold in a more interrupted fashion, over days and even months of gameplay as the player opts to take up other quests and/ or activities along the way. The player may choose to complete any number of other quest lines before completing the main quest, even though the main quest could be considered the central narrative thread, telling, as it does, the story of the Dragonborn's heroic journey from unnamed stranger entering the province as a captive, headed to the executioner's block, to Skyrim's savior.

The player's experience of the game is enriched through the various quest lines, each of which develops the emerging storyline, enlarges the player's understanding of the game's deep mythology, and contributes to the player-as-character's personal history and identity. The more time the player spends exploring the game world and engaging with its inhabitants, the more complicated become some of the choices confronting the player. For example, the Civil War quest line requires the player to choose sides. A player can simply opt to take up one side or the other in order to complete the game as quickly as possible. Or a player may opt to play through the game once as a member of the Imperial forces and a second time as a member of the rebel forces. Or a player may opt to take time to engage in conversation with all of the NPCs encountered in the game, learning from them their thoughts and opinions about the Civil War. As the player travels through the gameworld, chatting with a variety of characters from different races, occupations, economic classes, and political beliefs, they learn that opting to become a Stormcloak engaged in battle with the Empire requires the player to align themselves with a group of Nords who, in their desire to make Skyrim great again, hope to remove all non-Nords from Skyrim. Pro-Stormcloak characters may de-emphasize their racism by stressing the fact that the Civil War is about the freedom to worship their traditional Nord gods, currently outlawed by the Empire, or to free the Nords, who they see as suffering under Imperial control. They will argue that joining the Stormcloaks is the best way to restore traditional Nord values and customs to Skyrim. Pro-empire characters, however, may present the Empire as a necessary evil, providing stability and order. They may point out that Ulfric Stormcloak, the leader of the rebellion, is not a hero but a self-serving individual, interested only in power. They may argue that many good Nords serve in the Imperial army and that the stability provided by the Empire has allowed people from all over Tamriel (see below) to settle and thrive in Skyrim, regardless of race or creed. In this way, by taking time to explore the game world, the player encounters the deep mythology and history of the land and its characters and is able to use this information to determine game play actions that embed the player more fully in the game's story world.

As the above indicates, Skyrim, the province, and *Skyrim*, the game, are richly complex worlds. In his essay "Worlds," Mark J. P. Wolf observes that, like

> novels, narrative films, and television shows, many video games can be said to have a diegetic world, that is, an imaginary or fictional world in which game events take place, and where the game's characters live and exist.[5]

Skyrim offers the player the fictional world of Tamriel, an imaginary continent composed of nine provinces, one of which is Skyrim. Each province, inhabited by a specific race, is geographically, politically, and religiously unique. Skyrim, while home to the Nords, a race modeled after the real-life Vikings, has, by virtue of its stable political structures and thriving economy, become a land of opportunity for members of all of Tamriel's races. They bring with them the

---

5   Mark J. P. Wolf, "Worlds," 155.

history of Tamriel, including its various wars and disasters, so that, through its inhabitants, Tamriel's history is deeply embedded in that of Skyrim.

In this sense, then, while the gameplay is restricted to the province of Skyrim, the gameworld extends beyond Skyrim to all of Tamriel. At the same time, while the gameplay occurs within the "now" of Skyrim's present, game events propel the player into the deep past of the entire world and cosmos in which Tamriel and Skyrim exist. For example, the "Glory of the Dead" quest requires the player to explore Ysgramor's Tomb and recover Ysgramor's shield. In undertaking the quest, the player learns that the Nord, Ysgramor, known as the "First Harbinger," was the first human to settle in Skyrim. Undertaking the "Jagged Crown" and "Gauldur Legend" quests deepens this knowledge by revealing the ways that Harald, Ysgramor's heir, drove most of the other races out of the province in order to establish a Nordic empire. The desire to restore the glory of the Nordic past provides the rationale for the Stormcloak rebellion and the Civil War that shapes much of *Skyrim's* gameplay. The "now" of the game is deeply enmeshed in Skyrim's mythology, cosmology, and history.

The game's deep mythology emerges through gameplay as the player interacts with the gameworld. In this way, the game exemplifies what Henry Jenkins refers to as "narrative architecture." He points out that within

> an open-ended and exploratory narrative structure like a game, essential narrative information must be redundantly presented across a range of spaces and artifacts, since one cannot assume the player will necessarily locate or recognize the significance of any given element [...]. The game world becomes a kind of information space, a memory palace.[6]

Jenkins identifies two different types of narratives present in games: one generated by the player as they explore the game space and one that is pre-structured (written, as it were) and embedded in the game's design. In Jenkins' words, the

> embedded narrative doesn't require a branching story structure but, rather, depends on scrambling the pieces of a linear story and allowing us to reconstruct the plot through our acts of detection, speculation, exploration, and decryption.[7]

---

6 Jenkins, "Game Design as Narrative Architecture," 9.
7 Jenkins, 11.

Part of the pleasure of playing an open world game lies precisely in the non-linearity of the story embedded in the game. One must play *Skyrim* in order to reconstruct the narrative that informs the game, where playing the game involves not just taking up the game objectives but also engaging with the gameworld as fully as possible. While the game objectives provide pieces of the narrative, it is through talking to non-player characters, reading the hundreds of books and journals scattered across the gameworld, visiting shrines, and uncovering a variety of relics that the player comes to a fuller understanding of their place in the gameworld. To the extent that the player gains pleasure from story and not just from the gameplay (that is, completing the game objectives), the player will take time to make sense of the variable timeframes and timelines that exist in the game as information space and memory palace. In a sense, this is where the deep pleasures of playing an open world game lie.

2   *Skyrim* and Game Time

In his essay "Introduction to Game Time/Time to Play – An Examination of Game Temporality," Jesper Juul argues that most

> computer games project a game world, and to play them is therefore to engage in a kind of pretense-play: you are both "yourself" and you have another role in the game world. This duality is reflected in *game time*, which can be described as a basic duality of *play time* (the time the player takes to play) and *event time* (the time taken in the game world).[8]

He refines his analysis of time in video games in his book *Half Real: Video Games Between Real Rules and Fictional Worlds*. In this work, Juul renames event time as fictional time to better capture the ways in which games construct fictional worlds. As he states it, to

> play a game takes time. A game begins and ends. I would like to call this time *play time*. Play time denotes the time span taken to play a game [...] I propose the term *fictional time* to denominate the time of the events in the game world [...] Playing for two minutes can make a year pass in the fictional time/game world.[9]

---

8   Jesper Juul, "Introduction to Game Time/Time to Play," 1 (emphasis in original).
9   Jesper Juul, *Half-Real*, 142–143.

For example, one minute of play time in *Skyrim* equals twenty minutes of fictional time in the game world. That is, for every minute of play time (time in the real world outside the game) it takes to do something in the game, twenty minutes pass in the fictional world of the game. If the player engages in a game objective, such as searching through a dangerous location to find a hidden weapon, the way that play time maps onto fictional time will depend, in great part, on the player's skill at negotiating the space and defeating any obstacles encountered. It will also depend on whether or not the player takes time to explore the space, looking for hidden treasures, reading books collected along the way, or just waiting for fictional time to pass before continuing on. The player may also choose to replay certain sections in order to achieve different outcomes. In this way, completing the specific objective may involve thirty minutes of game time even as ten hours or more elapse in the fictional world of the game.

Juul compares play time/fictional time to the terms story/discourse as used in narrative studies. In this way, he offers a bridge between the old debate between ludologists and narratologists. In the early days of game studies, many scholars came to games from the fields of literary studies and narratological research – that is, narratology. As such, they tended to stress the storytelling and narrative aspects of games, thinking of them predominately as interactive fictions. Other scholars, including game developers and designers, in contrast, emphasized the game aspects and adopted the term "ludology" to distinguish their approach from that of the narratologists. In the past two decades, games and game studies have become more complex and nuanced. As Frans Mäyrä observes, debates

> can be useful in making even slight differences of opinion stand out more clearly. That is also true of the so-called "ludology-narratology debate" [...; even] the "ludology-narratology debate" has turned into discussion whether it ever really happened in the first place. [...] No-one actually seems to be willing to reduce games either into stories, or claim that they are only interaction, or gameplay, pure and simple, without any potential for storytelling.[10]

For example, reducing *Skyrim* to its basic gameplay fails to do justice to the pleasures offered by the rich story embedded in the gameworld. Thinking of it simply in terms of its basic story – that of the mythological hero, the Dragonborn, reborn in the player's character, who must fight the evil dragon

---

10   Frans Mäyrä, *An Introduction to Game Studies*, 10.

Alduin, the World-Eater, to restore balance to the world – does not even begin to capture the richly complex story world that emerges through playing the game. The player needs to engage with both play time and fictional time in the game in the same way that a reader or viewer, in order to have the richest reading or viewing experience, engages with both the story as told and the discourse, or the way of telling the story, in a written or media text (such as a movie or television series). A player who opts to consider the game "completed" as soon as the main quests are completed will end up with a relatively short play time and possibly very little fictional time as the focus is on play rather than story. A player who opts to take their time exploring every aspect of the world, talking to every non-player character and reading through the books scattered through the world, will chalk up a near infinite play time. To the extent that they complete the various quest lines, this player will have experienced both play time (discourse time) through gameplay and fictional time (story time) through exploration and discovery of the many narrative layers that form the foundation of Skyrim's history, big and small.

Fictional time is mapped onto play time when a player loads a game at a specific save point and then pulls up the game stats. For example, when I choose to save my current game, the screen provides me with information as to my character name, race, and level as well the date and time in the "real world." If I pull up the game stats, I learn that, within the game world, the day at which I saved the game is Middas; the time in the game world is 10:22 AM; and it is the 1st of Frostfall while in the real world it is 10:00 AM on July 10, 2019. I also discover that, in my current game, in the almost 31 hours of real time that I've been playing this particular one, 45 days have passed in the game world, totaling 1,080 hours of fictional time, of which I've spent 207 hours waiting and 63 hours sleeping. During the time spent waiting or sleeping, events continued in the gameworld. I might wake up and step out of my house, only to learn that, while I was sleeping, vampires attacked the town, killing some of my friends and companions. In this way, taking time to wait or sleep may impact the player's ability to complete specific game objectives.

Game objectives are linked to specific locations and characters within the game. In order to complete a specific objective, the player must learn about the place and the game objective related to it. This generally happens through conversation with one of the non-player characters in the game. A character may mention a place in passing or may give more specific information by means of direct conversation. For example, as the game begins, the player, caught up in a raid on the Stormcloak's encampment by the Imperial forces, is saved from execution by the appearance of a dragon. After escaping, they are directed to travel to Whiterun to inform the Jarl that dragons have returned to Skyrim.

In Whiterun, the player, now recognized as someone with skills and talents, is instructed by the Jarl to join the Whiterun forces and find and slay a dragon threatening the city. After helping the soldiers slay the dragon and learning, in the process, that they are the Dragonborn, with the ability to absorb the souls of dead dragons, the player receives a summons to visit the Greybeards, reclusive monks who will be able to instruct the player in the ways of the Dragonborn.

As one can see, much of the time spent playing the game is involved in traveling. In the sequences noted above, the player must travel from Helgen to the little town of Riverwood to Whiterun, the capital of Whiterun Hold, to High Hrothgar where the Greybeards live in seclusion. Players may travel by foot, horse, cart, or ship. For players intent on playing the game – that is, engaged in game objectives that will ensure the completion of the various quests – time spent traveling may be considered "dead time," necessary but not "real game play." For other players, spending time traveling can be almost meditative or contemplative. There is a slowness to travel that can be calming. One can stop along the way to visit a shrine or just enjoy the natural world in which one is passing.

3 Slowing Down and Speeding Up Time in *Skyrim*

One doesn't always associate video games with contemplation or meditation and yet games such as *Skyrim*, by virtue of their open world structure and emphasis on emergent and nonlinear play, provide the opportunity for what we might call contemplative or meditative play. It is possible to take a very linear, goal-oriented, approach to the game, tackling each quest line sequentially and in a serial manner, making progress towards finishing the game, if one measures such things by completing all the quests and activities. The game, however, builds a different approach to gameplay within its structure and story world. The game encourages contemplation. It does this diagetically through the many temples and shrines scattered throughout the gameworld. Early in the game, as the player travels the 7,000 steps up the mountainside to High Hrothgar, they pass a series of ten shrines, each set off to the side of the path. At some of them the player encounters characters who are meditating or praying next to the shrine. By stopping to talk to the characters, read the engravings, and contemplate each shrine, the player is able to learn about Skyrim's cosmology.

The game also uses game mechanics and game design to slow down time. With its vast expanses of marshes and swamps, enormous mountain ranges, and snow-swept glaciers, *Skyrim* is designed to evoke awe, amazement, and wonder. For those who stop and look up, the night sky is full of stars, moons, and, at times, Skyrim's version of the Aurora Borealis, flickering against the Northern skies. The player can sit and watch fish swim, or butterflies flitting by, or listen to the birds singing. When the player completes tasks for the jarls, they may be rewarded with land on which to build a house. The act of building the house can be very contemplative as it involves using a set of blueprints to design the house; collecting the necessary building materials such as lumber, clay, stone, and leather; and using the workbench and anvil to craft hinges, iron fittings, and nails. This task can take up days in the gameworld and hours in the "real" world. The floor plans include the possibility of designing a deck from which one can sit and watch the sun set or rise. Since it takes about 30 minutes for the sun to set in the gameworld, watching the sun set can occupy the player for about 2 to 3 minutes of inaction in terms of play time. In many ways, the act of chopping wood, building a house and its furnishings, planting and harvesting a garden, mining ore, digging up clay, smelting ore, forging new weapons and armor, enchanting objects, and/or creating potions can all be considered meditative actions, or, in my case, meditative stims, relying, as they do, on repetitious, quiet movements. They offer a respite from the many battles, big and small, with which the player must engage in order to meet the game objectives.

Of course, one's contemplation can always be interrupted by some threat that must be contained. It is, after all, an action-adventure game that assumes the player wants to engage in actions and adventures, what Jenkins refers to as a "lusory attitude" or "the orientation to play that makes gameplay possible in the first place."[11] The game's lusory attitude requires a very dangerous world. Because of the legacy of war, past and present, the world is filled with bandits and assassins who attack without provocation. There are the normal predatory animals such as wolves, mudcrabs, bears, and tigers. Since the game is an action adventure fantasy game, its world also includes mammoths, giants, dragons, trolls, vampires, ice wraiths, and giant spiders. The player's quiet stroll along the sea may be interrupted by a pack of ice wolves or wild horkers (Skyrim's version of walruses), attacking from behind. The player working in the garden may be interrupted by a giant, intent on killing the player's horse, cow, or chickens. The giant must be killed, and, because giants are very difficult to kill, violence may not just interrupt the player's tranquility; violence may lead to the player's death. The player sitting on the deck, quietly enjoying the sunset may be attacked by a dragon swooping down from the sky, or a

---

11   Jenkins, 63.

group of mages sending fireballs onto the deck. At these moments, the player has several options: enter into battle; refuse to fight and, therefore, die; or run away and live to contemplate another day. By choosing death, the player is able to re-enter the game at a previous save point, one that provides a return to quiet contemplation.

It is because *Skyrim* builds these possibilities of subverting ludic expectations into the gameworld that I, as an autistic player, can use the game as my great stim to manage the stresses and strains that result from navigating a neurotypical world with its noise and confusing expectations about communication and appropriate behavior. As I explore the gameworld, I discover areas in which I can wander without interruption by threats or demands to battle. I. can set aside my lusory attitudes in the frozen, ice-swept region along the sea that forms the Northernmost border of the gameworld. When I travel along the seashore filled with floating ice and look out across the horizon into a frozen, empty landscape that seems never to end, it is possible to slow down time. Many autistic people, as well as people with ADD or ADHD, describe their experience in temporal terms, as working in a different clock time than that experienced by neurotypical people. We might call this "neurodivergent time." I often describe my experience of occupying variable timeframes at any given time in the real world as being in a Jell-O cube filled with lightning. Wandering the game's frozen northern reaches, often beneath a sky filled with the northern lights reflected across two moons, I find that the Jell-O cube dissolves, and the lightning dissipates as time slows down and emptiness fills the noisy spaces.

I can wander in Skyrim's barren landscape as, simultaneously, a being in time and a being outside of time. As a player, I can enter the game and let real world time flow past me as I immerse myself in play time. I am outside real time. As a character in the game, I am a being in time, that is, in the game's time. I can enjoy the variable temporalities as Skyrim time is mapped onto time in the real world. But I am also outside play time, evading any game objectives or actions. I am in fictional time, wandering the snowy wastes alone, everything quiet except for the music that plays constantly as undertone to the game (and, if it gets too overwhelming, I can turn it off). The empty landscape is calming. It's all icebergs, glaciers, water, and windswept shores. My autism gives me the ability to get lost in watching the soap bubbles as I do the dishes or the light playing on the leaves on the tree outside my window. In the same way, I can spend time in the game watching the sun rise or set or the play of light on the snow-covered glacier. The game gives me a contemplative, meditative, stimming space from which I can emerge as a being in the world.

4   Time's Passage in *Skyrim*: Seriality and Simultaneity

Fictional time gives Skyrim its depth and richness. It also gives *Skyrim*, the game, its sense of what Jared Gardner calls "tensed time." In his essay on seriality and simultaneity, Gardner defines "tensed time" as "the time of the moving present, that time node at which a fixed but always-already lost past and an as-yet unreal future converge."[12] The player encounters this tensed time as they move through the game, always negotiating the lost past that informs the gameplay, the "now" of the game, and the as-yet unreal future that establishes the game's goals, objectives, and rewards. Games are immersive and interactive because they deploy both seriality, that is the way that events unfold through gameplay as a series of events that occur in the present tense of gameplay (play time) as well as in the present tense of the gameworld (fictional time). Quest objectives unfold in a serial or sequential manner, with each piece following from the previous one while leading to the next one until all of the quest objectives have been completed. While seriality is important to the game as a game, it is through simultaneity, or the sense that events unfolding in the fictional time of the game are occurring simultaneously, whether the player is present or not, that fulfills the experience of immersion in a storyworld with which the player interacts. In this way, games such as *Skyrim* provide an example of what Gardner observes as Einstein's discovery:

> After centuries of Newtonian physics, time was no longer absolute, nor was it any longer separate from the three dimensions that defined space. Einstein's special theory of relativity highlighted the paradoxical relationships between two seemingly contradictory models of time: *seriality*, as the model that corresponds with how we experience time; and *simultaneity*, as the model of time that emerges from Einstein's insights into relativity.[13]

When the player begins a quest objective, such as killing a vampire in their lair, the gameplay unfolds through a series of action that unfold in play time as well as fictional time. This time can be measured by clock time in the real world as well as in the fictional world of the game. When the player, having defeated the vampire and their minions, emerges from the lair, they will find that life in Skyrim has gone on as usual. Time in Skyrim has unfolded simultaneously

---

12   Jared Gardner "Serial/Simultaneity," 161.
13   Gardner, 161.

with the serialized events that the player encountered while completing the quest objective.

Successfully managing all of these variables requires the player to be, as Christopher Hanson points out, "simultaneously engaged in the immediate past, continuously processing the effects of actions or the environment in order to determine the now and its projection into the future."[14] In their essay "Time in Video Games: A Survey and Analysis," José P. Zagal and Michael Matias observe that a game

> can establish its own notions of cycle and duration that are potentially independent of cycles and duration in the real-world frame. [... G]ame-world days may be used to add atmosphere to a game.[15]

In *Skyrim*, day turns to night, days turn into weeks, weeks into months, and months into seasons. Weather changes from sun to thunderstorms to snow. The player may enter a space while it is day only to emerge, after completing the task at hand, to discover that it is night, or that what started as a sunny day has turned dark and gloomy, cold and snowy. Confronted by the passing of time, the player must process the effects of the action just completed (retention of the immediate past) while taking in the environment and the signs that time has passed in order to project their actions into the future and decide on the next course of action to be undertaken.

As the player completes a specific game objective, they are aware of how much time has passed in the real world, that is, how long it has taken to complete the required game action. In addition, as noted above, the game mechanics produce the sense of duration as well as diurnal and seasonal cycles, experienced when the player steps back into the gameworld and discovers that time has passed. Time's passage is also created through the ways the non-player characters (NPCS) move through the world and react to the player's actions, lack of actions, dialogue, and/or silence. In their article on artificial intelligence, Robin Johnson points out that the

> primary technique of game AI is based on moving NPCs through the game, including determining optimal paths with pathfinding algorithms. [...] NPCs increase the depth of the player's experience and increase the playability of the game.[16]

---

14   Hanson, 75.
15   Zagal and Mateas, 6–7.
16   Johnson, Robin, "Artificial Intelligence", 11.

For example, if the player remains motionless too long, the characters in *Skyrim* around them may sigh deeply, comment on the lack of action, and even sarcastically note that nothing is happening. Tired of waiting, they may take up conversations with other non-player characters or even wander off. If the NPC is a mercenary paid to accompany the player and assist them in battle, becoming bored and wandering off can seriously impact the player's success in a quest as well as finances because the player will need to find and rehire the mercenary. Because time passes in the game, locating the mercenary may be difficult as they are traveling back to the location where they were originally hired. The player must either try to follow them or wait for them to return to their original location.

While the sense of time passing alludes to time's linearity and seriality, other aspects of the game create the sense of simultaneity, an "in the meanwhile" of events happening all over Skyrim, whether the player is present or not. Once again, this sense is created through the adaptive AI as NPCs move throughout the gameworld. For example, entering a town at night, the player may encounter the townspeople heading home. Or the player may wander into the middle of a conversation, argument, or event to which they become a bystander, an observer. More pointedly, as noted above, the player may enter a town only to discover that, during their absence, vampires or bandits attacked and killed off some of their favorite NPCs, some of whom were waiting to assign the player specific quests, which are no longer available. As the player passes characters in the streets, inns, taverns, palaces, and along the roads, they will overhear snatches of conversation, some of which may be directed to them and some directed to other characters. It is important to pay attention since some of the overheard conversation includes vital information. Simultaneity maps play time onto fictional time as the "in the meanwhile" that provides a sense of aliveness and presence to the player as being-in-the-world also provides vital information for them to accomplish a game objective.

5   Suspended Time

Play time and fictional time are suspended when a game is paused and animated when the game is reactivated. As Hanson points out, pausing

> a game allows players to temporarily suspend play, effectively interrupting the game's flow of time. In our everyday life, time moves only forward, and we have no power to control its passage. However, games offer

us alternate temporalities that can be controlled and manipulated at our whim.[17]

Players can pause play in *Skyrim* for a variety of reasons. They may want to take a break from play or look up a strategy guide or walkthrough for assistance in navigating a particularly tricky game objective. Using the "pause" button superimposes the game's menu over the game action that has been frozen in time. The game is also paused every time the player pulls up the character menu. The player can do so at any time, including in the midst of a battle, and it allows the player to take advantage of items that might help them succeed at whatever they're attempting to accomplish. In this way, pausing a game

> complicates the bounded temporality of gameplay, potentially suspending and extending a game's temporality indefinitely. Pausing a game allows the player to manipulate the passage of time within a game, fragmenting it and thickening the present of a game as it is played.[18]

For example, while in the midst of a particularly challenging battle, the player might find their health or stamina is failing. The player can pause the game in order to drink a potion to restore their health and strength so as to defeat their opponents. Since many of the quests take place in labyrinthian spaces, filled with traps and obstacles, it is easy to lose one's way. The player may often find themselves turned around in a particularly dark and windy place. The only way out is to pause the game in order to bring up the map. There are also times in battle when success depends on a change of weapon or armor. In these instances, the player can bring up the character menu, pausing the game, in order to choose from among the weapons and/or the armor available to them, in order to ensure victory against their opponents. In these instances, the player suspends time in the gameworld in order to use time in the real world to strategize how best to accomplish the task at hand.

Skyrim provides other options to manipulate fictional or story time. The player may choose to wait and, by choosing the "wait" function, can control how much time they wish to wait. While pausing the game stops the diegetic time from unfolding, choosing the wait function merely allows time in the game to pass. In this way, noting that it is late at night, a player may select "wait" and choose the number of hours needed for the sun to rise in order to avoid

---

17   Hanson, 58.
18   Hanson, 60.

traveling a dangerous road in the dark. If, a player arrives in a town before the shops are open, the player may opt to choose "wait." The wait option provides the player with the current time in the game world. The player can then select how many hours to wait. So, for example, if the shops open at 8 and it is 5 AM, the player may choose to wait for three hours in order to purchase items necessary for their next quest. While the player can't wait while enemies are nearby, waiting while engaged in a game objective will restore the player's health and stamina. A variation of wait is "rest." Finding an unoccupied bed, the player may choose to rest. As in the case of "wait," the player can control how much time to spend resting. Unless one is a werewolf or vampire, one wakes "feeling refreshed." If the player/character has adopted children, the player wakes "filled with a mother's love" if gendered female or a "father's love" if gendered male. If married, the player/character wakes up filled with marital bliss.

If pausing a game suspends fictional time and waiting suspends action while allowing fictional time to elapse, saving a game freezes the game at a particular time in the gameplay as well as the gameworld. *Skyrim*, like many contemporary games, includes an autosave function in the game mechanics. The player is able to set some of the parameters for when the autosave function is triggered while others are set by the game system. For example, the game is automatically saved every time the player exits and enters a new location, including leaving or entering a town, home, inn, tavern, dungeon, cave, et cetera. In addition, the player can set the autosave to save while the player is resting, waiting, or opening up the character menu.

Since the autosave function may not always save the game at the most judicious times, a wise player saves the game often when engaged in game objectives that may result in death. When the character dies, the game resets to the last save point. At that point, the player is still alive and can take what was learned in previous attempts to play through the game section again. Saving often eliminates the loss of any progress made in the game up to that point; otherwise, the player may need to replay long sections of the game that led up to their untimely death. The player also has the option to reload any saved game. This allows the player to replay a section of game. Every time a section of the game is replayed or a new game is begun, it produces a new narrative that is both similar and dissimilar from what emerged in the previous encounter with the game world, game events, and game characters. As Hanson observes, when

> paused or saved, the temporality of a video game achieves a certain thickness: the present of the player's experience is extended. While the now of the game is frozen, the player enters an extended present in which she is

still invested in the now of the game but given the luxury of contemplating the best course of action from here.[19]

In other words, pausing or saving a game thickens the present of the game into one in which the player may more readily succeed in accomplishing the desired objectives or outcomes. Freezing time in the game extends it as an elongated present in which the player can make decisions, including searching out hints in an online strategy guide.

Finally, game time and fictional time overlap when the player is given the ability to manipulate time as part of the gameworld. For example, one of the shouts the Dragonborn learns and unlocks allows the player to slow time during battle. Time limits are included with some potions that allow for an increase in health, stamina, skills, or abilities for a limited time. In addition, while most of the game is played in the present, there are times when the player travels back in time as part of a game objective. In order to learn how to defeat the dragon king Alduin, the player must use the elder scrolls to enter the Time Wound, a gap in time. Entering the Time Wound, the player is transported back in time to a moment when ancient heroes battled a ferocious dragon. Watching the event unfold, the player learns what must be done to defeat Alduin in their own time.

6   Conclusion

In his essay about emergence, Joris Dormans observes that the "richest games with the highest replay value tend to rely on emergent techniques to create unending variations in their gameplay."[20] *Skyrim* offers a richly rewarding game experience because it combines elements of progression and emergence in a way that allows players to choose to emphasize either playtime or fictional time. According to John Sharp, games "of progression are those that lead the player along a set path as they move through the game" while games "of emergence are those games that in which the player moves through the game experience in more open-ended ways." In this way, games of progression "have narrower spaces of possibility due to the constraint placed upon player decision-making by the narrative progression." Games of emergence, on the other hand, "have broader spaces of possibility because the choices are more open, allowing greater player agency. The greater the space of possibility

19   Hanson, 89.
20   Dormans, Joris, 427.

a game has, the greater dimensionality a game has."[21] While *Skyrim* shares with other video games the basic game mechanics that allow players to manipulate time (pause, wait, rest, save, replay, repeat), it is in its open world and emergent play that it offers such a richly rewarding experience for players. Players who are more invested in play time may opt for the progressive elements of the game, found in the seriality and linearity of the various quests. Players like myself, who are more invested in fictional time, may opt for the emergent elements of the game, taking time to learn all the nuanced temporalities in the game, including the game's historical, cosmological, geographic, and mythological timescales. Players like myself, who take pleasure in the multiple games suspended in time, waiting to be animated at the touch of a button, enjoy the open world, sandbox aspects of the game. And, finally, autistic players like myself, who take comfort in the possibilities of slowing down our own internal time clocks, will prefer to spend time exploring, wandering, meditating, contemplating, and just sitting beneath the two moons shining against Skyrim's sublime Aurora Borealis.

References

Dormans, Joris. "Emergence." In *The Routledge Companion to Video Game Studies*, edited by Mark J. P. Wolf and Bernard Perron, 427–433. New York: Routledge, 2016.

Elder Scrolls V: Skyrim, Special Edition, Remastered. Bethesda Game Studios, Iron Galaxy, 2016.

Gardner, Jared. "Serial/Simultaneous." In *Time: A Vocabulary of the Present*, edited by Joel Burges and Amy J. Elias, 161–176. New York: New York University Press, 2016.

Hanson, Christopher. *Game Time: Understanding Temporality in Video Games.* Bloomington: Indiana University Press, 2018.

Jenkins, Henry. "Game Design as Narrative Architecture." Pdfs.Semanticscholar. Semantic Scholar. 9/21/2005. www.pdfs.semanticscholar.org. Downloaded on 29 July 2020.

Johnson, Robin. "Artificial Intelligence." In *The Routledge Companion to Video Game Studies*, edited by Mark J. P. Wolf and Bernard Perron, 10–18. New York: Routledge, 2016.

Juul, Jesper. "Introduction to Game Time/Time to Play – An Examination of Game Temporality." Jesperjuul.net. Jesper Juul. http://www.jesperjuul.net/timetoplay/ . Downloaded on 6/23/2019.

Juul, Jesper. *Half-Real: Video Games Between Real Rules and Fictional Worlds.* Cambridge: MIT Press, 2005.

---

21   Sharp, 97.

Mäyrä, Frans. *An Introduction to Game Studies: Games in Culture*. Los Angeles: Sage, 2008.

Sharp, John. "Dimensionality." In *The Routledge Companion to Video Game Studies*, edited by Mark J. P. Wolf and Bernard Perron, 91–98. New York: Routledge, 2016.

Tavinor, Grant. "Art and Aesthetics." In *The Routledge Companion to Video Game Studies*, edited by Mark J. P. Wolf and Bernard Perron, 59–66. New York: Routledge, 2016.

Wolf, Mark J. P. "Worlds." In *The Routledge Companion to Video Game Studies*, edited by Mark J. P. Wolf and Bernard Perro, 125–131. New York: Routledge, 2016.

Wolf, Mark J. P. "Time in the Video Game." In *Medium of the Video Game*, edited by Mark J. P. Wolf, 77–91. Austin: University of Texas Press, 2001.

Zagal, José and Michael Mateas. "Time in Video Games: A Survey and Analysis." In *Simulation & Gaming* 41: 6 (2010): 844–868. http://sag.sagepub.com.

CHAPTER 8

In the Forest of Realities: Impossible Worlds in Film and Television Narratives

Sonia Front

Abstract

In the twenty-first century, the postmodern multiplicity of temporal scales is accompanied by a new mode of temporality, established by the turn of the millennium and 9/11 terrorist attacks. The new time, in which the paradigms of linearity, causality, and succession are revised, is marked by the prevalent sense of threat and a sense of an ending, be it by terrorist means or as a result of climate change. In one strand of film and television narratives, the collapsed bridge between the centuries is symbolized by the temporal rifts brought about by the release of huge masses of energy, which produces the disruption of a single ontological reality and opens up impossible worlds, such as parallel universes, special zones or wormholes. On the basis of the series *FlashForward* and the film *Annihilation*, I analyze impossible spatio-temporalities and their influence on human time in an effort to advance the thesis that unlocking the scale of physics' temporality to human experience is physically and psychologically intolerable.

Keywords

Annihilation – Anthropocene – apocalyptic discourse – *FlashForward* – impossible worlds – spatio-temporal representations – temporal rifts

One of the characteristic features of postmodern changes in the culture of time has been the splintering of the sense of time into multiple temporal scales:[1] from the nanosecond of computer time to cosmic time, from genetic to geological time, from the microscopic scale of quantum mechanics to the macroscopic scale of relativity theory. There is no overarching concept of time that would unify all these scales to elucidate adequately the different layers

---

1 See Ursula K. Heise, *Chronoschisms: Time, Narrative, and Postmodernism* (New York: Columbia University, 1997), 38–76.

of reality that constitute the world. The multiplicity of scales persists in the twenty-first-century culture of time, but it is accompanied by a new mode of temporality, marked by what Frank Kermode has described as a "sense of an ending" in human collective consciousness. The end-times rhetoric of the last fifty years, encompassing a sense of historical exhaustion, conclusion, and the repetition of terms such as "late," "post," and "end" in reference to culture coincides with the sense of time passing in a different way in the new century.[2] Peter Boxall traces this postmillennial "recalibrated" time in novels by a wide array of writers, recognizing it as part of a broader trend.[3]

The new temporality, positioned between linear historical time and the ultimate end of humanity, was instituted by the turn of the millennium and 9/11 terrorist attacks.[4] Sociologically, the attacks launched a time different from the religious apocalypse in that it is marked by "the end of the world as we know it"; a rupture, in which "dramatic events reshape the relations of many individuals at once to history."[5] In the new mode, the paradigms of linearity, causality and succession are rewritten by the new technologies and global networks as well as by the temporal rift of the event (9/11) itself. This event rang down the curtain on the twentieth century and ushered a new social and cultural mood underpinned by trauma and anxiety. One of the consequences of 9/11, as Philip Tew argues, has been a surge in trauma narratives as a cultural response to a new global state of affairs. Roger Luckhurst also perceives 9/11 as a turning point at which pre-millennial "traumaculture" focused on individual experience was supplanted by a collective sense of trauma, "a broader post-September 11 traumatological culture, by a sociologically significant disposition that permeates both selfhood and artistic renditions of this perspective."[6] John R. Hall points out that apocalyptic discourse gained "a new currency with the attacks of 9/11" as an adequate manner in which to express the prevailing sense of crisis[7] and what Barbara Korte and Frédéric Regard have described as a sense of "precariousness" of human life, brought about by war on terror,

---

2 Peter Boxall, "Late: Fictional Time in the Twenty-First Century," *Contemporary Literature* 53.4 (2012): 681–682, 695.
3 See Boxall, "Late: Fictional Time in the Twenty-First Century."
4 Boxall, 698.
5 John R. Hall, "Apocalypse 9/11," in *New Religious Movements in the Twenty-First Century: Legal, Political and Social Challenges in Global Perspective*, ed. by Philip C. Lucas and Thomas Robbins (New York: Routledge, 2004), 216.
6 Philip Tew, *The Contemporary British Novel. Second Edition* (London: Continuum, 2007), 199. See also Roger Luckhurst, "Traumaculture," *New Formations* 50 (2003): 28–47.
7 Hall, 216.

abuses of technology and science, environmental exploitation, and so on.[8] Although these issues are not exclusive to the twenty-first century, they have come to the forefront of the concerns that characterize the present world.

In one strand of film and television narratives, the collapsed bridge between the centuries is symbolized by the temporal rifts produced by the release of huge masses of energy. The examples include a nuclear blast in *Dark* (Baran bo Odar, Jantje Friese, 2017–2020) and *Twin Peaks* (Mark Frost, David Lynch, 2017); terrorist attacks in *Continuum* (Simon Barry, 2012–2015), *Fringe* (J. J. Abrams, Alex Kurtzman, Roberto Orci, 2008–2013), and *Source Code* (Duncan Jones, 2011); a meteor hitting the earth in *Annihilation* (Alex Garland, 2018), a Large Hadron Collider experiment in *FlashForward* (Brannon Braga and David S. Goyer, 2009–2010), and so on. This energy brings about the disruption of a single ontological reality and opens up impossible temporalities – such as parallel universes, special zones, or wormholes – which can be scientifically valid because massive amounts of energy are needed to distort space-time. It follows, then, that the time of the new physics, normally imperceptible on the human scale, becomes visible around catastrophic events. They are symbolic of 9/11, which J. G. Ballard sees as an "attempt to free America from the 20th century."[9] The zero event inaugurates a new post-catastrophic reality and temporality, marked by the pervasive sense of threat and the imminence of the end.

Twenty-first-century narrative worldbuilding (in literature, film, and television) has been heavily influenced by science and the epistemic shifts in the understanding of space and time. Cinema and television are crucial platforms for the representation of scientific theories and for probing their philosophical implications, including emergent conceptualizations of time. Scientific discoveries – be they in geology, evolution, radioactivity, the expansion of the universe, or the laws of quantum mechanics – have always exerted influence on cultural temporal awareness and cultural understandings of time. Technological innovation based on science has played a great role in shaping and reshaping the human experience of time and space. By juxtaposing human time with the temporal scales of physics, normally severed from everyday reality, film and television narratives investigate their impact on the human perception of time and human temporality and identity. They explore the question of relativity and the subjectivity of time as well as the nature of

8   Barbara Korte and Frédéric Regard, "Narrating 'precariousness': Modes, Media, Ethics," in *Narrating "Precariousness": Modes, Media, Ethics*, ed. by Barbara Korte and Frédéric Regard (Heidelberg, 2014), 8.
9   J. G. Ballard, *Millennium People* (London: Harper, 2008), 139.

time and its relationship to human consciousness. They raise questions about the representability of time for subjects whose sense of identity is interlocked with displaced time frames that unhinge their customary being-towards-death and the temporal hierarchy of past, present, and future. The subjects negotiate their lived time and the sense of their selves as they attempt to situate themselves in a world governed by an unfamiliar temporal regime. The narrative thereby systematizes the human experience of time. It operates as a mediator between manifold frames of time; a bridge between physical time and human time, between the time of physics and cultural time. In this way, the narrative diffuses ideas from science into culture and propagates new temporal relations, unlocking new ways of seeing, incompatible with the traditional Newtonian framework. That framework no longer provides a legitimate paradigm within which to convey the human experience of time in the twenty-first century. According to physics, time and space are inextricably connected, mutually interdependent (whereby time impacts space and space impacts time), and, most importantly, relative. Yet capitalism, public institutions and social life still deploy the notion of absolute space and time although they are "a historical aberration, arising out of a specific articulation of reality in the eighteenth and nineteenth centuries."[10] That is why narratives that represent time as understood by the new physics are called "unnatural" or "impossible" in narratology.[11]

1   Impossible Temporalities

Jan Alber defines scenarios and events that are impossible (or "unnatural") as being so according to "the known laws governing the physical world; accepted principles of logic (such as the principle of non-contradiction); or standard human limitations of knowledge and ability."[12] He describes natural temporality as one in accordance with our "natural" or "real-world knowledge of time

---

10   Elana Gomel, *Narrative Space and Time. Representing Impossible Topologies in Literature* (New York, London: Routledge, 2014), 6.

11   Some notions of time proposed by physics are quite commonsensical: for example, the direction of experiential time as governed by the Second Law of Thermodynamics.

12   Jan Alber, *Unnatural Narrative: Impossible Worlds in Fiction and Drama* (Lincoln: University of Nebraska Press, 2016), 80. Alber views these logical impossibilities as compatible with the framework of possible worlds, which Leibniz defined as "those which do not imply a contradiction" (*Philosophical Papers and Letters*, vol. 2, ed. and trans. by Leroy E. Loemker [Dordrecht: Reidel, 1969], 513). This definition influenced the most common understanding of possible worlds as the worlds that obey the rules of the excluded middle and non-contradiction. Some theorists consider logical impossibilities

and temporal progression."[13] Even if the impossible features prominently in new scientific theories and these theories can result in the revision of the possible and impossible, Alber predicts that they will not impact the "cognitive parameters that we use to make sense of the real world" anyway.[14]

Rüdiger Heinze suggests that we need to differentiate between the actual physical laws connected with time and our assumptions about them.[15] Sometimes our intuitive notions of time are contradicted by actual physics; still, if a narrative undermines these intuitive assumptions, a common reader will regard it as unnatural. For instance, according to the deterministic equations of physics, there is no distinction between past, present and future, yet our intuitive assumption is that the past no longer exists while the future is yet to come. Heinze claims that the intuitive assumptions about time are decisive as they "determine our experience of the world and of life to such a degree that it is nonsensical to insist that temporality really functions quite differently."[16] The real-world knowledge is grounded in the Newtonian paradigm, and that is why the scenarios with impossible times and spaces seem unrealistic. Even so, they do not have to be unreal,[17] as many of them can be elucidated by means of modern physics: for example, narratives with parallel universes can be explained with the help of multiverse interpretation of quantum mechanics.

Impossible temporalities can also be elucidated by means of the notion of "paraspace," described by Samuel Delany as an alternative space, "a science fictional space that exists parallel to the normal space of the diegesis – a rhetorically heightened 'other realm'."[18] This space happens to be "largely mental" but always manifests itself materially.[19] Delany's notion of paraspace is similar to Brian McHale's concept of "the zone," which he characterizes as a kind

---

as violations of possible-worlds theory. For the discussion of the nuances see Alber, *Unnatural Narrative*, 29–32.

13   Jan Alber, "Unnatural Temporalities: Interfaces Between Postmodernism, Science Fiction, and the Fantastic," in *Narrative, Interrupted: The Plotless, the Disturbing and the Trivial in Literature*, ed. by Markku Lehtimäki, Laura Karttunen, Maria Mäkelä (Berlin, Boston: De Gruyter, 2012), 188.

14   Jan Alber, "Unnatural Spaces and Narrative Worlds," in *A Poetics of Unnatural Narrative*, ed. by Jan Alber, Henrik Skov Nielsen and Brian Richardson (Columbus: The Ohio State University Press, 2013), 64.

15   Rüdiger Heinze, "The Whirligig of Time. Toward a Poetics of Unnatural Temporality," in *A Poetics of Unnatural Narrative*, 33.

16   Heinze, "The Whirligig of Time," 34.

17   Gomel, *Narrative Space and Time*, 30.

18   Samuel R. Delany, "Is Cyberpunk a Good Thing or a Bad Thing?," *Mississippi Review* 47/48 (1988): 30.

19   Delany, 31.

of paradoxical space that juxtaposes incompatible worlds.[20] The zone/paraspace mediates colliding realities and shifting ontological levels. The ontological refigurations in the discussed narratives derive partly from temporal and spatial dislocations embedded in quantum mechanics. For Scott Bukatman, the subatomic level is the locale of "the ultimate paraspace, an imploded and 'violently acausal' realm in which the ontological status of the subject and universe is opened to question and positioned for redefinition."[21] The zone/paraspace thus constitutes the space where two discordant spatio-temporalities coincide, one superimposed upon another, which results in the renegotiation of the primary reality.

Impossible worlds have been central to spatio-temporal representation in twenty-first-century cinema and television narratives. Engaging science and innovative narrative design, they propose the reassessment of the human perception of reality, asserting that it is too limited, and they demonstrate the existence of other layers of reality or multiple realities. The most popular impossible topologies are parallel universes branching off the primary reality (*Fringe*, *The Man in the High Castle*); special zones, each of which is governed by its own logic system (Area X in *Annihilation*, Black Lodge in *Twin Peaks*, the tesseract in *Interstellar*); and a time continuum along which one can move back and forth (*Dark*, *FlashForward*). The impossible temporalities, as mentioned above, are produced by temporal ruptures, or zero events, which bring the scales of the new physics to the scale of human time. For example, a nuclear power plant explosion in *Dark* creates a wormhole in an underground tunnel in a cave that connects 2019 with 1986, 1953, and other years as well to confirm Einstein's statement quoted at the beginning of the show that the division between past, present, and future is an illusion. In *Fringe* the zero event is an attempt to break through the membrane between two parallel universes, which leads to a series of cracks in the fabric of reality and to the erosion of the laws of physics.[22] In *Source Code*, after a terrorist attack, a soldier (Jake Gyllenhaal) is repeatedly transported by means of a computer program into an eight-minute slice of the past to acquire information about the terrorists so as to preempt another attack, and, in so doing, he creates parallel universes with different versions of events.

---

20   Brian McHale, *Postmodernist Fiction* (London and New York: Routledge, 2004), 43, 44.

21   Scott Bukatman, *Terminal Identity. The Virtual Subject in Postmodern Science Fiction* (Durham and London: Duke University Press, 1993), 174.

22   On *Fringe*, see Sonia Front, "'There's More Than One of Everything' – Time Complexity in *Fringe*," in *Time's Urgency: The Study of Time XVI*, ed. by Carlos Montemayor and Robert Daniel (Leiden, Boston: Brill, 2019), 234–254.

The preoccupation of these television series and films with the complexities and malleability of time is conveyed not only on the thematic level but also on the formal one. The temporal devices that structure the narratives engage multiple timelines, flashbacks, flashforwards, flashes sideways (into a parallel universe), repetitions, and so on, to create a manifold range of narrative temporalities. The plot threads encompassing various layers of reality are often set in temporal contradistinction with one another, which implies discordant speeds, rhythms, and senses of time – of people, objects, and matter – that clash within the narrative and advance the plot. The viewer needs to navigate the constantly shifting temporalities in order to decipher the puzzle of the narrative and understand the complex temporal relations between the characters and events. These techniques of achronological narrative structure are not just "narrative gimmicks but rather are an intrinsic part [...], a core and consistent element of the narrative format."[23] They contribute to the portrayal of time as dynamic and inherent to the event, rather than as being an abstract entity that engulfs it. The narrative thereby operates as a semantic framework that helps the viewer to conceptualize and enact new notions of space and time.

Twenty-first-century narratives discussed in this chapter bring the various scales of time to the level of human time in an attempt both to demonstrate how people operate at the intersection of these scales and to ask questions about time, the relationship between the past, present, and future, and the nature of reality. In the two case studies that follow, I analyze impossible temporalities and the impact they exert upon human time in an effort to advance the thesis that opening up the scale of physics' temporality to human experience is physically and psychologically intolerable to the characters. The subject who faces that temporality experiences a loss of spatio-temporal coherence, which leads to the deconstruction and reshaping of subjectivity. Confronted with the discontinuity of identity, the individual either endeavors to regain that lost continuity that we persist in believing is essential to our being, or shifts to discontinuity without death, that is, a rebirth on a different plane, in the paraspace.[24]

23   J. P. Kelly, *Time, Technology and Narrative Form in Contemporary US Television Drama. Pause, Rewind, Record* (New York: Palgrave Macmillan, 2017), 159.
24   Bukatman, *Terminal Identity*, 281.

2   The Snapshot of the Future in *FlashForward*

In the series *FlashForward*, based on Robert J. Sawyer's novel (1999) under the same title,[25] the temporal rift takes the form of a blackout all humanity experiences for two minutes and seventeen seconds, during which people glimpse their future half a year from now. The narrative follows an FBI investigation led by Mark Benford (Joseph Fiennes) into the causes of the blackout. It appears to have been engineered by a group of evil-doers and accidentally amplified by a fictional counterpart of Large Hadron Collider experiment, called the National Linear Acceleration Project and based in Palo Alto, California. The main storylines concentrate on Mark's attempt to avert a future in which his marriage to Olivia (Sonya Walger) is to disintegrate and he is to die in an attack on the FBI building, while his colleague Demetri Noh (John Cho) struggles to solve his own murder so as to prevent it. There are several other threads devoted to characters from around the world who work either to fulfill their future or to preempt it. Their actions and interactions demonstrate that their lives are as intermeshed and entangled with other people's as pieces in a jigsaw puzzle. Through this, the series explores a contemporary experience of space-time that is dislocated by data flows and information networks.

To get a holistic picture of the memories of the future, the FBI team creates the Mosaic Project website, where everyone can submit a description of their flashforward. As people experienced matching visions of the same future, the pieces of tomorrow – with each person's vision constituting a small element in a greater whole of a vast mosaic – combine to represent a consistent, coherent and plausible future of individuals but also companies, organizations, and governments. The Mosaic Project thus becomes a unique database of human memories, a central repository that exhibits a circuitry of global interconnections and interdependencies. As the window washer-turned-preacher (Gil Bellows) explains, "Each one of us is unique but we are being stitched together to form a tapestry, something larger that can't be understood until we step back to see the whole thing."[26]

The narrative structure of the show contributes to illustrating this web of interconnections. The pilot starts with Mark emerging from a car wreck and looking at the multiple car crashes around him, a helicopter smashing into a

---

25   The novel is quite different from the series; therefore I do not include it in my analysis.
26   *FlashForward*, s. 1, ep. 11, "Revelation Zero: Part 1," written by Brannon Braga and David S. Goyer, aired March 18, 2010 on ABC, https://abc.com/shows/flashforward.

skyscraper, and people asking each other the question, "Is it an attack?"[27] He watches people dealing with the effects of the accidents, and the sequence ends with a long shot of the cityscape in turmoil. After that, the announcement "Four Hours Earlier" appears on the screen, and a set of characters and relations is introduced within the next six minutes. Mark, his wife Olivia, and his daughter Charlie (Lennon Wynn) start their day, and then this narrative thread forks out to other plotlines: Mark visits his friend and sponsor, Aaron (Brian F. O'Byrne), who talks about his dead daughter, Olivia calls her colleague Bryce (Zachary Knighton), who attempts to commit suicide, Charlie's nanny arrives at the house, Mark and Demetri engage in a chase during a case. Then, a sequence of cars in a traffic jam cuts to Mark's flashforward: a series of blurry, distorted images of Mark in his office rendered with some shots from his point of view and some from the outside, after which his consciousness – and everyone else's, as it turns out later – returns to the present. The viewer is returned to the starting point of the pilot episode: Mark emerging from a car wreck.

This blend of futurity and completion is intrinsic to the show's temporal structure and block-universe idea. After admitting their involvement in the time displacement, the physicists explain the mechanism of what happened. During the experiment, they created such extreme energies that "they sent shockwaves through the consciousness field and jolted all of humanity's awareness to a different place in space-time."[28] If we compare space-time to a pile of motion-picture frames stacked up, "now" is the currently illuminated frame. During the flashforward, it is the slice representing the moment half a year into the future that is illuminated, rather than the present one. This explanation is theoretically congruent with Einstein's idea of the block universe, in which past, present, and future are coexistent and immutable and share equal ontological status. The physicist Lee Smolin puts this deterministic view of the world as follows:

> Relativity strongly suggests that the whole history of the world is a timeless unity; present, past and future have no meaning apart from human subjectivity. Time is just another dimension of space, and the sense we have of experiencing moments passing is an illusion behind which is a timeless reality.[29]

27   *FlashForward*, s. 1, ep. 1, "No More Good Days," aired Sept. 24, 2009.
28   *FlashForward*, s. 1, ep. 16, "Let No Man Put Asunder," aired April 15, 2010.
29   Lee Smolin, *Time Reborn: From the Crisis in Physics to the Future of the Universe* (Boston, New York: Houghton Mifflin Harcourt, 2013), xxii.

During the flashforward, a future slice of time is superimposed upon the present slice, and its temporality clashes with the primary temporality, creating a mental paraspace for the characters. After that, the already seen future becomes relegated to the past: "It is a future that has already happened."[30] In other words, the virtuality of the future event is transformed into the actuality of the past event; the future is granted the full materiality and weight of the past. Yet, simultaneously, it is also what is to come; therefore, contingency and necessity, certainty and uncertainty, free will and determinism amalgamate in the paraspace of colliding realities. Some characters strive to preserve the openness and virtuality of the future. For instance, Olivia, a doctor, declares that the information about her patient gained from the flashforward is not "data," and she does not intend to incorporate it into her practice of medicine. However, she is forced to reconsider when her patient is about to die. She uses knowledge from the flashforward, which actually does save his life.

The future snapshot grants human consciousness access to a temporal perspective normally inaccessible to human perception and the result is that it cannot cope with this perspective. Human senses are able to process only one temporality at a time, and therefore the characters' contact with the present is lost for the whole duration of the flashforward. In consequence, the temporal rupture of the blackout is one of the most significant disasters in the history of the human race, as it entails unprecedented mayhem and death. The psychological consequences of the characters seeing the future, on the other hand, are such that it dismantles the present: "They see the future and the knowledge of it ends up destroying them."[31] Opening up the time of physics – here a possibility inherent in Einstein's block universe – to human experience is thus physically and psychologically overwhelming.

The event, "a black swan," that is, a high-impact event beyond human expectation and prediction that leads to global-scale consequences,[32] is highlighted as day zero and the plotlines that follow are designated as happening before or after the blackout. The time that follows the event is decidedly different from the time before; it is marked by anxiety, irrational behavior, and an obsession with the future that becomes so important that it overrides the present. People's reactions to the event depend on whether they support the idea of determinism or the idea of free will. Those who believe that the future is fixed lose all sense of agency and make decisions "based on what will happen, not

---

30   *FlashForward*, s. 1, ep. 2, "White to Play," aired Oct. 1, 2009. These are a German prisoner's (Curt Lowens) words.
31   *FlashForward*, s. 1, ep. 17, "The Garden of Forking Paths," aired April 22, 2010.
32   *FlashForward*, s. 1, ep. 4, "Black Swan," aired Oct. 15, 2009.

on what could," maintaining that one "can't be fighting fate."[33] Those who are convinced that people have free will and can forge their lives' trajectory argue that the "memories of events that haven't occurred yet" might be a possible future, the knowledge of which might enable the opportunity to change it.[34] The third group of people, those who did not have a vision, conclude they are going to be dead half a year from now, so they form a group called the Ghosts. Its subgroup, The Blue Hand, which signifies "the surrender to the inevitable," seek to at least control the manner in which they die; therefore during their meetings they create an opportunity for those who wish to commit suicide.[35]

The clash between fate and free will serves as the driving force of the show, which ends with the day pre-experienced in the flashforward and reveals that to some, the future ultimately comes to pass as predicted, while to others, it unfolds in a different way. Mark and Demetri, for example, manage to prevent Demetri's death. After many attempts to preserve her marriage and elude a new relationship, Olivia fails, resigned: "It's about this moment, right here. I fought it, and resisted it, and yet here it is. It happened. The future happened."[36] In some cases, different paths lead to the same outcome. Mark's and Demetri's colleague Al (Lee Thompson Young) has seen in his flashforward that he has caused a woman's death, so he decides to forestall this by committing suicide, hoping to "change the game" and prove that "the future is unwritten," and "our choices do matter,"[37] yet still, his colleague kills the woman in a car accident. Gibbons (Michael Massee) has created hundreds of small-scale flashforwards, and in each of them, he has seen a different possible future branching off each critical decision; however, in all his possible futures he dies on the same day. He theorizes:

> The point is once we've glimpsed [a possible future], the future wants to happen and gains weight – it's like atmosphere pressure bearing down, and if we want to escape that pressure, we have to do something drastic.[38]

Some flashforwards are misinterpreted: Charlie's babysitter Nicole (Peyton List) thinks she is being drowned when, in fact, she is being saved from drowning; Demetri's fiancée Zoey (Gabrielle Union) thinks she saw her and Demetri's

---

33   *FlashForward*, s. 1, ep. 3, "137 Sekunden," aired Oct. 8, 2009.
34   *FlashForward*, s. 1, ep. 1, "No More Good Days."
35   *FlashForward*, s. 1, ep. 7, "The Gift," aired Nov. 5, 2009.
36   *FlashForward*, s. 1, ep. 22, "Future Shock," aired May 27, 2010.
37   *FlashForward*, s. 1, ep. 7, "The Gift."
38   *FlashForward*, s. 1, ep. 17, "The Garden of Forking Paths."

wedding while it was his funeral; Mark's daughter hears "Mark Benford is dead," which turns out to be a guess, rather than a statement of fact.

Chaos theory has shown us that small changes in initial conditions produce considerable effects over time, and we might therefore assume that, because the characters' knowledge of the future constitutes a change in initial conditions, a different outcome would therefore occur than the one glimpsed in the flashforward. Yet in some cases, the flashforward seems to be to a future that considers the very flashforward in the past and its consequences to the human psyche. For instance, a police officer Janis (Christine Woods), who never wanted a baby, is pregnant in her vision, so now she wants to get pregnant. On a mission in Somalia, Demetri – who is afraid that he is going to die soon, as predicted – suggests he will be Janis's baby's father. This chain of events seems like a loop triggered by the consciousness shift itself. Ultimately, then, the future appears to be a blend of determinism and free will. The preacher argues that these visions are just snapshots, small portions taken out of context that need to be interpreted: "It's not fate versus free will, it's fate and free will."[39] It appears, then, that the multiplication of outcomes obviates the stasis of four-dimensional space-time.

The instability of the future is illustrated by the fact that the visions presented in the course of the season are always visually marked off from the present course of events. (Mark's snapshot is exceptionally unstable: it consists of hazy images stitched together by rapid cuts, which is explained by his being under the influence of alcohol). The characters experience the visions in the same way as they experience present events ("One second I was in the car, the next I was somewhere else"[40]), but the memories of those visions are marked off in a way reminiscent of the typical way in which memories of the distant past used to be presented in films and shows – with the memory framed by hazy edges, signifying their instability and provisional nature.[41] This strategy underlines a different ontological status for the future slice of time, which highlights provisionality and points to a multiplicity of possible futures. By acting upon that possible future, however, the characters transform the possible into the actual, and the image becomes stable.[42]

---

39   *FlashForward*, s. 1, ep. 11, "Revelation Zero: Part 1."

40   *FlashForward*, s. 1, ep. 1, "No More Good Days."

41   Now flashbacks are most often seamless shifts to past events in any type of film/show.

42   A similar strategy is used in the film *Source Code*, in which the protagonist repeatedly sees a hazy image of Chicago's sculpture Cloud Gate, which proves to be a flashforward of a possible future. With time the image becomes clearer until finally it becomes stable when the protagonist embarks on a course of events that leads him to the sculpture.

The future snapshots and The Mosaic Project are symptomatic of what Richard Grusin has termed "premediation," that is, the news media's propensity to anticipate future events, particularly after 9/11. He recognizes 9/11 as a "watershed moment, a sea change" that designates a transition from a culture of remediation – that is, thinking about the past – to the culture of premediation – that is, speculating about the future.[43] Grusin considers premediation as an essentially American response to 9/11, in which "the United States seeks to try to make sure that it never again experiences live a catastrophic event like this that has not already been premediated."[44] This demand "to see the future, not as it emerges immediately into the present, but before it ever happens" has replaced a cultural need for immediacy.[45] Similarly, FBI agents on the show strive not only to find those responsible for the blackout but also to establish whether there is going to be another blackout and preempt it.

While preempting the future half a year from now on the basis of 137 seconds' data – as it is attempted in *FlashForward* – might appear difficult, it is certainly even more so in the case of environmental despoliation that has spanned many decades and many generations and needs to be confronted with huge scales of geological time – as it happens in *Annihilation* to be explored in the following discussion – to predict its impact on environment in the present and in the future. Deep time encompasses a length of time that exists on a scale incompatible with human time. As *FlashForward* brings the huge scale of relativistic universe to the perception of the characters by means of a Large Hadron Collider experiment, involving the collision of enormous masses of energy, so does *Annihilation*, in which a forceful strike of a meteor leads to the encounter of human time with cosmological time and geological time.

3   Area X in *Annihilation*

Alex Garland's film *Annihilation* (2018), based on Jeff VanderMeer's novel *Annihilation* (2014), the first part of *The Southern Reach Trilogy* (2014), follows an expedition into a scientific anomaly, Area X, a zone that came into being about three years before when a meteor hit a lighthouse.[46] The ever-expanding

---

43   Richard A. Grusin, "Premediation," *Criticism* 46.1 (2004): 21.
44   Grusin, 21.
45   Grusin, 21.
46   The film introduces many significant changes from the novel, so I do not include the novel in my analysis. Garland reports that for him reading VanderMeer's *Annihilation* was a "dream-like experience." He read the novel only once and worked from the "memory of the book," hoping to adapt in this way the atmosphere of the novel. "'Annihilation'

zone is cut off from the rest of the world by a border called the Shimmer, a kind of iridescent rippling force field, a translucent membrane, which lets people and objects in but through which nothing returns. A government agency called the Southern Reach, created for the purpose of studying this scientific anomaly, regularly sends expeditions into its interior. The only survivor has been Kane (Oscar Isaac), who, some time after his expedition lost contact with the agency, appears by unknown means in the house he shares with Lena (Natalie Portman). A zombie-like other, he does not remember anything of the expedition and soon vomits up blood, after which he is taken to the Southern Reach medical unit.

Lena, a biologist, takes part in the next expedition, together with a psychologist, Dr Ventress (Jennifer Jason Leigh), a magnetologist, Anya Thorensen (Gina Rodriguez), a paramedic, Cass Sheppard (Tuva Novotny), and a physicist, Josie Radek (Tessa Thompson). After crossing the magnetic field, inside the paraspace, the team discovers an uncanny pristine ecosystem, consisting of forest and marshland filled with colorful birds and lush vegetation, and crystal clear air tinted by rainbow-colored, diluted Shimmer. Many of the plants and animals are mutations and oddities, duplications or corruptions of form: different species of flowers growing on one stem, duplicated deer with flowers on their branch-like antlers, anthropomorphic trees, a super-strong alligator with shark teeth, and a bear with a human voice. Garland enacts the otherness of the zone and the Shimmer by presenting them visually as phantasmagoric and dreamlike. The border assumes a "petrol slick" that looks "like a heat haze, but with prismatic effects," inspired by a Mandelbulb 3D fractal shape.[47] The hallucinogenic quality of Area X is meant to emphasize its otherworldliness, strangeness, unfamiliarity and spatio-temporal separateness, contributing to "a visceral, in-the-moment quality that often uses elements of surreal or transgressive horror for its tone, style and effects," characteristic of the genre of the New Weird, which *Annihilation* exemplifies.[48]

---

director Alex Garland chats with CNET about the upcoming film," https://www.youtube.com/watch?v=nYhT5Ey42gg; "Alex Garland 'Annihilation' Talks at Google," https://www.youtube.com/watch?v=w5i7idoijco.

47   Andrew Whitehurst, the film's visual effects supervisor, explains: "we unwrapped a Mandelbulb, made it into a wall, and then had multiple different Mandelbulbs running at different speeds, so you always have this churning shape that has some structure in it, which would ebb and flow continually." "Mandelbulbs, Mutations and Motion Capture: The Visual Effects of Annihilation," https://vfxblog.com/2018/03/12/mandelbulbs-mutations-and-motion-capture-the-visual-effects-of-annihilation/.

48   Jeff VanderMeer, "The New Weird: 'It's Alive?'," in Ann and Jeff VanderMeer, eds., *The New Weird* (San Francisco: Tachyon Publications, 2008), xvi.

The film does not reveal the origin of the zone, yet it suggests its alien provenance: in the beginning, a stream of light is directed from outer space towards the earth, then the stream flows towards the lighthouse before a meteor hits it.[49] The zero event of the meteor strike initiates an irreversible process that establishes a new temporality, characterized by a breakdown of capitalist linear time (understood as progress and as measure of productivity and economic growth), and replaces humanity's governance of the earth, splitting history into before and after. Area X, the self-contained, impenetrable segment of space-time, isolated from the outside world, is situated in total contradistinction to that world. Time in the zone is relative to the time outside: it flows at an accelerated rate, and, as a result, the processes of growth and transformation advance very fast. The inextricable connection of space and time, inherent in relativity theory,[50] is manifested in the zone in the shifts of space in relation to temporal processes. Space-time cannot be separated into space and time because the reconfigurations of space are accompanied by the reconfigurations of time, or, in other words, spatiality and temporality are mutually constitutive.

The zone exerts a transformative effect on the people who enter it. The scientists of the current expedition find a video left behind by the previous expedition in which Kane cuts open his companion's abdomen to reveal serpent-like insides twisting in his body. Soon the scientists themselves spot changes on their bodies: Josie's skin becomes tree-like; Anya's fingertips seem to be moving, and she starts to lose her sanity; Sheppard is killed by a bear, but part of her consciousness and her voice are imprisoned inside it; and Lena studies her blood under a microscope to discover that it has already mutated. It transpires that the transformation and appropriation by Area X begin at the moment the Shimmer is crossed. The physicist gradually theorizes what the Shimmer is: a prism that refracts not only light but also other physical properties, such as DNA. The zone demonstrates the ability to mutate matter and duplicate it, combine and recombine it across species and within them. People do not return from the paraspace because they are incorporated into other

---

49   VanderMeer's trilogy divulges more information. Area X, as the third part of the trilogy, *Acceptance*, explains, is a slice of space-time "transplanted from somewhere far remote" (Jeff VanderMeer, *The Southern Reach Trilogy* [Toronto: Harper Collins, 2015], 527). One of the scientists describes its origin in his report: "Area X has been created by an organism left behind by a civilization so advanced and so ancient and so alien to us and our own intent and our own thought processes that it has long since left us behind, left everything behind" (503).

50   See, for example, Sean Carroll, chapter 4, *From Eternity to Here. The Quest for the Ultimate Theory of Time* (Oxford: Oneworld, 2010).

organisms. This process constitutes a correction of the error in human biological make-up, which is foreshadowed by Kane's and Lena's conversation before his expedition, during which she argues:

> You take a cell, circumvent the Hayflick limit, you can prevent senescence. It means the cell doesn't grow old, it becomes immortal. Keeps dividing, doesn't die. We see aging as a natural process but it's actually a fault in our genes.[51]

The theme of mutations appears at the beginning of the film when Lena, a cell biologist, lectures at the university on cancer, which involves pathological cell divisions. Similarly, the aberrations produced by the Shimmer are termed "mutations," cancerous, "malignant, like tumors."[52] Additionally, although the government perceives the whole event as an environmental catastrophe, from the non-anthropological perspective, it can be construed the other way round: humanity is the cancer of the planet.

The complete reorientation of subjectivity causes for most characters an unbearable disintegration of the self when they are assimilated by the environment. Kane confesses on the tape, "My flesh moves like liquid. My mind is loose. I can't bear it,"[53] and he then commits suicide while his duplicate records the tape. While Kane is unable to shed his anthropocentric attachment to individuality and independence, other characters respond more open-mindedly to the inhuman, transformative power of nature, recognizing it as the price of survival; the price of surpassing human biological capabilities and achieving immortality. In the cave under the lighthouse, visually reminiscent of the brain, Ventress surrenders herself to the Shimmer. She realizes that the biotic intelligence has permeated her and that it will carry out annihilation: her mind, like other people's before her, will be fragmented and dissolved until no part remains.

In physics, annihilation takes place when a particle collides with its antiparticle and they annihilate each other while their mass is converted into energy. Accordingly, the purpose of the zone is not to obliterate human biology but to recreate it from the molecular level. It recycles Ventress's matter, which assumes the form of a vibrating fractal vortex that absorbs a few drops of Lena's blood. The drops multiply uncontrollably until a humanoid form

---

51   Alex Garland, dir., *Annihilation* (Paramount Pictures, Skydance Media, Scott Rudin Productions, DNA Films, 2018). The script was also written by Alex Garland.
52   Garland, *Annihilation*.
53   Garland, *Annihilation*.

emerges. When it turns into Lena's double and imitates each of her movements, she fights with it to maintain her anthropological privilege. She manages to outsmart it, handing it a grenade. Both the creature and the Shimmer are destroyed in the explosion, and Lena returns to the outside world. She is reunited with her supposed husband, who is, in fact, his alien double – a copy of his body without his consciousness. When he inquires whether she is Lena, she does not reply but her eyes glow, as his do, in the closing sequence of the film. This glow evinces that Lena has been infected by the Shimmer, which is corroborated by her transformed blood. However, after initial resistance, at this point, she perceives the transformation as a positive change. She attempts to get reconciled to the alterations within herself and to understand the rationale behind her relationship with the environment. This is demonstrated in the following exchange:

> Interrogator: What did it want?
> Lena: I don't think it wanted anything.
> Interrogator: But it ... attacked you.
> Lena: It mirrored me. I attacked it. I'm not sure it even knew I was there.
> Interrogator: It came here for a reason. It was mutating our environment, it was destroying everything.
> Lena: It wasn't destroying. It was changing everything ... It was making something new.

Although to the interrogator (Benedict Wong), the process of dismantling the duality between subject and object is tantamount to destruction, to Lena, it amounts to a positive metamorphosis. What is significant here is that the story is framed through a series of flashbacks that constitute Lena's testimony at the Southern Reach, which she delivers after her return from Area X. The narrative is thus told by a person who has already been mutated, which calls into question the subject position from which she is speaking.

Annihilation illustrates, as *FlashForward* does as well, how an altered spacetime may impact human psyche and body. The encounter with alien time leads to the scientists' memory loss and disorientation, and they preserve no memory of three or four days after crossing the border. The break with customary time leads to the loss of the sense of time: Lena believes she spent "days, maybe weeks"[54] in the zone, when, in fact, she was gone for almost four months. Additionally, the accelerated rate of time instigates disintegration of people's minds such as in the early stages of dementia, and leads to a total

---

54   Garland, *Annihilation*.

reconfiguration of human biology, which becomes intermingled with all forms of existence, mind, and matter. A process that would normally extend over thousands of years, in the alien spatio-temporality of Area X consumes only days, sometimes hours, and in the lighthouse – the center of the Shimmer – just minutes and seconds.

The sped-up rate of the transformative processes thereby allows human time to be set in relation to deep time. The hierarchy of being has been obviated; the individual is engulfed in the collective, and all steps of evolution are synthesized in each organism. Every organism thus encompasses what J. G. Ballard has called "archaeapsychic time," the temporality in which the boundary between the physical and the psychological is blurred, and in which the time of each epoch is compressed.[55] *Annihilation* establishes through this alternative temporal modality beyond capitalist modes of production and progress a posthuman vision that enables all species to coexist. In doing so, it exemplifies what Rosi Braidotti has called "zoe": "the wider scope of animal and non-human life," "the dynamic self-organizing structure of life itself" that denotes "generative vitality. It is the transversal force that cuts across and reconnects previously segregated species, categories and domains." For her, this zoe-centered egalitarianism is the crux of "the postanthropocentric turn."[56] The focus on zoe erases the differentiations between living and dying and establishes zoe as "a posthuman yet affirmative life-force."[57]

Annihilation represents broader scales of space and time, the deep time of the emergence of life, which integrates human time into the end of time, thereby adopting the perspective of the Anthropocene. Human time is juxtaposed with a prehistory that subsumes civilization and human consciousness. The transformation of the human species is brought about by a climate destabilized by destructive human habits. The environment is in bad condition, doomed to despoliation by humanity insensitive to the planetary crisis, to what Christophe Bonneuil and Jean-Baptiste Fressoz call the "shock of the Anthropocene."[58] The alien intervention allows humans to be upgraded into beings integrated into and united with the environment, experiencing a feeling of indissoluble connection. The film thereby enacts what Timothy Morton calls for: a reconfiguration of the concept of "nature," the undoing of the "habitual distinctions between nature and ourselves":

55   J. G. Ballard, *The Drowned World* (London: Harper Perennial, 2008), 44.
56   Rosi Braidotti, *The Posthuman* (Cambridge: Polity Press, 2013), 60.
57   Braidotti, 136, 115.
58   Christophe Bonneuil and Jean-Baptiste Fressoz, *The Shock of the Anthropocene: The Earth, History and Us*, trans. David Fernbach (London: Verso, 2016), 5.

It is supposed not just to describe, but also to provide a working model for a dissolving of the difference between subject and object, a dualism seen as the fundamental philosophical reason for human beings' destruction of the environment. If we could not merely figure out but actually *experience* the fact that we were embedded in our world, then we would be less likely de destroy it.[59]

The interrogation of the concepts of time and space in the film underscores the feasibility of new kinds of change that involve the transformation of humans from their complacent anthropocentrism to collectivity and relationality. Contrary to traditional tropes of alien invasion, in which humanity regains control, *Annihilation* offers a vision in which the human being does not hold the central position in the hierarchy of species but constitutes one of the many forms of existence, and human history represents only a fraction of the history of the universe.

The new space-time locked within the Shimmer forces the renegotiation of the human relation to space and time. Roy Scranton argues that we need to learn how to die in the Anthropocene, by which he means that we need "a new conceptual understanding of reality, a new relationship to the deep polyglot traditions of human culture that carbon-based capitalism has vitiated through commodification and assimilation."[60] In *Annihilation*, a retreat into archaea-psychic time and the displacement of the self to reach a new way of becoming lead to the abandonment of the destructive notions of space and time that contributed to the Anthropocene.

4   Conclusion

In the films and television series discussed here, the temporal rifts produced by the release of vast quantities of energy or other global-scale events open up realms in which customary conceptualizations of causality, linearity, temporality, and ontology break down. The realms are governed by a time that clashes with the primary reality's time and, as such, is beyond the capacities of the human organism. This temporality is shattering to the human body and psyche as it derails them from the trajectory of being-towards-death. The self becomes discontinuous with the world it inhabits. As the anchor of the subject

---

59   Timothy Morton, *Ecology Without Nature: Rethinking Environmental Aesthetics* (Cambridge: Harvard UP, 2007), 63–64. Emphasis in original.

60   Roy Scranton, *Learning to Die in the Anthropocene: Reflections on the End of a Civilization* (San Francisco: City Light Books, 2015), 19.

and the site of the interface with paraspace, the body is not able to cope with the temporal rupture; it becomes "the arena on which temporal dislocations are played out."[61] The psyche is not able to process clashing temporalities, either; therefore it responds – in the cases under discussion – with disorientation, loss of consciousness, and/or suicide. Confronted with an altered space-time, the characters struggle to sustain the integrity of their reality, and, by inference, their very selves. A more effective strategy, which allows for agency, is for characters to learn to navigate the tapestry of interwoven scales of time, and through this to learn to deal with time and space beyond human scales. To those who can master it, the transitional state generated at the intersection of various layers of reality opens up a multiplicity of exuberant existential possibilities and new modes of being. In *FlashForward*, knowledge of the future allows some to preempt it. In *Annihilation*, giving up one's individuality to be integrated into Area X as part of the ecosystem enables the subject to persist possibly for eternity.

Narratives showcasing temporal rifts that reveal multiple layers of reality are responding to the existential anxieties of the twenty-first century. These temporal rifts are symbolic of 9/11, which resulted in a breakdown of historical continuity. Many films and shows have allegorized 9/11, and, more widely, the apocalyptic spirit of the contemporary world, and fantasy and science fiction worlds have offered "alternatives to the frightening new reality,"[62] marked by fear of terrorism, consequences of capitalist exploitation, climate change, and so on. Spencer Kornhaber notes that in the alternate reality stories, "the appealing implication is not only that there is another place, but that we can understand what it is and how to get there." Significantly, the new visions that offer "imaginative comfort" are based on knowledge rather than faith.[63] Fiction has a momentous role to play as a tool through which we can imagine alternate realities that are not subjected to the dominant ideologies of our world. The narratives that I have dealt with here thus serve as a map of contemporary anxieties and a starting point from which to envision multiple futures and realities that invite us to imagine their impossibilities. Imagining them serves as a way to comprehend the contemporary world from future (im)possible locations of catastrophic aftermath. From that vantage point, the fictions provide insights

61   Alison Landsberg, "Cinematic Temporality: Modernity, Memory and the Nearness of the Past," in *Time, Media and Modernity*, ed. by Emily Keightley (London: Palgrave Macmillan, 2012), 86.

62   Susan Napier, *Anime from Akira to Howl's Moving Castle, Updated Edition: Experiencing Contemporary Japanese Animation* (St Martin's Press, 2005), xi.

63   Spencer Kornhaber, "Pop Culture is Having a Metaphysical Moment," *The Atlantic*, January 5, 2017, https://www.theatlantic.com/entertainment/archive/2017/01/the-oa -stranger-things-westworld-metaphysical-moment-alternative-realities-tv/511808/.

about the present world and might serve as a way of premediating the future
and animating social and political action.

References

Alber, Jan. "Unnatural Spaces and Narrative Worlds." In *A Poetics of Unnatural Narrative*, edited by Jan Alber, Henrik Skov Nielsen and Brian Richardson, 45–66. Columbus: The Ohio State University Press, 2013.

Alber, Jan. "Unnatural Temporalities: Interfaces Between Postmodernism, Science Fiction, and the Fantastic." In *Narrative, Interrupted. The Plotless, the Disturbing and the Trivial in Literature*, edited by Markku Lehtimäki, Laura Karttunen, Maria Mäkelä Berlin, 174–191. Boston: De Gruyter, 2012.

Alber, Jan. *Unnatural Narrative: Impossible Worlds in Fiction and Drama*. Lincoln: University of Nebraska Press, 2016.

"Alex Garland 'Annihilation' Talks at Google." https://www.youtube.com/watch?v=w5i7idoijco.

"'Annihilation' director Alex Garland chats with CNET about the upcoming film." https://www.youtube.com/watch?v=nYhT5Ey42gg.

Ballard, J. G. *The Drowned World*. London: Harper Perennial, 2008.

Ballard, J. G. *Millennium People*. London: Harper Perennial, 2008.

Bonneuil, Christophe and Jean-Baptiste Fressoz. *The Shock of the Anthropocene: The Earth, History and Us*. Translated by David Fernbach. London: Verso, 2016.

Boxall, Peter. "Late: Fictional Time in the Twenty-First Century." *Contemporary Literature* 53.4 (2012): 681–712.

Braga, Brannon and David S. Goyer. *FlashForward*. ABC, 2009–2010, 22 eps.

Braidotti, Rosi. *The Posthuman*. Cambridge: Polity Press, 2013.

Bukatman, Scott. *Terminal Identity. The Virtual Subject in Postmodern Science Fiction*. Durham and London: Duke University Press, 1993.

Carroll, Sean. *From Eternity to Here. The Quest for the Ultimate Theory of Time*. Oxford: Oneworld, 2010.

Delany, Samuel R. "Is Cyberpunk a Good Thing or a Bad Thing?," *Mississippi Review* 47/48 (1988): 28–35.

Front, Sonia. "'There's More Than One of Everything' – Time Complexity in *Fringe*." In *Time's Urgency. The Study of Time XVI*, edited by Carlos Montemayor and Robert Daniel, 234–254. Leiden, Boston: Brill, 2019.

Garland, Alex, dir. *Annihilation*. USA: Paramount Pictures, Skydance Media, Scott Rudin Productions, DNA Films, 2018.

Gomel, Elana. *Narrative Space and Time. Representing Impossible Topologies in Literature*. New York, London: Routledge, 2014.

Grusin, Richard A. "Premediation." *Criticism* 46.1 (2004): 17–39.

Hall, John R. "Apocalypse 9/11." In *New Religious Movements in the Twenty-First Century: Legal, Political and Social Challenges in Global Perspective*, edited by Philip C. Lucas and Thomas Robbins, 215–230. New York: Routledge, 2004.

Heinze, Rüdiger. "The Whirligig of Time. Toward a Poetics of Unnatural Temporality." In *A Poetics of Unnatural Narrative*, edited by Alber, Nielsen and Richardson, 31–44.

Heise, Ursula K. *Chronoschisms: Time, Narrative, and Postmodernism*. New York: Columbia University, 1997.

Kelly, J. P. *Time, Technology and Narrative Form in Contemporary US Television Drama. Pause, Rewind, Record*. New York: Palgrave Macmillan, 2017.

Kornhaber, Spencer. "Pop Culture is Having a Metaphysical Moment." *The Atlantic*, January 5, 2017. https://www.theatlantic.com/entertainment/archive/2017/01/the-oa-stranger-things-westworld-metaphysical-moment-alternative-realities-tv/511808/.

Korte, Barbara and Frédéric Regard. "Narrating 'precariousness': Modes, Media, Ethics." In *Narrating "precariousness": Modes, Media, Ethics*, edited by Barbara Korte and Frédéric Regard. Heidelberg, 2014.

Landsberg, Alison. "Cinematic Temporality: Modernity, Memory and the Nearness of the Past." In *Time, Media and Modernity*, edited by Emily Keightley, 85–101. London: Palgrave Macmillan, 2012.

Leibniz, Gottfried Wilhelm. *Philosophical Papers and Letters*, vol. 2, edited and translated by Leroy E. Loemker. Dordrecht: Reidel, 1969.

Luckhurst, Roger. "Traumaculture." *New Formations* 50 (2003): 28–47.

"Mandelbulbs, mutations and motion capture: the visual effects of Annihilation." https://vfxblog.com/2018/03/12/mandelbubs-mutations-and-motion-capture-the-visual-effects-of-annihilation/.

McHale, Brian. *Postmodernist Fiction*. London and New York: Routledge, 2004.

Morton, Timothy. *Ecology Without Nature: Rethinking Environmental Aesthetics*. Cambridge: Harvard University Press, 2007.

Napier, Susan. *Anime from Akira to Howl's Moving Castle, Updated Edition: Experiencing Contemporary Japanese Animation*. St. Martin's Press, 2005.

Scranton, Roy. *Learning to Die in the Anthropocene: Reflections on the End of a Civilization*. San Francisco: City Light Books, 2015.

Smolin, Lee. *Time Reborn: From the Crisis in Physics to the Future of the Universe*. Boston, New York: Houghton Mifflin Harcourt, 2013.

Tew, Philip. *The Contemporary British Novel. Second Edition*. London: Continuum, 2007.

VanderMeer, Jeff. "The New Weird: 'It's Alive?'." In *The New Weird*, edited by Ann and Jeff VanderMeer, ix–xviii. San Francisco: Tachyon Publications, 2008.

Vandermeer, Jeff. *The Southern Reach Trilogy*. Toronto: Harper Collins, 2015.

CHAPTER 9

"Out of Repetition Comes Variation": Varying Timelines, Invariant Time, and Dolores's Glitch in *Westworld*

Jo Alyson Parker and Thomas Weissert

Abstract

This article first explores how the multiple narrative timelines in the television science fiction series *Westworld* highlight the invariant narrative loops to which the android/gynoids "hosts" are subject. It then examines how host consciousness emerges once the hosts can remember their previous passages through their loops and thus begin to grasp their pasts and conceive of their futures. It concludes by considering how the fictional simulated beings of *Westworld*, as they escape from the unvarying timeloops to which they have been subject, demonstrate the role of time, memory, and narrative in the development of human consciousness.

Keywords

androids – consciousness – embedded narrative – glitch – gynoids – invariance – memory – metaconsciousness – narrative loops – self-narration – timelines – *Westworld*

• • •

Then, when are we? Is this now?

DOLORES in *Westworld*

∵

The HBO series *Westworld* (2016–ongoing), loosely based on Michael Crichton's 1977 film of the same name, presents a thought-provoking exploration of consciousness against the backdrop of a Wild-West theme park wherein human "guests" fulfill their fantasies by inflicting predations such as murder and rape on their android/gynoid "hosts." Co-created by Jonathan Nolan and Lisa Joy, the series flips the premise and emphasis of the original film, which focused on the plight of guests terrorized by malfunctioning hosts. Instead, the series focuses on – and viewer sympathy transfers to – the hosts, whose "lives" are defined by apparently unvarying narrative loops. After these loops are altered or terminated by, to use a series catchphrase, the "violent delights" of the guests, the hosts' memories are wiped, and they begin a passage through their loops anew. Their time is reset.

The first season of *Westworld* is a narrative about narrative in the most saturated sense. It concerns the writing of narratives as the park's co-creator Robert Ford and his fellow scriptwriters concoct the multiple interactive narratives that the hosts will enact. ("We sell complete immersion in a hundred interconnected narratives," proclaims Lee Sizemore, the head of Narrative and Design, in the pilot episode.[1]) Over the course of the season, the episodes skew narrative temporality by featuring varied timelines, thus toying with viewers' expectations for a coherent plot. And, importantly, it concerns the self-narrativizing that not only the fictional hosts do but also we humans do as both the foundation and the sign of our consciousness, of our plot of the self.

Through varied timelines in the overarching narrative and in the interactive embedded narratives that the hosts enact, *Westworld* highlights the invariant time that confines the programmed hosts and explores how random, human-caused deviations from machine-perfect behavior cycles lead to an accumulation of newly accessible memory traces that fuel and ignite the self-narration process necessary for and indicative of consciousness. The hosts' accession to consciousness, in effect, mimics humanity's own.

1 Narrative Loops and Varying Timelines

> I wake up every day, right here, right in Punxsutawney, and it's always February 2nd, and there's nothing I can do about it.
>
> PHIL CONNORS in *Groundhog Day*

∴

---

1   We use the episode titles to indicate the source of the passages. The appendix includes information about each of the episodes in season one and the order in which they appear, as well as a list of the key characters and the actors who play them.

The first season of *Westworld*, subtitled *The Maze*, takes place in what is essentially an immersive, interactive adult Disneyland that simultaneously replicates and deconstructs a mythic American West. It is populated with stock characters from Western lore, with the roles performed by the android/gynoid hosts, which the guests can use to satisfy their basest instincts – to "go black hat," in the words of one particularly predacious guest ("Dissonance Theory"). Charlotte Hale, the ambitious executive director of the company, sums up guest behavior best: "Most of the guests just want a warm body to shoot or to fuck" ("The Well-Tempered Clavier").[2] Thus the hosts are fixed in narrative loops that are designed to appeal to guests' fantasies and that may play out over one to several days before culminating in a climactic event. For example, Dolores Abernathy, at 30 years the oldest host in the park, enacts the role of the pretty and innocent rancher's daughter; she has been programmed, as she eventually realizes, "to be the damsel" ("Contrapasso") in need of rescue by a valiant guest. Teddy, the handsome young cowboy who "loves" Dolores, has been programmed to challenge the guests who will then vanquish him and violate her. Significantly, Dolores tells Teddy, "There's a path for everyone. Your path leads you back to me" ("The Original"). This time-worn metaphor proclaiming love's constancy becomes literalized: Teddy's loop will always lead him back to Dolores – from the meet-cute encounter in the streets of Sweetwater when he picks up the can that she has dropped and, later, to his "death" while fighting for her honor. After the hosts complete a passage through the loop, their memories are wiped and their bodies are repaired so that they can perform their scripted actions with the next set of paying guests. Although hosts can "kill" one another, their programming prevents them from harming a guest; they are pawns in a rigged game of which they are unaware.[3]

The first season of *Westworld* functions as a (mainly) self-contained narrative arc featuring a plot that plays out over the course of ten episodes.[4] This plot develops according to a complex nonlinear narrative structure that complicates the relationship between an overarching frame narrative taking place

---

2   We learn in the episode "Dissonance Theory" that, although hopeful storylines had been developed initially, the guests did not choose them.
3   Not only does the host programming protect the life of the guests, but the weapons themselves are rigged to be non-lethal to guests as well.
4   Prior to the airing of the first episode, "producers wanted to firm up their master plan for the entire series – all the way to the show's eventual finale" (Hibberd 2016), and it has been projected that the overall story may play out over five to seven seasons. Although some of the gaps in time that occur in the first season have been filled in during the second and third seasons, "Nolan wants fans to view *Westworld* seasons as self-contained movies with each subsequent season viewed as a sequel" (Schmidt 2018).

in the "real world" of the Delos Corporation, which controls the park and the hosts, and the scripted narrative loops that hosts enact and guests perturb. The real-world narrative concerns the attempt of Ford to bring to fruition a new scripted narrative despite the resistance of Delos authorities, who regard the aging scriptwriter as no longer useful. Significantly, as viewers discover toward the end of the season, Ford, while writing the new narrative, is simultaneously attempting to finish the job that he prevented his now deceased partner, Arnold, from completing over 30 years ago – that is, bringing the hosts to consciousness.[5] The other key real-world narrative features stand-alone scenes of what seems to be Ford's assistant Bernard interrogating Dolores about her thoughts and memories in an attempt to gauge whether she has achieved such consciousness. These interviews often appear at the outset of an episode as if to provide a context for what will follow.

The narrative loops that trap the hosts ostensibly operate as embedded narratives – stories within the frame story of the Delos Corporation and its internal power plays. These key embedded narratives include the Dolores/Teddy loop and a loop involving the saloon-and-brothel-keeper Maeve. Unlike a Disneyland animatronic loop that guests merely observe, however, the loops in Westworld are interactive, allowing guests to participate in the story. The loops consequently facilitate for viewers a metaleptic merging of narrative levels, so that they watch the inhabitants of the "real world" of Delos enter a scripted world to engage with its android/gynoid characters; in later episodes, viewers even watch these characters enter the real world to engage with its inhabitants. Two interactive narrative strands weave through the Dolores/Teddy loop, one involving the white-hatted, idealistic guest William, who begins to regard Dolores as something more than a programmed machine and to have feelings for her, and the other involving the mysterious guest the Man in Black, who repeatedly assaults both Dolores and Teddy.[6]

The Man in Black has also violently assaulted Maeve, as viewers discover when she begins to access memories of an earlier storyline loop in which she played the role of a homesteader with a young daughter.[7] Of course, the

---

5   See Lacko (2017) for a discussion of the quest-narrative structure of the first season's plot.

6   The Man in Black is a nod to the black-garbed malevolent malfunctioning android (played by Yul Brynner) in the 1977 film. In "The Adversary" episode, a defunct android in the basement is reminiscent of this android. Brynner's outfit in the film itself recalls the outfit he wears in the iconic 1960 Western film *The Magnificent Seven*.

7   The specific names used for the hosts are slippery because they derive from the role that hosts play when we first encounter them. "Maeve" is the name of the brothel-keeper, but the name goes with the role rather than the host. One of the initially puzzling features in the William storyline (one that viewers later discover is a crucial clue) is that the host

hosts' narrative loops are designed to function even despite the disruption of the guests' interactions with the hosts; as Theresa Cullen, the Delos Head of Quality Assurance, states, "The hosts are supposed to stay in their loops, stick to their scripts with minor improvisations" ("The Original") – improvisations that do not significantly alter the overarching unvarying loop.[8] Yet burgeoning host consciousness results in a merging of the embedded narratives with the frame narrative as the programmed objects stuck in their narrative loops attempt to become subjects controlling their own destinies. Viewers see this merging effect particularly in the case of the Maeve narrative strand: Maeve deliberately courts death, committing "suicide" by guest (and also by host), in order to wake up in the lab where hosts are repaired. When she eventually explores the facility with a human lab technician, she discovers to her horror heaps of "slaughtered" (deactivated) hosts, including Teddy.

Complicating the narrative structure even further, narrative strands that would seem to be occurring simultaneously actually take place in different time periods. In a very convoluted first season, two key reveals stand out: (1) the present-day Bernard is actually a host who has been modeled on the deceased Arnold, and it is actually Arnold himself and not Bernard interrogating Dolores in the past before the park opened; (2) Dolores's adventures with William took place over 30 years in the past, and the Man in Black is, in fact, an older, cynical version of the white-hatted idealist William. Although astute viewers guessed these reveals beforehand, the first plot twist does not occur until episode seven, the aptly titled "Trompe L'Oeil," and the second in the season finale, "The Bicameral Mind."[9]

Devoid of analeptic markers such as "30 years ago," the flashbacks do indeed serve as a *trompe l'oeil*, misleading many viewers (including us authors) to see a straightforward narrative timeline rather than a sedimentary layering of

---

known as the prostitute Clementine actually appears during those sequences in the role of Maeve. Most importantly, following Arnold's action from 30 years prior, Ford gives Dolores the additional role of "Wyatt" (a name itself a nod to the Old West); Wyatt is a mysterious cold-blooded killer from Ford's new narrative, a situation that leads to the somewhat schizophrenic nature of Delores's emerging consciousness.

8 Alex Goody (2019) discusses the fictional technology behind the hosts' loops: "As illustrated in the Story Line Template Builder from the Delos Handbook for New Employees (Delos Destinations 2018, 7), her story is mapped across a varied but ultimately limited number of combinations and it is through breaking from this loop that, whatever combination of the set pathways are played, always returns to the same place (restart and/or sleep), that Dolores wakes up in a different sort of labyrinthine space" (262).

9 Kim Renfro and Jenny Chang (2018) have located five different time periods covered in season 1. For a more extensive description of the events and their relation to one another, see also "A Chronological View of Westworld's Timeline (Season 1)" (2017).

multiple timelines. This trickery is abetted by the machine-like invariance of the hosts' narrative loops and the apparent agelessness of these mechanical entities. In each passage through her loop, Dolores appears exactly the same, she generally wears the same blue dress, and she essentially follows the same programmed behavior throughout her day. Stubbs, the head of security at Westworld, attests to Dolores's longevity: "She's been repaired so many times, she's practically brand-new" ("The Original"). His words reinforce the deceptive merging of varied temporalities enabled by the invariance of the hosts' appearance and their narrative loops. The closed and continuous topology of a narrative loop helps deceive the viewer.

It is not surprising if viewers tend to see events as succeeding one another in chronological order simply because the events are presented that way. In a dense and confusing sequence at the end of the third episode ("The Stray"), we see Dolores fending off an assault from the Man in Black, escaping from her ranch, and (following several rapid-fire edits) arriving at William's campfire and fainting into his arms. Viewers may conflate the two time periods, assuming that they are watching an earlier and a later event taking place in one evening rather than an event in the narrative present followed by another from 30 years prior. Viewers may confuse one of Dolores's later passages through the loop for an entirely different, earlier passage, and they may thus be tricked into thinking that William and the Man in Black are indeed two different men inhabiting the same epoch. Addressing an effect of nonchronological narratives, Elena Kalefanos points out: "Readers (listeners, viewers) make a fabula [that is, a story] by assembling in chronological sequence the events they discern in a representation" (Kalefanos 15). Yet the fabula viewers originally construct during the first season may be incorrect; a skewed temporal order may prompt the audience "initially to misinterpret causal relations among the reported events" (x–xi), according to Kalefanos. Viewers may mistake a chronological presentation for a causal succession.

In an oft-quoted line from the series, Ford likens the hosts' narrative loops to human routines: "And yet we live in loops, as tight and as closed as the hosts" ("Trace Decay"). Dolores herself says, "All lives have routines. Mine's no different" ("The Original") However, although a rudimentary equivalence exists between a host's loop and a human routine, a crucial distinction pertains: the differences among the repeating loops are lost on the hosts. We humans may be stuck in a daily 9-to-5 grind, but we remember previous jobs and project toward a future. We are aware of time's passage.[10]

---

10   In season 3, subtitled *The New World*, various hosts have escaped from Westworld to the real world, and it is suggested that there is little difference between host loops and human

To live in a host loop, however, entails no access to one's actions in the past because actual memories are presumably wiped: "The memories are purged at the end of every narrative loop" ("The Original"). Teddy, for example, has died a thousand times ("The Stray"), but he does not remember these deaths. In a new passage through her loop, Dolores can greet with a smile the Man in Black who has violated her repeatedly during previous passages, prompting his exasperated comment, "I've been coming here for 30 years, but you still don't remember me" ("The Original").[11] Indeed, before her (re)awakening, Dolores has no awareness that she has spent over 30 years in the park. Granted, it seems that the process of wiping hosts' memories may not be completely effective. After being shot during an extremely violent gunfight in the saloon, Maeve manages to see the arrival of the suited technicians who have come to clean up, and, in a subsequent passage through the loop, she remembers flashes of being hastily repaired. She then sketches a figure wearing a mysterious spacesuit and tries to hide it, only to discover a trove of nearly identical sketches, revealing to her the existence of prior, unremembered iterations of terror ("Dissonance Theory").[12] The ability to remember their pasts becomes crucial to the hosts' burgeoning sense of self.

Hosts are given backstories, implanted memories that help anchor them to their programmed existences, but these stories have no relation to what has happened to them while serving in the park over many years.[13] These "memories" are of *inexperienced* events, as opposed to those events that the embodied hosts actually undergo. Brothel-keeper Maeve has no memories – initially – of homesteader Maeve and of the traumatic murder of her daughter, but she has an implanted fictional memory detailing her voyage from England to the American frontier and the brothel. Thus hosts refer not to a real past but to one that is entirely fictitious. It is as if readers absorbed a novel and then appropriated its events to their memories of their own lives. Significantly, the hosts

---

loops. In the fifth episode, "Genre" (2020; written by Karrie Crouse and Jonathan Nolan and directed by Anna Foerster) the once-corporate-fixer/now-Dolores-copy Martin Connells, played by Tommy Flanagan, explains what the creators of a supercomputer called Rehoboam intended: "This is their god. [...] How they make the future. In order to do that, they watch everyone. Tell them what to do, where to live, who to love. Keep them in a loop."

11   It is presumably the fact that Dolores does not remember him and the apparent passion she shared with his younger self that occasions his vicious behavior toward her.

12   The time-looping film *Triangle* makes ample use of this trope, from notes to necklaces to bodies.

13   The one android in the first season that does not live in a loop is Bernard. He does, however, have an implanted backstory regarding his son's death that is intended to anchor him to the park.

behave like young children who, in the process of developing self-narrativizing abilities, operate at a level wherein their stories are "undifferentiated as to whose stories they are," in the words of developmental psychologist Katherine Nelson (2003, 31).

What the hosts have actually experienced in their pasts is at first inaccessible to them. Furthermore, although hosts may anticipate a future, they never progress beyond the limited span of the loop. When Teddy promises an incipiently conscious Dolores that they will be together "someday," she chafes against the meaningless temporal designation ("The Stray"). There is no someday for a machine whose existence is as regulated, predictable, and constrained to a limited timespan as that of a wind-up toy. Humans may subscribe to the optimistic belief in an overarching plan, saying, as Dolores does, "I know things will work out the way they're meant to" ("The Original"). But even if we believe "there is an order to our days – a purpose," we nevertheless regard ourselves as having agency as we move toward fulfilling that purpose in the future.[14] When Dolores makes such statements, however, they ironically reveal that she has no such agency in the invariant time in which she exists.

Ultimately, a meaningless existence of an endlessly repeating loop – with no true past upon which to draw and with no actual future for which to plan – is an existence devoid of consciousness. "None of this matters," says Maeve when she finally realizes that she is just a puppet of "the makers who pull your strings," compelled to repeat the endless ritual of welcoming guests to the brothel ("Dissonance Theory").

2   Glitching the Loop

> The idea of a 'glitch' strikes me as wonderfully eerie – a surrealistic kind of quivering as on a surrealistic kind of surface.
> SANDY in "The Turing Test: A Coffeehouse Conversation"

• • •

---

14   As theorists of the posthuman condition have reminded us, the notion of agency is a problematic one: "the presumption that there is an agency, desire, or will belonging to the self and clearly distinguished from the 'wills of others' is undercut in the posthuman, for the posthuman's collective heterogeneous quality implies a distributed cognition located in disparate parts that may be in only tenuous communication with one another" (Hayles 1999, 3–4). A claim for agency can occur only from a position of privilege. Certainly, the hosts' lack of agency would resonate for people not only who lived in past times but also who live in many situations today.

> Déjà vu is usually a glitch in the matrix.
>
> TRINITY in *The Matrix*

.:.

The main narrative arc of the first season traces the hosts' coming into consciousness. In the early episode "The Stray," Ford suggests that the hosts can in fact simulate human behavior sufficiently to fool an interrogator: "Our hosts began to pass the Turing Test after the first year."[15] For example, when he arrives in Westworld, William asks the woman who greets him (Angela in the present-day park, as viewers later discover), "Are you real?" She provocatively replies, "If you can't tell, does it matter?" ("Chestnut"). What seems to be at issue, however, is not simply to simulate human behavior but to replicate a human way of thinking that includes self-awareness and self-reflection. One of the participants in Douglas Hofstadter's "The Turing Test: A Coffeehouse Conversation," explains, "The really interesting things in AI will only begin to happen, I'd guess, when the program *itself* adopts the intentional stance toward itself!" (1981, 88; Hofstadter's emphasis). Similarly, Ford notes that Bernard is "a machine who knows its own true nature" ("Trace Decay"). Hofstadter elsewhere discusses "genuine artificial intelligence" (Somers 2013) that would truly replicate the way in which humans think, as opposed to the machine-thinking that occurs, for example, in AIs like IBM's Watson or DeepMind's chess-playing AlphaZero computer program. Indeed, what Hofstadter refers to as artificial intelligence might be more accurately termed artificial *consciousness* – in the sense of human consciousness.[16]

In *Westworld*, the explanation for the androids/gynoids coming into consciousness is the now mainly discredited Bicameral Mind theory, put forward

---

15   In the second-season episode "Reunion," during a sequence set in the past wherein Arnold attempts to persuade the Delos Corporation to invest in Westworld, a group of hosts effectively convinces Logan Delos, the owner's son, that they are human.

16   AI thinking is not like human thinking. In discussing Watson's thinking process, David Ferrucci, leader of the Watson team at IBM, says, "Did we sit down when we built Watson and try to model human cognition? [...] Absolutely not. We just tried to create a machine that could win at *Jeopardy*. [...] It's *artificial* intelligence, right? Which is almost to say *not-human* intelligence. Why would you expect the science of artificial intelligence to produce human intelligence?" (Somers 2013; Somers's emphasis). According to Kissinger et alia, AlphaZero "became the best chess player in the world [...] by playing neither like a grandmaster nor a preexisiting program. It conceived and executed moves that both humans and human-trained machines found counterintuitive" (2019, 24).

in Julian Jaynes's *The Origin of Consciousness in the Breakdown of the Bicameral Mind* (1976). As Ford explains in "The Stray," "Arnold built a version in which the hosts heard their programming as an inner monologue, with the hopes that in time, their own inner voice would take over," thus "bootstrapping consciousness."[17] What is significant here is not Jaynes's theory but the idea of hearing the inner monologue – of narrating a self.

Viewers discover in the season finale that at least one host, Dolores, through her interviews with Arnold, had achieved consciousness 34 years before the "present-day" events. When Arnold discovered the breakthrough, however, he realized to his horror that Westworld would be "a living hell" for conscious hosts, who would be aware of their exploitation but powerless to stop it. Resolved to "break the loop before it begins," he had Dolores shoot the other hosts, himself, and then herself ("The Bicameral Mind"). But corporate greed triumphed over ethical considerations. After Arnold's death, Ford repaired the hosts, removed all apparent remnants of consciousness from them, and allowed the park to open.[18] Nevertheless, at some point, a change of heart occurred in Ford. In the present-day timeline, in conjunction with his promised "new narrative," he has begun upgrading the hosts with what he refers to as "reveries" – the ability to access past memories of their actual experiences, thus enabling them to develop a self-narrating, future-projecting consciousness.[19] The seeming verbal trigger for accessing memories is a line from *Romeo and Juliet* that Dolores's father whispers to her and that she later whispers to Maeve: "These violent delights have violent ends" ("The Original" and "Chestnut"). This Shakespearean insight presages what will happen once the hosts remember all the violence they have endured in countless passages

---

17   It is beyond the purview of this paper to explore how convincing the Bicameral Mind theory is or how (un)likely the *Westworld* technology is. For discussions of the Bicameral Mind theory in regard to *Westworld*, see Johnson (2017), Konstantin (2017), and Lacko (2017). Responding to the plausibility of the *Westworld* conscious hosts, John Searle tells an interviewer: "A computer program is defined purely in terms of symbol manipulations. [...] They are not sufficient for mental human contents" (Doty 2016). Stephen Beckner points out, "It's a given that the show's biggest conceits are completely unrealistic. [...] [W]hen it comes to how the brain works, we don't even know what we don't know, much less how to replicate it in Artificial Intelligence at a human level" (2017, 50). As Steven Pinker has noted with regard to human consciousness, "We understand what it means for a device to respond to a red light or a loud noise – we can even build one that does – but humans are the only devices in the universe that respond to danger, praise, English, and beauty" (1997).

18   Jones (2016) provides a provocative discussion of *Westworld* in light of Marxist theory.

19   How can supposedly erased memories be restored? One could liken this process to the deleting of files from a computer hard drive. When a file is erased, it is merely removed from the index but not overwritten. As such, it is still retrievable with the right software.

through their loops – essentially, the guest-instigated variations in their seemingly invariant existences.

This access to memory, however, initially manifests itself as a "glitch" whereby past experiences, devoid of timestamps, disconcertingly intrude into the present day of the hosts and consequently disrupt the timeline for the viewers.[20] Maeve, for example, begins to have vivid memories of herself as a homesteader (playing with her daughter, attempting to protect her, and watching the Man in Black murder her before turning on Maeve). She initially thinks that these memories are only dreams. How can this "real" past conform to the memory implants that currently control her brothel-loop trajectory? When glitching, Dolores may experience a jarring memory that she has previously been in a particular place and performed a particular action but that the outcome was different. These differences account for our own possible confusion because they are often presented in rapid, almost jerky, succession. Although Dolores remembers her journey with the white-hatted William, past and present timelines conjoin so thoroughly that she cannot fathom that their journey took place over 30 years ago. In fact, during a final confrontation with the Man in Black, whose earlier acts of violence she has now begun to remember, Dolores asserts poignantly that she is waiting for someone who will come for her – that is, she is waiting for the young man whom the Man in Black once was.

Unlike human memories that change upon review and fade with time, the machine-perfect memories of the hosts do not, such that the hosts initially cannot tell the difference between their experience of remembering and their experience of the now. After seeing/remembering the early days of the park when she slaughtered the hosts, Dolores asks William, "Then, when are we? Is this now?" ("Trace Decay"). Dolores's inability to anchor herself in time is reminiscent of the Zen koan "When is now?" – a koan that Susan Blackmore references in her book on consciousness, along with others such as "Who am I?" and "What is this?" (2017, 133).[21] Katherine Nelson's comment on the developing narrative competence in children is also germane: "The not now begins to be filled with specific semi-plots but they are not ordered

---

20   Carol Erwin (2019) touches on the interplay of past and present in the reveries: "The reveries are critical for the Hosts because they blend past and present, and that blending gives the Hosts the knowledge and the language they need to resist the hegemonic narrative; they realize that their past nightmares are present day realities" (121).

21   The latter questions also are pertinent to the Westworld hosts as they struggle with questions of who/what they are and what constitutes reality. Blackmore notes, "I have spent a long time sitting with these questions and finding that the ordinary sense of being a self having a self having a stream of experiences disappears." Certainly, Westworld prompts viewers to consider what "illusions" (to use Blackmore's term) constitute our own sense of self.

among themselves" (2003, 31). As viewers struggle to figure out a timeline in the series, to understand what event has led to another and determine the proper temporal sequence of events, they may find themselves asking similar questions.[22] Viewers, too, are glitching, made aware of the importance of a continuous narrative – not only in our viewing experience but also in our own sense of self.[23]

Bernard's comment to Theresa highlights the sedimentation of memories: "The ability to deviate from programmed behavior arises from a recall of past iterations" ("Trompe l'Oeil"). The implanted memories meant to anchor the hosts to their identities create a false sense of self and provide a simulation of the passage of time rather than the actual sensation. Self-awareness, however, comprises the experiences that one has had over time, including, in the case of the *Westworld* hosts, understanding the living hell they have endured.[24] Indeed, Ford suggests that suffering helps facilitate consciousness and that the hosts needed "[t]ime to understand your enemy" – specifically, humans ("The Bicameral Mind"). No wonder Theresa cautions that "[t]he concerns with the reveries is that the hosts will remember some of their experiences and act on them" ("Trompe l'Oeil"), avenging themselves upon those who have caused their pain.

A recurring image and puzzle throughout the first season is the *Westworld* maze. In one of Arnold's conversations with Dolores, he says,

---

22   It is perhaps not insignificant that Jonathan Nolan wrote the short story "Memento Mori" upon which his brother Christopher based his film *Memento*. With its backwards narrative structure, the film puts viewers in a position of the protagonist, whose loss of short-term memory entails that he lives a fragmented existence (see Parker 2004). In some sense, too, the protagonist's plight is similar to that of the hosts' in that he exists in a limited timespan that ends when all his memories are erased.

23   Calling *Westworld* "a fable about time," Glòria Salvadó-Corretger and Fran Benavente (2019) argue that the series presents "a game in which the spectator must be able to go back in narrative time to reconfigure the story and thus try to find answers to questions about existence and identity raised by the temporal complexity articulated over the long duration of the series" (2). In the meantime, the spectator is subject to a "powerful ontological bewilderment" (4). Although Salvadó-Corretger and Benavente also focus on the connection between the temporal structure of the series and the hosts' and viewers' sense of self, their focus is on the temporal twist and the ontological doubts prompted in the spectator rather than the hosts' mimicry of a human accession to consciousness.

24   This is not to say that our human sense of self is necessarily authentic. Our memories are faulty, spotty, even totaly wrong at times. As the monster in Shelley Jackson's landmark hypertext novel *Patchwork Girl* (1995) states: "So, within each of you there is at least one other entirely different you, made up of all you've forgotten ... and nothing you remember ... More accurately, there are many other you's, each with a different combination of memories." (This statement occurs in the lexia sequence "story/séance/she goes on.") The difference for the *Westworld* hosts is that the *only* memories to which they have access, at least initially, are false.

> There's something I'd like you to try. It's a game, a secret. It's called The
> Maze. If you can find the center, then maybe you can be free.
>
> "Dissonance Theory"

Then, while traveling with William, Dolores meets up with her own self, who
tells her, "We must follow the maze" ("Contrapasso"). In the final episode,
Arnold tells Dolores,

> Consciousness isn't a journey upward but a journey inward. Not a pyra-
> mid, but a maze. Every choice could bring you closer to the center or
> spiraling to the edges, to madness.
>
> "The Bicameral Mind"

The Man in Black mistakenly believes the maze is a real site in the park, the
one puzzle he has not yet solved in 30 years. It is instead, however, a symbol
for host consciousness, imprinted on the inside of host scalps in what seems to
be a graphic aspiration.[25] Teddy claims that "[t]he maze is the sum of a man's
life, the choices he makes" ("The Adversary"). Thus consciousness, according
to *Westworld* philosophy, entails a narrative sense of self – an awareness of
past choices that have led to present circumstances. And, as we discuss in the
next section, this *Westworld* philosophy dovetails fairly neatly with ideas about
human consciousness.

3   Narrating the Self

> Perhaps consciousness itself *is* the concentrated cumulative essence of
> all those transformative processes that lumped together we call time.
>
> FREDERICK TURNER, "Time's Speech in *The Winter's Tale*"
> (Turner's emphasis)

What consciousness is, how it emerges, or why it does so are not at issue
here. The emphasis here is instead upon the processual connections among

---

25   Goody (2019) discusses the maze in terms of Jorge-Luis Borges's "The Garden of Forking
      Paths" and Gilles Deleuze. See, too, Salvadó-Corretger and Benavente (2019), who argue
      that "[t]he complex temporal structure of Westworld emerges in the form of a labyrinth"
      and that this "convoluted structure is no more than a pretext to prolong the story, a mere
      narrative device" (11).

temporal awareness, self-narration, and consciousness, demonstrated in season one of *Westworld* through the alternation between varying timelines and invariant time. The disorienting time-scheme occurring throughout the season reinforces these connections in that viewers' struggle to make sense of the overall story they are watching reflects the hosts' struggle to make sense of their own stories. Ultimately, in order to achieve consciousness, Maeve and Dolores must stop glitching and acquire the ability to reconstruct a timeline by ordering the events of their entire existences and thus narrating a coherent cause-effect plot of the self. "Your memories are your first step to consciousness," Bernard tells Maeve. ("The Bicameral Mind").

The authors' 2019 article on time-loop fiction ("Eternal Recursion, the Emergence of Metaconsciousness, and the Imperative for Closure") uses the term "metaconsciousness" to delineate certain characters' capacity to transcend the loop that traps them and to recall past passages through the loop; metaconsciousness enables these characters to effect better outcomes with each passage through the loop so as to eventually break out of the loop itself.[26] Similarly in *Westworld*, in order to achieve consciousness, the hosts themselves must achieve metaconsciousness of their own loops, gaining a transcendent perspective of their prior passages through them. Dolores's seeing herself in a crowd ("Dissonance Theory") and talking to another version of herself ("The Bicameral Mind") signify her burgeoning transcendent perspective.

A sign that the hosts have broken the scripted temporal invariance of their loops occurs when they can actually project forward in time and begin to narrate their own stories, purposefully varying their loop existences. In an early episode, Dolores muses: "I think when I discover who I am, I'll be free" ("The Stray"). She thus equates her sense of self with freedom from the loop-life she only incipiently apprehends. She has "imagined a story where I didn't have to be the damsel" ("Dissonance Theory"). Only once she can comprehend most of her experiences – her earlier accession to consciousness, her killing the other hosts and Arnold, her sojourn with young William, her violations by the Man in Black – can she begin to narrate a self that is not the damsel. Bernard's comment to Maeve reinforces the notion of the cause-effect plot of self: "How can you learn from your mistakes if you can't remember them?" ("The Bicameral

---

26   We use the term "metaconsciousness" in a more specific way than the general meaning of consciousness about consciousness. Nevertheless, our usage corresponds to this general meaning as these phrases from Philip Merlan might suggest: "enlarged consciousness," "intelligizing all intelligibilia," and "a consciousness of a higher order" (1963, 84).

Mind").[27] Over the course of season one, the voice in Dolores's head begins as Arnold's, evolves into an admixture of their dual voices, and finally ends as her own voice. To achieve consciousness, she must realize that the voice she imagines she hears from without is undeniably her own, and it must become the voice she hears from within, enabling her to follow her own directives. Although this process invokes the Bicameral-Mind theory that underpins the *Westworld* AI-consciousness explanation, it is in keeping with more recent speculations about consciousness as well. As philosopher John Bickle has hypothesized, for example, the self comprises a "continuous quiet inner speech produced by activity in the human brain's language regions" (2003, 199).

Maeve's accession to consciousness also reinforces the connection between consciousness and a narrative sense of self. Episode 5 ("Contrapasso") ends with her waking in the real-world lab for an impending memory wipe followed by her reinsertion back into her brothel loop. The following episode ("The Adversary") begins with her dying yet again but then waking up in the lab and asking, "Now, where were we?" She has transcended her loop-bound existence and can move forward in real time. Now with access to her memories and control of her time, she asserts, "Time to write my own fucking story." Bickle points out with regard to human consciousness: "continuous self-constructing linguistic productions" create "a self – not only in causal control of important, and behavioral events but also aware of exerting this control" (199). Significantly, in beginning to self-narrate, Maeve can even take charge of unawakened hosts: she narrates their actions, as if she were an author, and those hosts then carry out her directives.[28] Maeve even begins to understand her own programming: "There are some elegant formal structures, a kind of recursive beauty, but complex, like two minds arguing with one another" ("Trace Decay"). The "two minds," we might suppose, are Arnold's, the original programmer, and her subjective will. In parallel to the voice inside Dolores's head, the programming in Maeve is transitioning to her control. A provocative sequence in the episode involves her altering her own programming so that she has optimum intelligence and less loyalty, yet another sign that she is now writing her own story.[29]

---

27   This line echoes a question asked by Leonard, the memory-challenged protagonist of *Memento*: "How can I heal [...] how am I supposed to heal if I can't feel time?"

28   In the second season, we are introduced to the concept of the "mesh network," whereby hosts can exchange data among themselves, which may account for Maeve's control of unawakened hosts.

29   As we discover in the final episode of season one, Maeve's awakening and plan to escape the park have themselves been programmed. By the end of the episode, however, she finally goes "off script" definitively as she scraps the escape plan and returns to the park to look for her daughter. Lester del Rey's short story "Helen O'Loy," published in *Astounding*

In the felicitously titled *I Am a Strange Loop* (2007), Hofstadter offers a succinct explanation of what a machine would need in order to be endowed with an "I." He had read how the Stanford Artificial Intelligence laboratory had developed a robot vehicle ("Stanley"), which "drove all by itself across the Nevada desert, relying just on its laser rangefinders, its television camera, and GPS navigation" (190). Stanley could not be an "I," however, because

> [t]his would require the vehicle to have a full episodic memory of thousands of experiences it had had, as well as an episodic projectory (what it would expect to happen in its "life", and what it would hope, and what it would fear), as well as an episodic subjunctory, detailing its thoughts about near misses it had had and what would most likely have happened had things gone another way.
>
> HOFSTADTER 2007, 191

To be an "I," a machine (whether Stanley, Dolores, or Maeve) must remember its past, plan its future, and assess its actions – the very things that make each of us an "I."

When viewers reach the final configuration of the plot in the season finale, they can construct the timeline that may have evaded them for most of season one.[30] Just as the hosts are able to make a cause-effect plot of the self, viewers can now comprehend the connections among various events and "make a fabula," to borrow Kalefanos's words. It is, of course, not the ultimate fabula. The second and third seasons of the series fill in additional gaps, and the second even provides a take on prior events from yet another perspective. But, for the time being, viewers can stop glitching. They have finally been given enough information to achieve metaconsciousness over the plot of season one.

More importantly, perhaps, viewers can now grasp why a cause-effect plot of the self is significant. Despite its fantastical story of consciousness-achieving androids and gynoids, *Westworld*'s focus on self-narrativizing, as we have suggested, aligns with current understanding of the integral relations among consciousness, narrative, and time. Daniel Dennett has called the self "a center of narrative gravity" and likened us conscious beings to "virtuoso novelists" who

---

Science Fiction in 1938, may be one of the first examples of a story wherein a gynoid achieves consciousness. Interestingly, Helen, the gynoid, does so through absorbing narratives that she then imitates.

30   In season two, a similar confusion of timelines occurs. The newly aware host Bernard deliberately scrambles his memory at a certain point in the plot arc. As viewers process each episode, they once again are glitching as they struggle, like Bernard, to figure out, "Is this now?"

"try to make all of our material cohere into a single good story. And that story is our autobiography" (1992, 114).[31] Gary Fireman et alia have noted:

> The stories we tell ourselves and others, for ourselves and others, are a central means by which we come to know ourselves, and others, thereby enriching our conscious awareness.
>
> FIREMAN ET AL. 2003, 3

Indeed, "Narrative does not merely capture aspects of the self for description, communication, and examination; narrative constructs the self" (5), as *Westworld* cleverly illuminates.

The creators of *Westworld* have provided viewers a laboratory in which to study consciousness. The hosts are given access to an-all-too perfect set of their own experiences whose time markers have been removed but each of which is completely indistinguishable from the experience of the now. The hosts begin with the voice and programming of their programmers. As they struggle to re-sequence these experiences into a temporal order, they create their own autobiography and develop their own inner voice and program. In setting up these fictive hosts with these impossible conditions and focusing on the hosts' construction of the self through narrative, *Westworld* deconstructs the process of self-making that we humans take for granted: as conscious beings, our accession to self-awareness occurs simultaneously *with* self-awareness.

Furthermore, through highlighting the initially discontinuous situations of the hosts, the series deconstructs the continuity of consciousness over time. According to Valerie Gray Hardcastle, "we want to understand our world and our selves as meaning something, as stories, as things with plots, with beginnings, middles, and ends" (38). Whereas hosts such as Dolores and Maeve achieve consciousness through a painstaking process of figuring out how the beginnings, middles, and ends fit together, as humans, we adhere to the notion of a continuous self persisting across time. It's all we know. Susan Blackmore argues that such thinking may be "the grand delusion of consciousness": "So we jump to the false conclusion that at every waking moment we must be a conscious self, experiencing a stream of conscious experiences" (2017, 132).

---

31   See Brandon 2016 for a critique of Dennett's position. Brandon argues that Dennett ignores the issue of embodiment. The link between consciousness and embodiment is beyond the scope and purpose of this paper. Nevertheless, *Westworld* offers a provocative take on embodiment in that certain host minds may have the capability to exchange bodies with other hosts, a circumstance that becomes especially exacerbated in the third season, when multiple Dolores-minds inhabit a variety of different bodies.

Nevertheless, it is a delusion to which our self-narrativizing selves subscribe. Through its poignant narrative of entities attempting to escape their invariant time loops and its structural temporal variations, *Westworld: The Maze* helps us understand that we *are* our own story – a story that we are continually narrating and rewriting in light of enhanced apprehension of our "reveries."

Acknowledgements

Jo Alyson Parker would like to thank the Saint Joseph's University Board of Faculty Research and Development for granting her a Michael J. Morris Grant for 2018–19, which facilitated her work on this project and her participation at the 2019 ISST Conference. Both authors would like to thank Sonia Front and Richard Fusco for their insightful comments on earlier drafts and the two anonymous referees for their helpful suggestions.

Appendix

Principal Actors and Their Roles in Westworld *Season 1* (The Maze)

Ben Barnes: Logan Delos
Ed Harris: William/the Man in Black
Luke Hemsworth: Ashley Stubbs
Louis Herthum: Peter Abernathy
Anthony Hopkins: Dr. Robert Ford
Sidse Babbet Knudsen: Theresa Cullen
James Marsden: Teddy
Thandie Newton: Maeve Millay
Simon Quarterman: Lee Sizemore
Talulah Riley: Angela
Angela Sarafyan: Clementine Pennyfeather
Tessa Thompson: Charlotte Hale
Evan Rachel Wood: Dolores Abernathy
Jeffrey Wright: Bernard Lowe/Arnold Weber

Season 1 Episodes Referenced

"The Original." S01, E01. Directed by Jonathan Nolan. Story by Lisa Joy, Jonathan Nolan, and Michael Crichton. Teleplay by Jonathan Nolan and Lisa Joy. HBO, October 2, 2016.
"Chestnut." S01, E02. Directed by Richard J. Lewis. Teleplay by Jonathan Nolan and Lisa Joy. HBO, October 7, 2016.

"The Stray." S01, E03. Directed by Neil Marshall. Teleplay by Daniel T. Thompsen and Lisa Joy. HBO, October 16, 2016.

"Dissonance Theory." S01, E04. Directed by Vincenzo Natali. Teleplay by Ed Brubaker and Jonathan Nolan. HBO, October 23, 2016.

"Contrapasso." S01, E05. Directed by Jonny Campbell. Story by Dominic Mitchell and Lisa Joy. Teleplay by Lisa Joy. HBO, October 30, 2016.

"The Adversary." S01, E06. Directed by Frederick E. O. Toye. Halley Gross and Jonathan Nolan. HBO, November 6, 2016.

"Trompe L'Oeil." S01, E07. Directed by Frederick E. O. Toye. Halley Gross and Jonathan Nolan. HBO, November 13, 2016.

"Trace Decay." S01, E08. Directed by Stephen Williams. Teleplay by Charles Yu and Lisa Joy. HBO, November 20, 2016.

"The Well-Tempered Clavier." S01, E09. Directed by Michelle MacLaren. Teleplay by Dan Dietz and Katherine Lingenfelter. HBO, November 27, 2016.

"The Bicameral Mind." S01, E10. Directed by Jonathan Nolan. Teleplay by Lisa Joy and Jonathan Nolan. HBO, December 4, 2016.

References

Beckner, Stephen. 2017. "Out of the Loop, Lost in the Maze: The Stealth Determinism of *Westworld*." *Skeptic*, January 17.

Blackmore, Susan. 2017. *Consciousness: A Very Short Introduction*. Second Edition. Oxford: Oxford University Press.

Brandon, Priscilla. 2016. "Body and Self: An Entangled Narrative." *Phenomenology and the Cognitive Sciences* 15: 67–83. DOI 10.1007/s11097-014-9369-8.

"A Chronological View of Westworld's Timeline (Season 1)." 2017. *r/Westworld: Subreddit for the HBO Series* Westworld. https://www.reddit.com/r/westworld/comments/5gqaau/a_chronological_view_of_westworlds_timeline/.

del Rey, Lester. 1938. "Helen O'Loy." *Astounding Science Fiction*. December.

Dennett, Daniel. 1992. "The Self as a Center of Narrative Gravity." In *Self and Consciousness: Multiple Perspectives*, edited by F. Kessel, P. Cole and D. Johnson, 103–115. Hillsdale, NJ: Erlbaum, 1992. Danish translation, "Selvet som fortællingens tyngdepunkt," *Philosophia* 15 (1986): 275–288.

Doty, Meriah. 2016. "'Westworld' Premise Refuted: Why the Robots Can't Be Conscious." *The Wrap*, November 15.

Erwin, Carol. 2019. "The Frontier Myth of Memories, Dreams, and Trauma in *Westworld*." In *Reading* Westworld, edited by Alex Goody and Antonia Mackay, 119–140. Palgrave Macmillan.

Fireman, Gary D., Ted E. McVay, Jr., and Owen J. Flanagan. 2003. Introduction. In *Narrative and Consciousness: Literature, Psychology, and the Brain*, edited by Gary D. Fireman, Ted E. McVay, Jr., and Owen J. Flanagan, 3–13. New York: Oxford University Press.

Goody, Alex. 2019. "The Theme Park of Forking Paths: Text, Intertext and Hypertext in *Westworld*." In *Reading* Westworld, edited by Alex Goody and Antonia Mackay, 255–275. Palgrave Macmillan.

Groundhog Day. 1993. Screenplay by Danny Rubin. Dir. Harold Ramis. Perf. Bill Murray, Andie McDowell, Chris Elliot, Stephen Tobolowsky. Columbia Pictures.

Hardcastle, Valerie Gray. 2003. "The Development of Self." In *Narrative and Consciousness: Literature, Psychology, and the Brain*, edited by Gary D. Fireman, Ted E. McVay, Jr., and Owen J. Flanagan, 37–50. New York: Oxford University Press.

Hayles, N. Katherine. 1999. *How We Became Posthuman: Virtual Bodies in Cybernetics, Literature, and Informatics*. Chicago: University of Chicago Press.

Hibberd, James. 2016. "Westworld Has Already Figured out the Next 5 Seasons." *Entertainment*. September 18. https://ew.com/article/2016/09/08/westworld-plan/.

Hofstadter, Douglas R. 1981. "The Turing Test: A Coffeehouse Conversation." In *The Mind's Eye: Fantasies and Reflection on Self and Soul*, edited by Douglas R. Hofstadter and Daniel C. Dennett, 69–92. New York: Bantam Books.

Hofstadter, Douglas R. 2007. *I Am a Strange Loop*. New York. Basic Books.

Jackson, Shelley. 1995. *Patchwork Girl; Or A Modern Monster, by Mary/Shelley & Herself: A Graveyard, a Journal, a Quilt, a Story & Broken Accents*. CD-ROM. Watertown, MA: Eastgate Systems.

Johnson, Jeremy. 2017. "The Philosophy of Westworld." *Vocal: Futurism*. https://omni .media/the-philosophy-of-westworld.

Jones, Eileen. 2016. "The Android Manifesto: Finding Marx in Westworld." *In These Times*. December 19. http://inthesetimes.com/article/19728/karl-marx-in -westworld-android-manifesto-trump-false-consciousness.

Kafalenos, Emma. 2006. *Narrative Causalities*. Theory and Interpretation of Narrative. Columbus: Ohio State University Press.

Kissinger, Henry A., Eric Schmidt, and Daniel Huttenlocher. 2019. "The Metamorphosis." *Atlantic Monthly*, August, 24–26.

Konstantin, Rayhert. 2017. "The Philosophy of Artificial Consciousness in the First Season of TV Series 'Westworld.'" *Philosophical Sciences* 151(5): 88–92.

Lacko, Ivan. 2017. "On the Path to Sentience: Post-Digital Narratives in 'Westworld.'" *World Literature Studies* 9(3): 29–40. http://www.wls.sav.sk/wp-content/uploads/ WLS3_2017_Lacko.pdf.

The Matrix. 1999. Screenplay by Lana and Lilly Wachowski. Dir. Lana and Lilly Wachowski. Perf. Keanu Reeves, Laurence Fishburne, Carrie-Anne Moss.

Memento. 2000. Screenplay by Christopher Nolan. Dir. Christopher Nolan. Perf. Guy Pearce, Carrie-Anne Moss, Joe Pantoliano. Newmarket.

Merlan, Philip. 1963. *Monopsychism, Mysticism, Metaconsciousnes: Problems of the Soul in the Neoaristotelian and Neoplatonic Tradition*. The Hague: Martinus Nijhoff.

Nelson, Katherine. 2003. "Narrative and the Emergence of a Consciousness of Self." In *Narrative and Consciousness: Literature, Psychology, and the Brain*, edited by Gary D. Fireman, Ted E. McVay, Jr., and Owen J. Flanagan, 17–36. New York: Oxford University Press.

Parker, Jo Alyson. 2004. "Remembering the Future: *Memento*, the Reverse of Time's Arrow, and the Defects of Memory." *KronoScope: Journal for the Study of Time* 4(2): 239–257.

Parker, Jo Alyson and Thomas Weissert. 2019. "Eternal Recursion, the Emergence of Metaconsciousness, and the Imperative for Closure." In *Time's Urgency*. The Study of Time 16, edited by Carlos Montemayor and Robert Daniel, 131–156. Leiden: Brill.

Pinker, Steven. *How the Mind Works*. New York: Scientific American Modern Classics, 1977.

Renfro, Kim and Jenny Chang. 2018. "An Essential Timeline of Every Important Event in 'Westworld.'" *Insider*. June 25, 2018. http://www.thisisinsider.com/westworld-timeline-spoilers-2018-4.

Salvadó-Corretger, Glòria and Fran Benavente. 2019. "Time to Dream, Time to Remember: Patterns of Time and Metaphysics in *Westworld*." *Television & New Media*, 1–19. https://journals.sagepub.com/doi/pdf/10.1177/1527476419894947.

Schmidt, Patrick. 2018. "How Many Seasons Will Westworld Last? Possibly as Many as 6." *Fansided*. April 22. https://fansided.com/2018/04/22/westworld-how-many-seasons-hbo/.

Somers, James. 2013. "The Man Who Would Teach Machines to Think." *The Atlantic*, November. https://www.theatlantic.com/magazine/archive/2013/11/the-man-who-would-teach-machines-to-think/309529/.

Turner, Frederick. 2018. "Time's Speech in *The Winter's Tale*: How the Past Can Be Undying." *KronoScope: Journal for the Study of Time* 18(1): 17–28.

CHAPTER 10

Time in Variance and Time's Invariance in Richard McGuire's *Here*

Arkadiusz Misztal

Abstract

The article examines the interplay of time's variance and time's invariance in Richard McGuire's graphic novel *Here* by studying narrative and visual constructions of hereness as an intrinsic part of places in relation to their ever-changing timescapes. It argues that the intensity and ingenuity of McGuire's narrative resides in the seamless bridging of time and space that generates multiple, heterogeneous dimensionalities that become simultaneously present. By engaging the reader's interaction, *Here* makes full use of imaginative modality and mobility to construct a narrative space that becomes temporally charged, a space in which the past and the future meet in the moving now.

Keywords

comics – deixis – frame – *Here* – hereness – imagining – invariance – modality – nowness – Richard McGuire – simultaneity – timescape – timelines – variance

1   Introduction

Published 25 years after the game-changing comic strip "Here" (1989), which revolutionized narrative possibilities of the comics medium, Richard McGuire's graphic novel *Here* (2014) continues to expand these possibilities in the deceptively simple story of a corner of a room and of the events that take place in that space over the course of hundreds of thousands of years. Comprising 304 pages of interconnected and overlapping storylines, the novel spans thousands of years of American life, moving back and forth in time but maintaining a single vantage point, a corner of the living room, set in McGuire's childhood home in Perth Amboy, NJ. In its continuous flipping back and forth in time, *Here* traces an impressively wide temporal horizon.

Some of the panels go back as far as the primordial eras when life on Earth began, while others spring forward to a distant future when humanity no longer exists. The in-between is filled with hundreds of moments capturing the lives of Native Americans, European settlers, American families from different centuries and decades, all of these moments taking place in the space between a cozy fireplace and a window.[1] Instead of chronological progression, *Here* conflates these moments in a nonlinear narrative to create temporal palimpsests in which multiple dimensionalities become simultaneously present.

By exploring the coexistence of continuity and discontinuity, change and stasis, McGuire's book discloses temporal dimensions of "hereness" and encourages the reader to reflect on the narrative constructions of places in connection to their ever-changing timescapes. A timescape perspective, as Barbara Adam (1998) has shown, enables the shift from explicit space and implicit time to "the complex temporalities of contextual being, becoming and dwelling" (11), which as I argue, lie at the core of *Here*. In constructing his graphic narrative, McGuire masterfully traces temporal features of our living by acknowledging the complex rythmicities and tempos, repetitions and contingencies as well as radically different timescales in our constructions and articulations of "hereness." With the focus centered on the dynamics of visual simultaneity, this article examines these constructions by looking into the framing devices that introduce multiple and overlapping temporalities of *Here* by means of visual and verbal juxtapositions. Subsequently, it discusses the ways in which McGuire's narrative seeks to resolve the tensions between heterogeneous temporalities and timeframes by encouraging a time shuffling kind of reading capable of integrating temporal consistencies and constants. Drawing on Hegel's presentation of deixis and sense-certainty, it argues McGuire's narrative temporally charges the spatial deictic referent "here" by appealing to the productive dynamics of imagining and generates multiple dimensionalities that tap into the workings of memory as well as individual and collective experience of time.

2   "Now, Where Were We?": Time Bending, Tabular Reading, and the Two-Dimensional Architecture of the Page

The book-length project provided McGuire with space to conceive, develop, and try out new ideas and possibilities as he reworked his 1989 strip by setting

---

1   This space is not always restricted to a living room; in the course of the narrative it becomes a stage on which, for instance, different configurations of the earth mass are presented.

it in a new key. With its wider historical scope and range of characters, *Here* devotes more attention to the remote past and future of his New Jersey home-town than the strip, which is loosely organized around Billy, who is born on the first page, grows up in the house, leaves and later visits his childhood home, and eventually dies. The narrative trajectories and storylines of *Here* are far more complicated as McGuire explores the varied timescapes of the "City by the Bay," as Perth Amboy calls itself, in the context of geological events and his-tory of Earth timeline. Located at the Terminal Moraine, a ridgelike accumula-tion of glacial debris that extends across central New Jersey to the western side of the state, Perth Amboy is variously revealed as a site of a prehistoric ocean, a hunting territory where dinosaurs roam, a primeval forest with lashed vegeta-tion, and as *Lenapehoking*, land of Lenni-Lanape Native Americans who occu-pied this part of the North American continent for thousands of years before European colonization. *Here* carefully registers Native Americans' presence and features a number of panels in which the characters speak the language of the Lenni-Lanape Indians. It recreates also Lenni-Lanape's first encoun-ters with the Dutch in 1624, which paved the way for the settlement of Perth Amboy by Scottish colonists around 1681. The colonial period in McGuire's book is, in turn, centered around William Franklin, the last royal governor of New Jersey and illegitimate son of Benjamin Franklin.[2] *Here* also incorporates McGuire's family history presented against the backdrop of their house, which is constructed in 1907, inhabited, refurnished, and redecorated many times. In the course of the narrative the house undergoes fire (1997), a burglary (1997), a partial collapse (2015), and a flood (2111); its final destruction in an atomic explosion occurs in 2313.

Here differs most conspicuously from the original black and white strip with respect to the book's visual composition made up of vibrant hues and colors, reminding one of both "Edward Hopper's moody, light-struck realism" (Garner 2014) and the lush color tones of the American landscape painter George Innes, who between 1864 and 1870 lived and worked in Perth Amboy.[3] Working

---

2   In 1775 Benjamin Franklin visits town and spends a night in his son's mansion. Giving William the full picture of the American Revolution and the war's development, Franklin urges his son to join the patriot cause, assuring the young man that General Washington would wel-come him with open arms. William stubbornly refuses his father's advice and some months later is forcefully evicted from the governor's mansion and exiled to Connecticut where is placed under guard. These events and historical details (along with hundreds of others) find their way into the novel as some of its central subplots.

3   Innes's residence and studio were located at the corner of Convey Boulevard and Smith Street; they were built by Marcus Smith in exchange for Innes's work *Peace and Plenty*. Smith and his wife Rebbeca were the *spiritus movens* behind the progressive artist colony in Eagleswood, which, apart from Innes, hosted other prominent individuals such as Louis Tiffany, William

FIGURE 10.1   From Richard McGuire, *Here* (London: Hamish Hilton 2014)
REPRODUCED BY KIND PERMISSION OF THE WYLIE AGENCY

in color, McGuire also perfected a technique that involved the preparation
of monochrome drawings, which were subsequently scanned, then digitally
textured and colorized. Many of the drawings are directly based on photographs that the author collected or came across during his research. *Here*'s
composition, as McGuire himself admits, draws on many diverse sources such
as Frank O. King's comic strip *Gasoline Alley*[4] or the work of Tadanori Yokoo,

---

Page, and Henry Thoreau. It is worth noting that *Here* makes even a more explicit reference
to Innes in a sequence of panels from 1870, in which the painters George Innes and William
Dunlap, a native of Perth Amboy, are joined into a composite figure of an artist, who sketches
on the grounds of William Franklin's onetime mansion. Intent on capturing the beauty of
the site, the artist ignores his handsome, red-haired muse, who complains, "Why don't you
want to paint me?"

4 *Gasoline Alley*, introduced in 1918, was the first comic strip in which characters aged in real
time. The fictional town of Gasoline Alley is modeled after King's hometown, the village of
Tomah in western Wisconsin, and many of its characters are based on the citizens of the
town. For a comprehensive survey of contemporary American graphic fiction and McGuire's
place in it, see Gardner 2012.

FIGURE 10.2   From Richard McGuire, "Here," *RAW* 2(1): 69–75
REPRODUCED BY KIND PERMISSION OF THE WYLIE AGENCY

a celebrated Japanese artist and graphic designer, in particular his *Waterfall Rapture* book (1996), comprising thousands of postcards that feature images of waterfalls.[5] Yokoo's book made McGuire realize that "one strong idea – the persistence of a single pictorial moment, without start or end – could provide the basis for an entire book" (Smith 2014, 55). Without undermining the importance of these influences and inspirations, I should note that McGuire recognized the creative possibilities of this strong idea in his 1989 comics strip and, as I have argued elsewhere, *Here* consciously and directly draws on "Here" to the extent that the visual and narrative DNA that McGuire developed in his strip shapes and sustains also the body of the novel.[6]

Much like its black and white predecessor *Here* explores the direct relation between layout and narrative structure to depart from both chronological progression and the traditional up-down and left-right reading of comics. By invoking year-labeled windows of time that float freely into each frame of action, McGuire introduces, to use Chris Ware's expression, "a third 'in-and-out' direction of overlapping palimpsest of framed historical space" (Greenberg 2014) in which the past meets the future in the present. Beginning with the fourth page, the relations between panels become increasingly complex as McGuire skillfully shapes images and various elements of layout not only to create "a deft, relaxed sense of space" (Mouly 2018) but also to transform a page into a field where images enter into dialogue with each other in *presentia* (Groeensteen 2007, 147). Apart from the linear transition between two adjacent panels, *Here* invites the reader to consider translinear interrelations between panels on different parts of the page. The reading for sequence is thus complemented by "tabular reading," or what Gérard Genette has called, "a global and synchronic look" across the page (Mikkonen 2017, 33).[7]

It is only through this global and synchronic look that the reader can fully appreciate the dynamic quality of juxtapositions, a quality also characterizing McGuire's books for children. In one of them, *Night Becomes Day*, two frames face each other on each spread and generate a dynamic widescreen effect when the nearly square volume is held open (Smith 2014, 25). The composition of *Here* is clearly much more complex as it introduces its own logic of spatial connectivity (Mikkonen 2017, 46) that challenges even the most experienced eyes and minds of graphic fiction readers. As Ware puts it, *Here* slices space "into pictures, and then shuffles it all up – past, present, and future hopelessly

---

5   I'm grateful to Raji Steineck for helping me to recognize and appreciate Japanese inspirations in McGuire's work.

6   For a detailed discussion of the comic strip and its relation to *Here*, see Misztal (2020).

7   For tabular reading, see Mikkonen 2017, 36–37 and 64–66.

intermingled – taking time out of page and placing it squarely back into the consciousness and more importantly, the control of the reader" (Ware 2015, 6). The reader therefore must work out a new relation between "a before" and "an after," between "now" and "then," as they not only coincide, but also coexist in the narrative. If the "yearly" time stamps, as well as other time props and cues are explicit, they do not, however, act as temporally prescriptive. While they guide the reader as he or she moves in space and time, they do not turn panels into "quasi-punctual" units of time. Instead they create the page layout in which the embedded frames, while focused on one point in space, a corner of a room, become in the reader's mind a temporal meeting point: a site where the future and the past meet the present.

This heightened sense of temporal fluidity and plasticity is particularly prominent in the digital version of *Here*, which develops and expands the original concept behind the 1989 strip. As McGuire admitted in an interview, he hit upon the idea of free-floating time windows for the strip after his first encounter with the computer operating system *Windows* in the late 80s.[8] This idea has been further refined in the digital edition of *Here* by making the graphic narrative truly interactive. In the interactive version, "where time and space are nothing but finite" (McGuire 2015), the reader encounters not only animated gifs but also clickable panels, which are not connected to the backgrounds. The reader can shuffle pages by clicking the date or can follow a single story by clicking through panels. By making the dates and panels interactive, the digital edition encourages the reader to rearrange and reorganize the order of the pages and thus create new combinations and connections. The reader can essentially read the book almost in any order, and each time can come away with a new understanding of the story.[9]

And yet this interactivity remains deeply indebted to the more traditional logic of the visual. In order to infuse his graphic narrative with a sense of directness and immediacy, McGuire, interestingly enough, turned his attention to photography. Attempting to capture "the moods and events of daily

---

8 Windows 1.0, developed as a graphical operating environment for IBM PC s, made its debut in 1985. Instead of typing MS-DOS commands, the user could point and click to access the windows. Its successor, Windows 2.0, released two years later, offered a more elegant and easier to use interface. It added keyboard shortcuts, desktop icons, and resizable, overlapping windows.

9 It is worth noting that the Polish translation of *Here* (2016) seeks to capture these dynamics by abandoning the traditional book-binding format and offering a box containing 154 non-paginated sheets stacked one on another. Instead of reading from cover to cover, the reader is expected to reshuffle, re-arrange the sheets as he or she tries out new possible combinations and patterns.

life, passing by in a perpetual present tense, [h]e found a congenial visual resource for *Here* in amateur snapshot photographs, which he mined for fragmentary narratives, gestures, and outfits he never could have made up" (Smith 2014, 55). Furthermore, some of the panels appear to be inspired by the late nineteenth-century chronophotographic experiments of Étienne-Jules Marey and Eadweard James Muybridge.[10] *Here*, testifying to McGuire's powers of attention, playfully invokes the chronophotographic explorations of tension between stillness and motion, between single and multiple views, presenting, for instance, a sequence of panels depicting a wild bird flying into the room and chasing away a female figure. And yet if photography has helped McGuire to develop the specific two-dimensional architecture of the page in *Here*, architecture that "allows for the transgression of temporal continuity and spatial consistency" (Corsten 2012, 93), the book's visual composition does not fall prey to photographic indexicality that freezes the action in a sequence of snapshots.[11] McGuire is well aware that graphic fiction as a multimodal medium reconfigures time and space in different ways than photography or cinema: comics activate the mobile concepts of frame and window through panel and thus can effortlessly transcend indexical attachment to the real world.[12]

As Greg Smith (2013) reminds us, the comics panel, as a component part of an artistic whole, can act as a still image, a shot, and a capture of duration (232). *Here* makes full use of the inherent mobility of the panel and the visual logic of the comics page to frame both a stasis and an unfolding of time. The

---

10   The former, a French biophysicist, used photographic images to study force by arresting individual instants of time and superimposing them on each other. Marey aptly called this method "chronophotography," literally "the photography of time." Muybridge likewise employed cameras in his detailed studies of movement, which paved the way for motion pictures. For a detailed discussion of these experiments, see Braun 1992 and Solnit 2004.

11   Another important photographic influence that left its mark on *Here* is the photobook *Evidence* (1977), a game changer and a seminal example of conceptual photography. Comprising 59 images taken from public corporate archives, carefully and intentionally deprived of original contextualizing aspects, *Evidence* presents enigmatic and visually stunning records of human activity.

12   It needs to be noted that the relations between the comics panel and chronophotographic image are much more complex than "straight borrowings." While comics aesthetics was shaped by experiments in chronophotography and inherited its techniques, the new codes and techniques were frequently deployed, as Scott Bukatman (2012, 64) argues, to parodic effect as the comics aimed to upset "the rationalist impulse to map the moving body's navigation of graphed space" (37) by introducing its own spatiotemporal *illogics*.

FIGURE 10.3   From Richard McGuire, *Here* (London: Hamish Hilton, 2014)
REPRODUCED BY KIND PERMISSION OF THE WYLIE AGENCY

comics' direct relationship between layout and narrative structure enables
McGuire to arrange his panels in such a way that they form multiple (and
often unexpected) interconnections with each other. The sparse textual elements, brief dialogues, and terse phrases compel the reader to formulate links
and connect the images by shuffling them around. Consider, for instance, the
temporal relations in Figure 10.3, where the déjà vu moment is set against
the time window from 1995 as well as the panel from 1352 that features a nude
female Native American taking a swim in a river. As the flipping back and forth
between panels overrides their initial fixity, the panels operate more as slices
of duration rather than points in time or snapshots.[13]

---

13   For a detailed discussion, see Misztal (2020).

3   "Where Did the Time Go?!": The Imaginal Leaping and
       Construction of Hereness

Despite the photographic-like realism of its composition and detailed, painstaking research into Perth Amboy's recent and ancient history, conducted in
the course of his work on *Here*, the book, as McGuire explains, "is not about
his house or hometown but an abstract or symbolic idea of home" (Smith 2014,
55). To put it succinctly, *Here* explores the hereness of home and the togetherness of families as projected against the massive sweep of time, which makes
generations, cultures, and entire civilizations disappear. The impulse of *Here*
is, however, not directed at documenting and preserving traces of their existence but appears to be more concerned with pondering the relationship
between history and memory and with reflecting upon the peculiar human
ways of occupying time and space.

   Here, as I argue, draws attention to our constructions of hereness and
nowness as intrinsic parts of the changing timescapes. Comprising complex
rythmicities and tempos, consistencies and constants, repetitions and contingencies, these timescapes, when taken together, reveal distinctive features
of our living on Earth in different historical periods. As Frederick Turner
remarks, "every new family is an act of the imagination, the making of a new
'here'" (Turner 2020, 14). McGuire's work recognizes this act in its detailed
portrayal of family life and parent-child relationships across generations and
cultures, emphasizing little rituals and small moments (birthday celebrations,
Christmas dinners, New Year's Eve parties, entertaining guests, housecleaning
chores, etc.) taking place within intimate domestic space. By bringing to the
fore the experiencing of the "here," McGuire's novel reminds us that even the
simplest and seemingly neutral spatial indicators are anthropocentric; they
express distance as measured from self. This anthropocentric perspective, as
Yi-Fu Tuan points out, is clearly reflected in English in the close interrelation
between personal pronouns, demonstrative pronouns, and adverbs of locations: "*I* am always *here*, and what is here I call *this*. In contrast to the here
where I am, *you* are *there* and *he* is *yonder*" (Tuan 2001, 47; emphasis in original). While "here" can be semantically defined in contrast to "there" it does
not, in our daily experience, necessarily entail the former. To use Thomas
Morton's expression, a "here" can be so cool that it "does not warm itself up
with references to 'there'" (Tuan 2001, 127).[14] Yet the "here" in McGuire's hand

---

14   In Morton's view the hermit's life is characterized by this quality: "It is a life of low defi-
      nition in which there is little to decide, in which there are few transactions or none, in
      which there are no packages delivered" (qtd. in Tuan 2001, 127).

does not remain frigid as it becomes a hallmark of transformation of space, revealing the intimate connectivity among people, their times, and the places they create.

Hereness thus indicates a form of "centeredness" as rooted in the person and linked with one's definite location and immediate presence. If "here," as Turner points out, "always already acknowledges the limits of my perspective," it also allows us to transcend them by being open to the possibility of going over and looking at the world from the other person's side, a new here:

> The fundamental quality of anything that has existence ("stands for," to translate the Latin) is that the world is different for it than it is for anything else, whatever its location. It is the center of its own world. It is the place where things that are pointed at or pointed out are pointed *from*.
>
> TURNER 2020, 13; emphasis in original

This deixic quality inherent in the act of pointing is one of the hallmarks of poetic sensibility, which through its construction of a new here transcends the limits by "turning everything in the world into a 'you,' and then shift[ing] the poet's 'here' into the point of view of the entity that is addressed" (Turner 2019). These dynamics, as I argue, underline also the visual poetry and narrative lines of McGuire's work, which in its articulation of hereness explores various modes of pointing.

To study these dynamics, I resort, in what follows, to Hegel's (1977) critical perspective on "hereness" developed in *Phenomenology of Spirit*, in particular in the sections where he characterizes the human comprehension of objects and places by paying close attention to the use of spatiotemporal terms such as demonstratives and indexicals. While Hegel acknowledges their discursive dimension and function, he ultimately approaches these terms as manifestation of a conceptual scheme that provides intelligibility of a single here and now within a larger spatiotemporal context. By capturing the non-linguistic dimension of pointing, Hegel seeks, first, to move beyond the simple reference between an I and a pointed-to object and then to analyze "thisness" and "hereness" as a set of categories that allows us to articulate the meaning of the objects as objects of consciousness. This theoretical approach is helpful in illuminating the specific deixicality of McGuire's graphic narrative, especially in its relation to the productive dynamics of imagining.

The spatial deixis, as signaled by the title and many textual and visual references, plays a prominent role in *Here*, but its function is rarely ever only indexical. The fixed point of view, namely that of a house corner, illustrates that pointing as a momentary reference to an object is incomplete. As Hegel has argued, pointing in spite of its apparent directness and immediacy cannot grasp an object in its concreteness and essentiality. A deictic act generates a multiplicity of spatial and temporal points, which create complex spatial and temporal relations by means of indexical markers such as "This," "Here," "Now," etcetera, but it cannot articulate a "rich" knowledge of objects and the truth of our immediate relation to them. As Ludwig Siep (2014) writes, pointing in Hegel's presentation of the experience of sense-certainty fails to

> capture anything enduring, anything that persists independently of the act of pointing. It can only indicate a spatiotemporal position, but without one knowing whether what 'is meant' is a point, a surface, etc.
>
> SIEP 2014, 76

Thus even the gesture of non-linguistic indication (what Hegel calls "ostending") does not capture a singular "This" but reveals a subdivided sensible manifold, "a simple plurality of many Heres" (Hegel 1977, 64):

> The *Here pointed out*, to which I hold fast, is similarly a *this* Here which, in fact, is *not* this Here, but a Before and Behind, an Above and Below, a Right and Left. The Above is itself similarly this manifold otherness of above, below, etc. The Here, which was supposed to have been pointed out, vanishes in other Heres, but these likewise vanish.
>
> HEGEL 1977, 64; emphasis in original

If, on the one hand, pointing lays bare a sense of fixedness and stability that I experiences,[15] it signals also "a conceptual scheme that goes far beyond the simple reference of this I to that object" (Siep 2014, 75). Within this scheme, pointing discloses the truth of sense-certainty as the most immediate relation of thought to reality by approaching the Here or the Now as "a movement which contains various moments" (Hegel 1977, 64). The intelligibility of these moments is, as Hegel maintains, a function of "Hereness" or "Nowness" as an abstract universal.[16]

McGuire's work does not aim at this kind of philosophical abstractness but instead, as I argue, more specifically and directly recognizes and celebrates

---

15   In turning around from a tree to a house, my "Here" itself does not vanish but "abides constant in the vanishing of the house, tree etc. and is indifferently house or tree" (61). As Hegel points out, "this vanishing of the single Now and Here [...] is prevented by the fact that I hold them fast. Now is day because I see it, 'Here' is a tree for the same reason" (61).

16   Hegel (1977) thus distinguishes "a Here of other Heres" that "in its own self is a 'simple togetherness of many Heres'; i.e., it is a universal" (66).

hereness as a movement comprising various moments. *Here*, as a narrative contemplation on time in variance and time's invariance, visualizes the coexistence of tensions and consistencies and invites the reader to actively negotiate the former without undermining the latter. In acknowledging our embodied condition, pointing in McGuire's narrative reveals both a dynamic manifoldness of "Heres" and a sense of immobility that we experience when forming our minds' relationships with outer reality. Confronting the flow of time, we find ourselves at the same Here, at the same point from which we approach the reality. But this motionlessness is overcome when we shift our attention from what has been moved to what moves, from a gesture to what animates the gesture. In doing so, we discover that the eye's movement lacks the power to unify the panels, and, if we are to make sense of all the juxtapositions that *Here* so abundantly offers, we need to appeal to the faculty that does not merely connect points in space or moments in time but is capable of establishing a relational and rational order among the panels by transcending their spatial juxtaposition and temporal succession. In other words, the eye's visual simultaneity must be complemented by a noetically inflected type of instantaneity, namely the one made available through our imaginative explorations of the hiatus among panels, images, and captions. The intensity and ingenuity of McGuire's narrative reside thus in the seamless bridging of time and space that engages the reader's interaction and generates multiple dimensionalities that tap into the workings of memory as well as into individual and collective experience of time.[17] This interaction activates also imaginal modality that animates all of the temporal gestures that the reader performs as he or she is growing more and more clearly aware that the spatial deictic referent "here" becomes temporally charged in the course of reading the novel.[18]

By encouraging a constant reshuffling among the panels, *Here* prompts the reader to move beyond the immediate panel sequence and consider simultaneous relationships between the panels. The stasis of a panel is in this way overcome by the reader's leaping forward and backward in time as she or he attempts to actualize and bring into being dormant relationships and patterns. The imaginal leaping is not meant to generate random oscillations between the captured moments, nor does it seek to replace the spatial deictic referent with the temporal one. McGuire's work does not reduce "here" to a mere substitute for "now" but rather unveils "heresein," to use Turner's (2019)

---

17   As Ian Gordon (2019) puts it, "*Here* is not only an innovative comic, but a plea for historical imagination and the use of memory in shaping ourselves" (282).

18   Graphic fiction, as Ware (2006) points out, is much like a musical score; it is the reader that brings this graphic music alive.

phrase, in its specific temporal horizon defined by locality, centeredness, and interiority.[19] More precisely, through its narrative strategies that stimulate imaginative mobility *Here* compellingly shows how my being always and irrevocably here "entails also the creation of a double temporal horizon ('backward' and 'forward') within which I am the permanently moving now" (Castoriadis 1977, 325).

McGuire's narratives appeal thus to the productive dynamics of imagining, in particular to its topographical power "involving *naturally* and *artificially* articulated fields, backgrounds, and foregrounds within which images emerge, are formed, and are determinately placed, and in the context of which the mind can engage in imaginative movement, in imaginative work and play" (Sepper 2013, 8; original emphasis). This generative and transformative mobility is not limited to mental imagery but emerges, as Sepper (2013) argues in his study, as an essential feature of the imaginatively engaged mind. Its operations are more attuned to "making and remaking, contextualizing and recontextualizing appearances" rather than to "envisioning and fixing them in mind" (8). This imagining power can, for instance, be applied in sport drills to improve players' sensory-motor skills and develop their strategic thinking by means of make-believe scenarios that restructure certain aspects of space and time of an initial situation and produce a new field that remains in relation with the original one: "What was initially taken as part of a larger field comes to be taken – imagined – for its own sake, and this part in its turn becomes the encompassing field for the partial activity" (89). The imaginative restructuring brings out both spatial and temporal shift as

> [o]ne can define within a field a subfield that can be temporarily inhabited for its own sake, after which one fits it back into the original field. *This creates a backward and forward movement, a "rolling" back and forth of imagined fields.*
>
> SEPPER 2013, 89–90; emphasis added

This "rolling" as McGuire's narrative shows, can be even more dynamic and complex in literary works that self-reflexively examine the flexibility and amplitude of imaginative thought as involved in articulating various ways of inhabiting space and time. The mobility, signaled also in Hegel's presentation

---

19   It is worth noting that Heidegger's concept of dwelling necessarily implies the notion of a world: here is where one dwells and here is constituted by a horizon of possibilities. This horizon, which constantly changes and recedes as we approach it, can be perhaps best understood, as Ricoeur (1991) suggests, as an inexhaustible capacity: "In each experience there is something there but also something which is only potential" (453).

of hereness,[20] does not only generate the oscillation between the presented moments but is also capable of generating more complex forms of here and now. Apart from conveying the noetic construction of imagined time, these forms embrace also the collective articulation and evaluation of time.

The transition from noetic time, the realm of human consciousness, to sociotemporality, to use J. T. Fraser's categories from his model of nested temporalities,[21] animates McGuire's narrative as it extends its time horizons beyond the limits of an individual consciousness and introduces a truly global, cosmic perspective that transcends the timespan of communities, cultures, and civilizations. These wide temporal horizons allow also McGuire to bring in the Anthropocene perspective from which our individual or communal "heres" are seen as an integral part of the Earth System, and which acts as a persistent reminder of human capacities and responsibilities for shaping future outcomes. By reaching back to primordial times (some of the panels go back to 3,000,000,000 BCE) and simultaneously foreshadowing the end of our solar system as we, human beings, know it ("In eight Million years, with its fuel supply running low, our Sun will start to swell in size, becoming a red giant.... engulfing the orbits of Mercury, Venus, and our Earth" [McGuire 2014]), *Here* creates a narrative symphony of moments, in which the near and the most distant futures and pasts meet in the moving now and in which the clichéd everyday phrases such as "Where did the time go?!" or "Now, where were we?" (McGuire 2014) acquire unexpected urgency and poignancy. On the one hand, the small moments and rituals projected against the massive sweep of time become deeply immersive and overwhelmingly personal, even nostalgic (especially when bringing memories of those "moments when we were all together in the same room" [McGuire 2014]). On the other, McGuire's narrative symphony elicits a more cogitative response by inducing a moment-by-moment meditation on the complexity of time and our relationship to it. Reflecting on the flow of time from our "heres," we confront yet another enigma of our existence as we become aware of both time in variance and time's invariance.

---

20   Hegel acknowledges this mobility as inward movement necessary for the initial constitution and preservation of intuition. The imaginative dynamics of a back and forth movement are likewise indispensable in Kant's schematism and play a prominent role in Fichte's presentation of imagination, especially in his formulation of the concept of *Schweben*. For a lucid discussion of the efforts to conceptualize these dynamics as intrinsic features of imagination in the Western thought, see Sepper 2013 and Brann 1991.

21   J. T. Fraser's model of nested temporalities that I draw on here provides a useful conceptual framework for classifying various temporalities and timescapes in McGuire's narrative. See, for instance, Fraser 1975 and Fraser 1987. For a brief but lucid overview of the model, see Ostovich's contribution to this volume.

References

Adam, Barbara. 1998. *Timescapes of Modernity: The Environment and Invisible Hazards.* London: Routledge.

Brann, Eva T. H. 1991. *The World of the Imagination: Sum and Substance.* Savage: Rowman & Littlefield.

Bukatman, Scott. 2012. *The Poetics of Slumberland: Animated Spirits and the Animating Spirit.* Berkeley: University of California Press.

Castoriadis, Cornelius. 1977. "Radical Imagination and the Social Instituting Imaginary." In *The Castoriadis Reader,* edited by David Ames Curtis, 319–337. Oxford: Blackwell Publishers.

Corsten, Rikke Platz. 2012. "Thirty-Two Floors of Disruption: Time and Space in Alan Moore's 'How Things Work Out'." In *Crossing Boundaries in Graphic Narrative. Essays on Forms, Series and Genres,* edited by Jake Jakaitis und James F. Wurtz, 93–103. Jefferson: McFarland.

Fraser, J. T. 1975. *Of Time, Passion, and Knowledge: Reflections on the Strategy of Existence.* New York: Braziller.

Fraser, J. T. 1987. *Time: The Familiar Stranger.* Amherst: The University of Massachusetts Press.

Gardner, Jared. 2012. *Projections: Comics and the History of Twenty-First-Century Storytelling.* Stanford: Stanford University Press.

Garner, Dwight. 2014. "While Stuck in a Corner, an Artist Bends Time." *The New York Times.* December 23, 2014. https://www.nytimes.com/2014/12/24/books/here-richard-mcguires-new-graphic-novel.html.

Gordon, Ian. 2019. "Bildungsromane and Graphic Narratives." In *A History of the Bildungsroman,* edited by Sarah Graham, 267–282. Cambridge: Cambridge University Press.

Greenberg, Julia. "5 Reasons to Read the Time-Travelling Graphic Novel *Here.*" *Wired.* September 12, 2014. https://www.wired.com/2014/12/here-graphic-novel-richard-mcguire/.

Groeensteen, Thierry. 2007. *The System of Comics.* Translated by Bart Beaty and Nick Nguyen. Jackson: University Press of Mississippi.

Hegel, G. W. F. 1977. *Phenomenology of Spirit.* Translated by A. V. Miller. Oxford: Oxford University Press.

McGuire, Richard. 1989. "Here". *RAW* 2(1): 69–74.

McGuire, Richard. 1994. *Night Becomes Day.* New York: Viking.

McGuire, Richard. 2014. *Here.* London: Hamish Hilton.

McGuire, Richard. 2015. *Here Interactive.* New York: Penguin. itunes.apple.com/us/book/here.

McGuire, Richard. 2016. *Tutaj.* Translated by Krzysztof Uliszewski. Warsaw: Wydawnictwo Komiksowe.

Mikkonen, Kai. 2017. *The Narratology of Comic Art.* London: Routledge.

Misztal, Arkadiusz. 2020. "All Times, One Place, and All At Once: Time Shuffling in Richard McGuire's 'Here' and Here." *Image & Narrative* 21(1): 72–81. http://www .imageandnarrative.be/index.php/imagenarrative/article/view/2443/1942.

Mouly, Françoise. 2018. "The Profound Mundity of Richard McGuire's 'My Things'." *The New Yorker,* 18 May 2018. https://www.newyorker.com/culture/culture-desk/ the-profound-mundanity-of-richard-mcguires-my-things.

Pedler, Martyn. 2012. "'3X2(9YZ)4A': Stasis and Speed in Contemporary Superhero Comics." In *Crossing Boundaries in Graphic Narrative,* edited by Jake Jakaitis and James F. Wurtz, 177–187. Jefferson: McFarland & Company.

Ricœur, Paul and Mario J. Valdés. 1991. *A Ricoeur Reader: Reflection and Imagination.* New York; London: Harvester Wheatsheaf.

Sepper, Dennis L. 2013. *Understanding Imagination: The Reason of Images.* New York: Springer.

Siep, Ludwig. 2014. *Hegel's Phenomenology of Spirit.* Translated by Daniel Smyth. Cambridge: Cambridge University Press.

Smith, Greg M. 2013. "Comics in the Intersecting Histories of the Window, the Frame, and the Panel." In *From Comic Strips to Graphic Novels: Contributions to the Theory and History of Graphic Narrative,* edited by Daniel Stein and Jan-Noël Thon, 219–238. Berlin: De Gruyter.

Smith, Joel. 2014. "From Here to Here: Richard McGuire Makes a Book." *Five Dials* 35: 54–55.

Solnit, Rebecca. 2004. *River of Shadows: Eadweard Muybridge and the Technological Wild West.* New York: Penguin.

Sultan, Larry and Mike Mandel. 1977. *Evidence.* Greenbrae: Clatsworthy Colorvues.

Tuan, Yi-Fu. 2001. *Space and Place: The Perspective of Experience.* Minneapolis: University of Minnesota Press.

Turner, Fredrick. 2020. "Metric and Variation: The Tempo of Poetry, the Poetry of Time." *KronoScope* 20(1): 7–16.

Ware, Chris. 2006."Richard McGuire and 'Here': A Grateful Appreciation." *Comic Art* 8: 5–7.

Ware, Chris. 2015. "Chris Ware on *Here* by Richard McGuire – a game-changing graphic Novel." *The Guardian.* December 17, 2015. https://www.theguardian.com/ books/2014/dec/17/chris-ware-here-richard-mcguire-review-graphic-novel.

Yokoo, Tadanori. 1996. *Waterfall Rapture Postcards of Falling Water.* Tokyo: Shinchosha Company.

PART 3

Measuring Time's Variance

⁛

CHAPTER 11

Variance in Time Morphologies in Production and Consumption of Incense in Medieval Japan

Vroni Ammann

Abstract

Time was variable in medieval Japan, with plural morphologies of time coexisting simultaneously. Not only can they be found in different realms of medieval life, but research on this question suggests even several time conceptions occurred in one and the same symbolic form. This paper aims to support the thesis by examining temporalities inherent in clocks and specifically in incense clocks, and it focuses on the symbolic form of economy for a first delineation of economic time. Incense, closely connected to the religious realm, was used as a medium for measuring time in a rather secular function. While it is an important good in the context of religious life, incense plays the role of a traded good in economic calculations or as a medium for artistic expression as well. A temporal analysis of a set of incense instructions, in its function as a manual for mixing incense as well as an expression of noble scholarship, informs how people measured time and scheduled their life. In contrast to other medieval documents, incense instructions were written without conscious thought towards time; therefore, the analysis here is focused mostly on chronographic and chronopolitical information.

Keywords

cyclical time – eleventh- to fifteenth-century Japan – Gofushimiin shinkan takimonohō – incense – incense clock – linear time – medieval Japanese economy – symbolic forms – time morphology – Tōdaiji

1 Introduction

In this paper I aim to connect economic temporal thinking and incense by means of the incense clock as well as a medieval document serving as an instruction for incense mixtures. First, I will discuss the function of different clocks and their temporality, focusing on incense clocks and showing a more detailed approach with the example of the incense clock used during the spring ceremony of Shunie 修二会 at Tōdaiji Temple 東大寺.[1] This discussion is followed by the outline of a developing economy, facilitating the consumption of exotic goods like incense, and highlighting linear temporal thinking required for the planning and scheduling of trade. The analysis is based upon the morphologies of time developed by Maki Yūsuke (2017) and adapted by Raji Steineck (2017), specifically a differentiation of linear, cyclic, and circular time. The analysis of examples drawn from the Gofushimiin shinkan takimonohō 後伏見院宸翰薫物方 (the incense formulae of the imperial house Go-Fushimi), a document written at the beginning of fourteenth century, will conclude the investigation on incense and its relation to time. I will apply the terminology used by Roland Harweg (2009, 1–59) on how time is measured in a text (chronography) as well as that used by Günter Dux (1989, for example, 49–65) and Eviatar Zerubavel (1976, 88–93) on how time is organized (chronopolicy).

2   Circular and Cyclic Time

In history, points are often made by the formulation of binary oppositions, like the famous good-bad, West-East or, in terms of temporality, linearity and circularity. These two temporalities then are singled out and unite under their wing several other time morphologies. Griffiths (2005, 57), for example, mentions deviant morphologies, but he does not take them into account in the following discussion.[2] Maki, like others, does not differentiate between circular and cyclic time, but still he tries to contrast 4 different morphologies in one diagram. The two reversible time morphologies in his work are called *hanpukutekina jikan* 反復的な時間 (oscillating time) and *enkantekina jikan* 円環的な時間 (circular time) (Maki 2017, 195; Steineck 201, 24–26). Both forms are entirely reversible and therefore highly abstract; as such a world would consist of no memory, no development. and no history. Cyclic time, however, a

---

1   東大寺 "Eastern Great Temple": Buddhist temple complex in Nara and center of Kegon school of Buddhism. It was founded after a series of disasters, and, after several relocations, the imperial court finally settled in Heijōkyō 平城京 (today's Nara), and it was commissioned by the imperial couple Shōmu 聖武 and Kōmyō 光明. Although it was founded earlier (738), Tōdaiji was not opened officially until 752, celebrated with the eye-opening ceremony for the large bronze statue Buddha Vairocana (Tsutsui 2018, 7).

2   Griffiths determines the time scape of most societies through history to be cyclical and cites examples of the Hopi image of time as self-contained wheel, of time moving in cycles, and of the ancient Greeks, where time itself was imagined as a circle. All these morphologies combined to something not moving in a straight line are then positioned against linear time, a "rare and recent idea."

morphology not discussed in either authors' works, contains a linear vector, and one could imagine this as the more adaptable morphology to everyday life, picturing time to flow in a spiral: circular when seen from above, oscillating between two points translated on the vector of time when seen from the side. Hence, I propose to use the concepts of oscillating and circular time in an abstract context. While a worldview like this is imaginable, the sources we are using suggest consciousness about differences between last years' first month and the current one. This differentiation would not be possible in a circularly moving flow of time. Circular time sets development repeatedly to zero, and a progression only is visible in the simple counting of numbers (examples here may be the living through thousands of lives in Buddhist belief, or praying a high and abstract number of prayers for the redemption of all during the spring festival shunie). Circular time rather belongs in abstract and tautological thought of the religious symbolic form (Cassirer 2010), while cyclic time with even small changes over the months or years is close to concrete human experience, even more so than the modern linear understanding of time. Griffiths never bothers to speak of linear or cyclic time schemes, but he calls the time of modern Euro-American culture dead time, which is disembodied from the life of the people, "the moment struck dumb by the striking clock, the deadening character of routines, schedules and endlessly counted and accounted time" (2016, 54–55). This dead time is contrasted with the living time of indigenous peoples, measured by tree calendars and insect and bird clocks.[3] Such a clock only works with the expectation of ever repeating events, working as the "alarm" which sets off action. The exact hour in a countable timeline is of little importance and does not support these actions and their outcome in a helpful way.

While such natural clocks are not considered in the following part, I would like to stress their usefulness as agents of the environment, strongly connected to the task for which they are used. Other than "living clocks," the following devices usually do not take into account the sensible ecosystem with its multitude of factors, but only rely on a few, or even one (water, sunlight, burning incense). Clocks, in this sense, are simplifications of processes, in their development moving away from their origin and dropping their connection to the measured element. Digital clocks from our modern life show this tendency in extremis, even shedding representational material and expression, to tick away time in a sequence of abstract numbers measured in the invisible world of an atomic energy-level change. The clocks analyzed here are somewhere in

---

3   Many thanks to the anonymous reviewer, who, amongst other important changes, suggested Griffith's article.

198

between and allow for an interpretation of the clock users' view on time in a macroscopic environment.

3   Clocks and Temporality

Besides observing the position of the stars or our sun, individuals in medieval Japan were using several different timekeeping devices, each one specified through their function for certain phases of day (sunlight necessary for a sundial) or year (temperatures over freezing point, as the sometimes huge clepsydra often were built outside), some even relatively independent from the environment and weather. Also, a clock allowed for more accuracy than approximations like "let us finish work at sundown." A court official probably needed to know about the hour of a specific day for planning a meeting of high-ranking nobility or for the organization of an imperial ritual conducted by a large part of the court personnel. It was necessary for such a person to follow a schedule, to write a structured diary, and to wake up by the sound of a bell in the early morning to manage the work of dozens of individuals into a functioning organism. Clocks and diaries therefore played an important role in the management and organization of the imperial court, and certainly the same was true in important Buddhist temples and other centers of power, assembling crowds of different social strata in one place.

Sundials have been used very early in human history and were mentioned in Greek and Babylonian documents well before the beginning of the Christian calendar. Bedini (1994, 8) cites the Zhōulǐ 周禮 (Rites of Zhou, a Confucian classic supposedly composed in the first century CE), mentioning an "installation of a straight pole to observe its shadow," followed by a reference to the Zhōubì suànjīng 周髀算經 (Arithmetical classic regarding the Zhou gnomon), supposedly composed in the first century BCE and describing the use of a gnomon during the time of the Western Zhou dynasty (eleventh century–770 BCE), hence placing the appearance of the sundial in East Asia into the millennium before Christ. Certainly the intricacies of later sundials developed over time, but the key time morphologies remained the same. Sundials measure time in an absolute way, which means that they show the hour during the course of a day and acknowledge the changing length of days by the shift of the shadows' position over the year, but without relating the measured moments to the temporal position of the observer in a linear flow of time of a larger time scale. Even if a sundial is using the relative movement of earth and sun for its operation, it seems, in itself, motionless. This impression stems from the fact that a sundial is measuring time cyclically, reflecting the same conditions of the annual cycle one year after the other, bound to the changes the solar system may undergo over time.

For the measurement of time without starlight, new developments, like the water clock or the incense clock, were necessary. The water clock, or clepsydra, measures time periods by means of successive water tanks, and it was used in Egypt and Babylonia "centuries before early Shang period" (Bedini 1994, 10), which places its appearance before 1800 BCE. Bedini (1994, 10–11) lists several sources that describe the use of a simple form of water clock in early China at least from the year 2356 BCE. The Nihon shoki 日本書紀 (Chronicles of Japan, 720) mentions the first water clock to be introduced to Japan in the year 660 by the crown prince, the later Tenji Tennō (626–672).[4] The hours were announced by drums or bells and with this introduced abstract time to the people. Their rhythm of life now was regulated by an external source.[5] As with sundials, clepsydrae were improved and extended over the centuries, but the principle remained the same: water flows steadily from one bowl into the next, while the water surface or a floater marks the elapsed time on a scale on the inside of the tank. It measures duration, as opposed to the sundial, which measures and shows points in time. This means that, if one marks the starting and ending point of a given length of time, the visual representation on the incense clock always will be the same, as the ember wanders over the plane. This is not the case with sundial, as the length of the fractions of time on the dial changes constantly, depending on the sun's position during the year. The measurement of duration is absolute, and it cannot be related to external moments in time, with the exception of using additional devices that count the time segments. In this function and with a sufficient supply of water, the water clock in principle is measuring time in a circular, unchanging manner. Other clock types, like the lamp clock or the sand clock, operate according to the same pattern, measuring duration instead of points in time, allowing for the measurement of daily tasks and scheduling rather than connecting the observers' time to specific moments in history.

In medieval Japan, a clock's function was not reduced to measuring time. The command over time was used by the elite as a tool of power (Griffiths 2005, 60), and all types of clocks subsequently found their way to the Japanese centers of power. They also were used as gifts to the wealthy: Oda Nobunaga, for example, a key figure in the power structure of feudal sixteenth century

---

4   Original source: Nihon shoki, scroll no. 26. https://zh.wikisource.org/wiki/日本書紀/卷第 廿六. Accessed 10 Oct. 2019, 19:13.

5   On the announcement of time, see Hashimoto 1978, 99–102; on abstract time see Maki 2017, 124.

Japan, was well known to be interested in foreign technology and collecting mechanical clocks. Clocks in this context were normatively used for showcasing power rather than measuring time.[6]

Clocks have been used for different things than timekeeping, a fact that alerts us to the difference of clock time and lived time altogether. Lived time, as Canales (2016, 117) explains, citing the philosophers Bergson, Husserl, and Heidegger, differs by the concept of "personal equation" from clock time. It is the difference between time assessed by a person and time marked by a machine (or in our context, by a device). This concept separates "world time" and "lived time" and makes the differences between the nature of the world and the world conceived by human consciousness tangible (Canales 2016, 117) The incense clock, the subject of the next section specifically measures time in a context, whereas time personally experienced may significantly diverge from measured time, for example, during meditation, or while reading and writing.

4   Incense Clocks

The incense clock has rarely been described in Japanese documents over the centuries and has received much less attention than other timekeepers. It is possible to find incense clocks in museums nowadays, but these exhibits usually were built in Edo period or later. It is an object one finds regularly in the personal belongings of monks, still in use for measuring the monks' time of meditation and prayer. Other than Tōdaiji incense clock for example, these clocks usually are called *jōkōban* 常香盤,[7] and they take a distinctive form different from other incense clocks. Observed from the side, they look like a heavy dumbbell, with a square surface looked upon from above. On the lower end, they are equipped with one or two small drawers, which contain the utensils for incense preparation. The upper container is protected by an elaborate latticed lid and is intended to be filled with ash. By using a mold, the user forms and presses a pattern from powdered incense. This allows the user to burn the incense gradually, similar to burning an incense stick (see figures 11.1 and 11.2). Plaques usually tag the points of passing time by displaying the 12 characters of the Chinese zodiac. Other than the equinoctial 24 hours of modern timekeepers,

---

6   Drawn from Angelika Koch's presentation "Diplomatic Devices: Clocks and Clashing Time(s) in late 16th and early 17th Century Japan," given at the conference "Time in Variance" conference. Also see Angelika Koch (2020), 80–82.

7   Incense seal burner. *Jōkō* is synonymous for *hyakkokukō* 百刻香 or *kōin* 香印, all expressions describing a pattern made from powdered incense and pressed in from on an ash bed, used for the measurement of time; see Jinbo 2004, 366, 412.

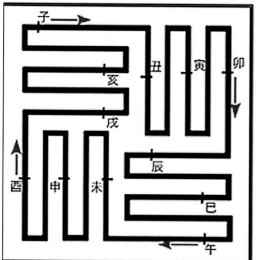

FIGURES 11.1 AND 11.2   The schematic representation of a common pattern (the Buddhist swastika) used for incense clocks and reference plaques manually positioned to indicate the change of the hour

PHOTOGRAPH TAKEN JUNE 15, 2019 BY THE AUTHOR OF HER OWN INCENSE CLOCK

the system adopted from China after the Taika Reforms (645–650 CE) consists of 12 double hours of the circadian period (one day and one night), represented by the names of the 12 earthly branches for differentiation and characterization. Even if the hours first were equinoctial (or equal), there is evidence that religious use increasingly propagated temporal hours with changing length during the year from the tenth century on (Steger and Steineck 2017, 12–13; Hashimoto 1978, 25–27; Miyake 2010, 168). Temporal daytime hours at summer solstice, for example, measure about 144 minutes, while at winter solstice their duration shortens to 95 minutes. The plaques on the surface of the incense clock therefore were not installed permanently in order to allow for adjustment.

The size of the clock and, accordingly, the size of the incense seal on top, as well as the pattern itself, define the measurable time; it ranges from a few hours to a few days. The incense clocks still available in today's specialized shops, for example, burn for about three modern hours; the *jikōban* 時香盤 (time incense burner or incense clock) of Tōdaiji on the other hand lasts for about a day. Morphologically, the clock reflects a segmented cyclic understanding of time with a consciousness regarding the flow of time and its changes in one year, divisible in repeated segments. The linearity inherent in cyclic thought shows the consideration of annual events, but its focal function never exceeds the cycle of a year. The huge incense coils that one may find in Chinese temples, burning for several months due to their impressive size, serve to exude fragrance on a long-time basis rather than measure time for the sake of linear continuation.

5   The Incense Clock of Tōdaiji Nigatsudō

The *jikōban* of Tōdaiji stands close to the central sanctum in the Naijin 内陣 (inner chamber) of the Nigatsudō 二月堂 (hall of the second month), its name already indicative of the important annual spring ritual of the Shunie. The ritual had been held during the second month of the Chinese lunisolar calendar, but nowadays it takes place between the 1st and the 14th of March. First performed by the monk Jitchū 実忠 (?–824)[8] as one of the most important rituals of Tōdaiji, Shunie was held annually without interruption from 752 CE. It was conducted so that one could repent one's transgressions before Buddha (*keka* 悔過), but nowadays visitors are mainly attracted by the spectacular big torches, scattering their burning embers from the porch of the Nigatsudō over the crowd at 7 PM every evening, as well as the drawing of water ceremony (Omizutori お水取り) performed at the climax of the ritual at day 12 (Nara National Museum 2016, 4). Still, the eleven monks of the Rengyōshū 練行衆 (those who practice continuously) follow a rigid regime during these days, and for their work they are highly revered by the crowd.[9] Each of these secluded men fulfils the specific tasks at defined moments during the day (Satō 2009, 65–70). The routine of the Shunie ceremonial can be gleaned from the Tōdaiji shunie gyōji shidai jikokubyō 修二会行事次第時刻表 (festival program of the Tōdaiji shunie) (Horiike 2008, 13),[10] and it is described in a task-oriented table by Satō (2009, 83).[11] These tasks have to be performed during specific time slots, the *rokuji* 六時 or "six Buddhist hours,"[12] measured by the *jikōban*, as table 11.1 below indicates:

---

8   Buddhist monk of Kegon school and pupil of Roben 良弁 (689–773, founder of Tōdaiji).

9   The number of monks fluctuated during the centuries from only a few to about 20 men. Nowadays each of the four appointees (shishiki 四職) is supported by an aide (only the Dōtsukasa 堂司, the head of the hall as grand marshal of ceremonies depends on two aides), as they lead the other members of the Rengyōshū through complex tasks.

10   Further extensive information regarding the ceremony from December 16, when the election for the members of the Rengyōshū is announced, until March 18, when the eternal lamp will be extinguished, can be found on pages 98 to 161.

11   The pages following contain a schedule of the days preceding the main ceremonial phase.

12   Dejitaru Daijirin デジタル大辞林. 2012. Second edition. "六時". Accessed on www .japanknowledge.com, 28 August 2019, 12:44.

TABLE 11.1   Rokuji, or the six Buddhist hours

Buddhist hour	Transliteration	Translation	Zodiacal system	Modern time
日中 (中時)	*nitchū*	noon, mid-day (also daytime)	hour of the horse	11 AM to 1 PM; "during day"
日没 (中時)	*nichimotsu*	sunset	hour of the rooster	5 PM to 7 PM
初夜 (大時)	*shoya*	nightfall	hour of the dog	7 to 9 PM
半夜 (小時)	*han'ya*	or chūya 中夜: midnight (also night-time)	hour of the mouse	11 PM to 1 AM; "during night"
後夜 (大時)	*goya*	early hours	hour of the tiger	3 to 5 AM
晨朝 (小時)	*jinjō*	sunrise	hour of the rabbit	5 to 7 AM

The six hours are distinguished into three groups of "big," "medium," and "small hours." These specifications do not indicate their length, but they mark their importance. The two "big hours," *shoya* and *goya*, therefore both demand main repentance prayers with the most repetitions, while the prayers scheduled for "small" and "medium hours" are short or their count of repetition is reduced, but admittedly these prayers can be finished in the already starting hour following the big ones (Satō 2009, 95–96). In contrast to the present Western system of measuring the hour in a linear progression from moment to moment, the hours in the double-hour system are measured in durations. This task-oriented approach to scheduling a busy day corresponds to the importance of the Buddhist rites: monks rather have to fulfil these tasks than stick to a specific starting and ending point in their daily schedule; also, it allows for a certain flexibility regarding unforeseen interferences or personal speed.

The incense clock standing in Nigatsudō dates back to early Edo period, when the hall was destroyed in a fire and reconstructed in 1667 (see figure 11.3). Its shape has been the same for all the clocks preceding this one: two lacquered wooden boxes, one above the other, one filled with ash, the other with instruments, standing in front of the inner sanctum of the hall of the second month.[13] Its dimensions are about 35 cm in width and depth, and both boxes add up to about 16 cm in height. On the surface of the flat-pressed ash, one can observe an accurate w-shaped form out of powdered incense, with six plaques demarking the six Buddhist hours.

---

13   Ueno Shūshin, interview during a research stay in Nara in March, 2019.

FIGURE 11.3   Tōdaiji *jikōban*
                PHOTOGRAPH TAKEN MARCH 13, 2019 BY THE AUTHOR

The responsibility for the handling of the incense clock lies with the Minami no Shū no Ichi 南衆之一 (head of the southern group), a member of the Rengyōshū. Available research literature holds that the incense clock of the Nigatsudō was used as time measurement device for ensuring the accuracy of the schedule without the sun in the line of sight (see, for example Bedini 1994, 166), but during my visit I was informed that now a modern wristwatch is used for the timing of the procedure.

But why has incense been used for measuring time in the first place? This question obtains significance especially when one is standing in front of the hall, where no fragrances can be discerned anymore in the fragrant air of the festival. The amount of incense burners in the inner sanctum of Nigatsudō alone – Satō depicts nine incense burners in her layout plan (2009, 78) – as well as the huge bowl for visitors' incense offerings in front of the western opening fill the air with strong odors. This overwhelming mixture of fragrances coming from sources other than the incense clock suggests that time has not been measured olfactorily but visually by one watching the incense clock's ember drawing closer to the next plaque. The casual manner in regard to the nonexistence of a recipe for incense powder equally suggests that the use of incense in this context is of a rather symbolic nature. Therefore, the reasons for using incense at all remain in the realm of esoteric Buddhism and pragmatism: incense arrived in Japan together with Buddhism and is an essential component of memorial services.[14] The Lotus sutra for example declares in a passage that a person holding a Buddhist memorial service and using objects like *hana* 華 (flowers), *kō* 香 (incense), *yōraku* 瓔珞 (personal ornament), *dōban* 憧幡 (hanging banner), *sōgai* 繪蓋 (silk canopy), *kōyu* 香油 (incense oil) or *sotō* 蘇燈 (butter lamp), will be the source of countless and boundless benefits for others (Matsubara 2012, 16).[15] It seems only logical that monks combined the symbolic power of incense with the convenient burning attributes and used it in the already known form of a time measurement device. Incense clocks usually being artful objects, the Tōdaiji *jikōban* is an inornate, well-worn black box, standing in the right corner of the Naijin, easily carried around if required, which indicates as well a pragmatic approach rather than one of enjoying incense in a refined and tasteful way, such as a noble or a monk might do in a different context. Other than this special timekeeper, incense clocks have been used in the palace, government offices, Buddhist temples, or scholars' studies mainly for measuring the time of meditation or learning in a stimulating environment, and while incense was used as a medium for measuring time, incense materials also have been used for their agreeable fragrance and for medical purposes. The reference for incense (*kō* 香) in old sources often is inseparably connected to the reference of drugs or medicine (*yaku* 藥), incense ingredients in some cases not only being one but also the other. These versatile properties converted incense and incense materials into sought-after commodities.

6   The Temporal Aspect of Incense as a Commercial Commodity

In this section, I will follow the trading routes of three important materials used for incense. The long-term character of international trade with mainly durable goods plays a role in the adoption of a more linear understanding of time and triggers society to see time as a commodity itself.

Historical trade with goods like incense usually is described in a very general way, and it is difficult to find information on specific aromatic ingredients, let alone any declarations on market volume, amounts, or traders. Also,

---

14   Ariga goes as far as to say, that incense is indispensable for Buddhist services (1995, 8).
15   Original source Myōhōrengekyō 妙法蓮華經: http://www.amtb.org.tw/pdf/08-01gcbeta
      .pdf, page 88. Accessed 10 Oct. 2019, 19:05.

documents on incense clocks specifically are rare. However, documents like the Gofushimiin shinkan takimonohō, or the Kunjū ruishō 薫集類抄 (a type selection of collected works about incense)[16] provide a general overview of different ingredients and instructions on how they should be mixed and used in a detailed, almost fastidious way. Concentrating on *nerikō* 練香 or kneaded incense, these texts were portrayed as the authoritative instruction for the enjoyment of incense, written for a select group of distinguished and refined people. It is not coincidental that the authorship of the Gofushimiin shinkan takimonohō is attributed to a Japanese emperor. Incense texts often were conveyed in secret from generation to generation, as "profound and secret teachings" of certain families (Go-Fushimi Tennō 1971, 19/564). The *makkō* 抹香 (incense powder) for the Tōdaiji *jikōban* for example is not mentioned in writings, and instructions on how to handle the clock were orally transmitted from monk to monk without an instructive document whatsoever. In regard to the incense clock powder, the consensus seems to be that a person may choose the ingredients freely while mixing incense for his or her personal device. Still, one online source provides information regarding Buddhist and Shintoist custom ingredients: both use *jinkō* 沈香 (aloeswood), *byakudan* 白檀 (sandalwood) and *chōji* 丁子 (clove). Buddhist monks add *shikimi no ki* シキミの木 (Japanese star anise), while in the Shinto context one adds *nemu no ki* ネムノキ (Persian silk tree) to the select ingredients above.[17] The blog entry further instructs one to use these ingredients as one likes, and it omits quantities. By personal observation of the *jikōban* incense seal of Tōdaiji, I deducted from its warm yellow color and fragrance that the powder mainly contained sandalwood.

Incense ingredients had to be imported to Japan, as only Japanese star anise and Persian silk tree grow on Japanese territory. Aloeswood was the main ingredient even before the shift from Okō 御香 to Kōdō 香道 in the sixteenth century. The differences between these two practices are considerable, and each required a different set of materials for the enjoyment of incense: Okō practitioners produced singular mixtures of incense and blended unique compositions by using a choice out of about 60 different kinds of incense ingredients,

---

16   Written and compiled by Fujiwara Norikane 藤原範兼 (1107–1165) from 1163 to 1165.

17   The art dealer Iwano Art gives additional information about rare or unfamiliar antiquities and art pieces in a continuous blog. Such blog entries may spark an interest in goods Iwano Art at times has on offer. The category here reads "Useful information regarding antiquities and art pieces" and contains an article titled "Has the time really been measured by fragrance? Let's try and look for a jōkōban!", a short text with a few unreferenced pictures highlighting incense clocks (Iwano Bijutsu Kabushiki Gaisha いわの美術株式会社). https://iwano.biz/column/kotto/kotto_method/0331-kohdokei-new.html. Accessed 6 June, 2019, 13:47.

while Kōdō contest participants were to distinguish aloeswood during the famous and refined incense contests in Edo period Japan (Hayakawa 2007, 9). Aloeswood is a resin-saturated wood of the family of Aquilaria trees, resinous as the result of an infection with fungus. The quality of aloeswood depends on factors like the species of tree, the geographic origin, the part of a tree, the incubation period, as well as the harvesting method.

Aloeswood first was found on beaches in considerable amounts, as the heavy and durable pieces of wood were washed ashore over the centuries. Singularly famous pieces of aloeswood with unique and recognizable fragrances were known as *meikō* 名香, adorned with a distinctive name mentioned through history. As the story goes, the famous *meikō* Ranjatai 蘭奢待[18] was chosen as a present for empress Suiko 推古 (554–628) in the sixth century, as fishermen noticed an aromatic fragrance when they were burning driftwood for warmth.[19] A moderate request for the fragrant wood developed, and when findings of aloeswood on the beach became scarce, Japanese consumers depended on the import of fresher material harvested directly from the plant. As trees from the Aquilaria family only grow in tropic areas, these long trading routes have to be considered an important factor regarding the acquisition of incense ingre-dients. Cultivation of aloeswood only started very recently, and, in contrast to agrarian incense ingredients, the seasonality of produce therefore did not impact on the availability. However, the increasing scarcity of collected wood, the slow and rare development of infected wood, and the specific knowledge necessary for acquiring freshly developed aloeswood from living trees certainly led to more demand than traders were able to supply.

Different qualities of aloeswood were imported from different countries: high quality from Cambodia, medium quality from South Vietnam, and lower quality from the Middle East, South Asia, and the island of Hainan, China (Minagawa 2014, 126–127). Sandalwood, an aromatic yellow wood growing in tropical rain forests, was used not only for woodcarving, incense, and incense oil, but also as *zukō* 塗香 (rubbing incense), a form of perfume in powdered

---

18   The characters of the name contain the characters of Tōdaiji (東大寺), insinuating the close bond of this relic and the temple. The large piece of aloeswood measures 156 cm in length and weighs 11.6 kg; amongst many other considerably sized aloeswood pieces, it was stored (and still is) in the Shōsōin repository of Tōdaiji (Shōsōinhōko 正倉院宝庫) since the middle of the eighth century. Several famous leaders, like Ashikaga Yoshimasa (1435–90), Oda Nobunaga (1534–82) and the emperor Meiji (1852–1912) are said to have cut pieces from Ranjatai. Information derived from the Shōsōin homepage: http://shosoin.kunaicho.go.jp/ja-JP/Treasure?id=0000012162. Accessed 10 August, 2019, 15:10.

19   Original source *Nihon shoki*, scroll no. 22. https://zh.wikisource.org/wiki/日本書紀/卷第廿二. Accessed 3 Oct. 2019, 18:40.

form for the cleansing and odorizing of the body. Sandalwood was imported from India (Jinbo 2004, 411–412). Clove grew exclusively on the islands of the Moluccas in Indonesia until the eighteenth century, when it was successfully cultivated on the Zanzibar archipelago in Africa (Jinbo 2004, 388).

Incense and drugs originating from India, Persia, and Southeast Asia irregularly reached Guangzhou via sea and land travel on the silk road routes from the eighth century on; other ports were Quanzhou and Yangzhou (Matsubara 2012: 19). Minagawa mentions the province of Hoan Chau (today Xu Nghe) in Vietnam as another place of trade for aloeswood but does not elaborate on the nature of the trade or the location of a port (2014, 127). The port of entry in Japan then was Dazaifu in Northern Kyūshū, a port in Western Japan not only closer to China and Korea by distance but also sociologically, compared to the capital cities in central and Eastern Japan.[20] With the beginning of the Heian period (794–1185), trade between Tang China (618–907) and Japan increased, and the Dazaifu Kōrokan 太宰府鴻臚館, a Japanese branch office for the reception of Chinese traders, was established. The institution was subsequently also used by Song-Chinese traders (960–1279). When Yuan China (1279–1368) came to power, offices for overseas trade were established – for example the office in Quanzhou in 1277. A trading network developed between these offices and the East coast of India through Southeast Asia (Matsubara 2012, 29–45). Even if this development facilitated the access to incense material considerably and channeled the trade volume of rare material to eager and solvent consumers in Japan, the travel routes for incense ingredients still were very long and dependent on the seasons and good weather. These factors heavily influenced schedules and deals between traders and consumers; a certain flexibility and patience was asked from all trading participants. This time-consuming aspect of rare exotics and non-domestic trade needs to be taken into consideration regarding availability at certain points of time (e.g. annual ceremonies, funerals), price, and further processing of incense materials.

While in the eighth century the "chief-handlers" of commerce most likely were lower-ranking bureaucrats (Farris 1998, 327) in a "multi-centered" economy with a multiplicity of official and unofficial markets (Farris 1998, 313), a hierarchy in terms of first access to goods developed. The most important temples were prime consumers and therefore first buyers not only because of

---

20   Amino Yoshihiko was the first historian to put peripheral regions, as well as the sea, between the nations' territories, into the center of trading maps, emphasizing the importance of unregulated sea routes and ports regarding trading connections to other countries. For further reading on the subject, see, for example, "Rethinking Japanese History" (Amino 2012).

their sufficient funds, but also because Tōdaiji, for example, served as the organizational institution operating the transactions with the traders at the port. Tōdaiji, like other temples, offered a secure space for a marketplace close to the temple where traders avoided the taxes of other authorities (Segal 2011, 79). This development again shortened the time for incense material to arrive at its destination, at least for Tōdaiji as the consumer. From the thirteenth century or earlier, Buddhist monks started to serve as money market agents in their function of collecting monetary donations, while at the same time local markets became the most important sites for the use of cash money in the countryside (Segal 2011, 67–68). This also applies to the Takimonouri 薫物売り or incense seller, a profession shown in a beautifully illustrated scroll from 1500, the Shichijūichiban shokunin utaawase 七十一番職人歌合 (poetry contest with the subject of artisans on 71 panels). Every panel, divided by a short paragraph, includes two professions with a poem each. The female incense- seller on panel number 60 is sitting next to the drug-seller on the floor, frozen in the act of measuring incense. The poem reads: "It is interesting to observe / how in the humid evening air / countless fragrances are prepared."[21] The accompanying passage tells us about people enjoying a moonlit evening and hearing the incense mortar sounding through the noise of the voices, even if nothing is sold anymore. Unfortunately, it is not possible to deduce whether the incense-seller is a travelling merchant or if she comes to sell at a stall, both ways of selling common in medieval Japan. We therefore cannot extract more information from this scroll about how incense was sold exactly.

A travelling merchant was able to quickly move across the political borders between the provinces, independent of the regulations of Kamakura or Kyoto appointees (Segal 2011, 83). This liberty and swiftness of trade must have applied to a travelling incense-seller as well, shortening the time again for incense to arrive. The description of lengthy travels and the troublesome acquirement of incense material, though abbreviated over the centuries' developments of economy, still may show how consumers of exotics had to plan their purchases if they wanted to produce incense and show off their abilities at scheduled incense contests. Linear time is inherent in economic considerations, and an awareness of linear time is a crucial requirement for the success of any market agent: consciousness of developments from the past into the future, planning for the achievement of future aims, scheduling of transactions with the aim of profit, or the situational adaption in case of shortage or excess – all these examples bear traits of a linear understanding of time and cannot function in

---

21   Original: 六十番　薫物売(たきものうり)「随分此香ども／選り整へたれば／この夕
暮のしめりにおもしろき」(Iwasaki 1993, 122). Own translation.

a circular concept of time. Circular time makes little sense in economic con-
templations, where repetition without progress means stagnation and failure,
while cyclic time allows at least for some progress and development in a limited sense, compared to strictly linear time.

Linear time however has not been constituted solely by the arrival of a market economy; it always has been part of a cyclic understanding of time. While
other concepts lost their importance in this increasingly economic environment, the concept of linear time grew in popularity. Still, all the other temporal
concepts continued to play their role in other areas, coexisting and intermin-
gling with newer and more advantageous concepts in regard of socioeconomic
developments. The concept of symbolic forms creating bubbles of different
temporalities helps us to understand the possibility of simultaneously existing
temporal morphologies, as for example linear time has much less impact in
religious or artistic realms.

A supporting perspective on the increasing importance of a linear morphology of time is the close link of economic thinking to the incipient reification of
time during the institutionalization of trade with foreign countries. This was
effected by the installation of traders' offices and the arrangements for a regulated market volume and content. With the exchange of commodities and later
the use of money in growing civilizations and an augmented and differentiated
production, time started to be a factor regarding the scheduling of processing
material into end products. This link is not only intuitively perceptible, but
also factual, as Maki explains: Compared to the concrete time of earlier non-
monetary societies, time now is measured as well as exchanged with material, detached of all personal relations. It is homogenized as a resource, which
can be earned, wasted, saved, and so on. This process disregards personal life,
where time certainly is qualitatively differentiated and heterogeneous. This
shift from quality to quantity is not limited to time but also refers to material. In non-monetary societies personal needs restrict quantity to a certain
degree and emphasize quality, but in a monetary economy the focus lies on
quantitative profit, where limits are non-existent (Maki 2017, 298–303). This
argument highlights the process of changing time regimes, a process mirroring the changes of a societies' experience. Different time morphologies lose
their relevance in some realms of life but not in others; they even may gain in
importance. Especially in times of change, this transformation is accentuated
by conflicts in all areas, for example shown during the development of night
work and the resulting higher wages during night shifts, as work during this
time impacts on the well-being and private life of a worker.

Incense not only measures time, but it is also part of a market economy
dictated by schedules, as well as the treating of time as an asset, and we can

expect documents about incense to mirror the variance of coexisting time concepts in their respective realms, as I discuss below.

7   Incense Instruction

This section investigates the variance of time by means of text analysis, focusing on the symbolic form of economics. Other symbolic forms like religion, art or administration, in which incense certainly plays an important role, will be mentioned but not tracked further for the sake of clarity. For analysis I turn to the incense instruction Gofushimiin shinkan takimonohō written or commissioned by Go-Fushimi Tennō 後伏見 (1288–1336) in the early fourteenth century. Texts like the Gofushimiin shinkan takimonohō, usually called kōdōsho or "writings about the way of incense," serve to educate the reader regarding the correct attitude towards the production and consumption of incense, and provide information about the origin of incense consumption and about the contemporary approach. The Gofushimiin shinkan takimonohō contains the author's opinions and detailed instructions on how to mix fragrances in different ways, and it can be divided structurally into three different text types: commentary, formulae with the mixing ratio, and instruction. The commentaries are descriptive and explain the reason for this document, focusing on the connection of certain fragrances and seasons, as well as on the bond between incense and Buddhism. The formulae usually are offset by title and indention, as well as visually by gaps between the incense ingredients and their quantity. The text is structured in a peculiar way, as first the author comments on incense in general and then provides a secret formula for *kurobō*, one of the six fragrances.[22] In between this formula and the other fragrances' formulae, a large section is dedicated to instructions for preparing the incense ingredients and the composite additives. This part then is followed by a general choice of recipes for each of the six fragrances.

Chronographically, the Gofushimiin shinkan takimonohō is absolute, relative to our present only implicitly via our knowledge of the reign of Go-Fushimi Tennō (1298 to 1301), but without chronometric information regarding the

---

22   The Mukusa no takimono 六種の薫物, produced and enjoyed since the seventh century: *baika* 梅花 (plum blossom), *kurobō* 黒方 (black formula), *jijū* 侍従 (lord stewart), *rakuyō* 落葉 (falling leaves), *kikka* 菊花 (chrysanthemum flower), *kayō* 荷葉 (lotus leaf). *baika* is the fragrance related to spring, *kayō* to summer, *kikka* to autumn, *rakuyō* to winter, *jijū* to love and *kurobō* to fortunate events (Kōdō Bunka Kenkyūkai, 2015: 25). See the next example for more information on the interpretation of the different fragrances in the source.

composition date. This disconnection fits its function as a manual, as recipes are generally valid independent of their publication date or the time of the user. Units may be temporally viewed regarding their application in the respective present, necessitating translation to more modern units and containing information about the time of composition as well, but the units used here do not limit the timeframe in which this document was written, as their use spans centuries. The main focus in the analysis of the Gofushimiin shinkan takimonohō is to show 1) a variance in time, as well as 2) defining economic thought as a marker of temporality.

In the beginning of the text we find a commentary about the connection of incense to Buddhism – how incense is the companion of Buddhist gods, constantly present in the religious realm:

薫物は仏の御世菩薩聖衆の沈檀の匂ひよりはじまりて。から国にも是をまなびうつせり。我国につたはりし事は。

GO-FUSHIMI TENNŌ 1971, 19/562

[The knowledge about] incense begins with the aloes- and sandalwood fragrance of the Buddhist gods, which was learned and imitated in ancient China too. [This knowledge] reached our country.[23]

In this beginning, fragrance connects the mythological past with the historical present. As the fragrance of the gods should be copied by incense, it serves as a medium not only permeating the air but also transcending time, chronographically reaching from a time that is absolute (what Harweg [2009, 2, 18] calls mythographic) and unconnected to the present of an observer into a time that is relational and historiographic with a continuous link to the present. (Also see Steineck 2018, 175–176). The concept of multitudinous lifespans to live through in Buddhist thought illustrates in a very concrete way cyclic time consciousness: circular in its movement, but not endlessly repetitive, going towards a destination. Incense and the knowledge about it are sacred and valuable, its value raised by the information about the way incense came to Japan. Buddhism and incense belong together, incense being indispensable for religious action. Therefore, the value of incense is defined not only by supply (or its shortage) but also by its religious necessity.

A very different example in the same paragraph of the document connects each of the six kinds of fragrance to seasons:

---

23   The translations of these parts of the Gofushimiin shinkan takimonohō are my own.

春は梅花。むめの花の香に似たり。夏は荷葉。はすの花の香に通へり。
秋は落葉。もみぢ散頃ほに出てまねくなるすゝきのよそほひも覚ゆなり。冬
は菊花。きくのはなむら々々々うつろふ色。露にかほり水にうつす香にことな
らず。小野宮殿実頼の御秘法には長生久視のか香なりとしるされたり。黒
方。冬ふかくさえたるに。あさからぬ気をふくめるにより。四季にわたりて身
にしむ色のなづかしき匂ひかねたり。侍従。秋風蕭颯たる夕。心にくきおり
ふしものあはれにて。

GO-FUSHIMI TENNŌ 1971, 19/562

Spring is *baika*. It is similar to the fragrance of plum blossom. Summer
is *kayō*. It is like the fragrance of Indian lotus. Autumn is *rakuyō*. It calls
forth the colors of autumn foliage and is evocative of the rustling pampas
grass. Winter is *kikka*. The chrysanthemum is a flower with magnificently
ephemeral colors, and it cannot be distinguished between fragrance waft-
ing through morning dew and incense transferred by water. The secret
recipe of Saneyori from Ono Palace lets us know of a scent of longevity: *kurobō*. In winter there is freezing chilliness, which inspires serious
feelings. This fragrance seems to freeze the body while going through the
seasons, and it serves as a fragrance of retrospection. *Jijū* is an evening of
cooling love. The season is detestable, and sometimes it evokes an awareness of impermanence.

Incense here is linked to seasons and feelings conjured by the thought of the
flow of time. Two different interpretations are possible: one is the seasonality of consumption, a view conveyed by delimiting, implicit-material temporal
expressions, exhibiting a practical, or prescriptive dimension of chronography.
A different interpretation emphasizes the symbolic nature of the fragrance,
expressed through the seasons. In this case, time expresses typological attributes of incense, thetically affirming the temporality of its characteristics.
These interpretations are not exclusive, and they may both apply. The seasons
also show cyclic time as markers of a year cycle, constantly repeating their succession. This passage further shows the duality of permanence and impermanence. Longevity is achieved by one's applying the right blend of fragrance,
and the context here implies not only deceleration of aging but even stagnation by freezing, bordering on permanence. Knowledge about the impermanence of things, however, is expressed through the aesthetic concept of *mono
no aware* 物の哀れ (awareness of impermanence).

Both examples are drawn from the descriptive commentary, and they display generally a more archaic worldview. The author mentions this explicitly
in the following sentences:

今の世には其方思ひ々々なり。名ある程の方もさま々々なれども。[...]是
皆時に随て昔の人は知けれど。今の世にさらに聞えず。

GO-FUSHIMI TENNŌ 1971, 19/562

I am studying these formulas, even if there are many new formulae
nowadays. [...] These [the six fragrances] were well known by people of
earlier times, but now they are unheard of.

The author here refers to knowledge from at least a few centuries ago, also to
be found in the Kunjū ruishō, a comprehensive document compiled in mid-
twelfth century. Even if many different ways of using incense have developed
recently, the author feels nostalgic admiration towards the old ways and, ignoring the new incense manuals, feels drawn to the old recipes.

The examples above show expressions of the authors' perception of time
in a personal and emotive syntax; the next examples are drawn from the prescriptive and practical instructions and formulae in the same text, devoid of
emotion and typecasting an instructional text or recipe. The first example is
concerned with the roasting of shells, an ingredient quite commonly used in
almost all mixtures:

凡三四日ばかり是を炙り。但貝の体にしたがひて遅速ある也。是を折てこ
ゝろみるに。おれやすくてそのくちき黄なり。是を持てそのご期とす。こが
れぬれば尤不宜也。

GO-FUSHIMI TENNŌ 1971, 19/563

Roast them about three or four days but subject the tempo to the size of
the shells. Try to break them when they are brittle, and the opening is
getting yellow. Then the right amount of time has passed. It is really not
advisable to apply *amadura* if the shells are burned.

On the quantitative level, the exemplary instruction here provides formal
chronographic expressions: determined ("three to four days long"), as well as
undetermined ("as long as it takes the shells to become brittle") measurable.
This is a common feature of the instructive part of the document, as it serves
to temporally organize the sequence of steps. One could argue that the intuitive awareness of material changing irreversibly into different material by the
use of fire may show signs of linear thinking. On the qualitative level, time
here is assessed in an explicit way, and affirmed in its existence ("then the right
amount of time has passed") in a rare chronothetic expression to be found in

such instructions. Time usually holds the function of organization and is recognized as something existing, but the pondering about time concepts is not a
primary feature in texts like the Gofushimiin shinkan takimonohō.

The next example shows economic thinking and a consciousness about
supply problems:

> 梅花。ぢん。八両。丁子。三両三分。せんかう。一分二朱。かい。三両。か
> んぞ。一分。白だん。三分。ざかう。二分。くんろく。一分。
> 梅花のすくなくわかちたる定。
> ぢん。二両二朱。[...]かんぞ。四分。
>
> GO-FUSHIMI TENNŌ 1971, 19/568

Baika formula: aloeswood: 8 *ryō*; clove: 3 *ryō* 3 *bu*; lower quality aloeswood: 1 *bu* 2 *shu*; shell: 3 *ryō*; Chinese liquorice: 1 *bu*; sandalwood: 3 *bu*;
musk: 2 *bu*; frankincense: 1 *bu*. This is the instruction in case you want to
mix a small amount of *baika* incense: aloeswood: 2 *ryō* 2 *shu* [...] Chinese
liquorice: 4 *bu*.

The sentence between the two recipes provides instructions on how to proceed if a consumer either does not have enough ingredients or does not want
to produce too much of the same incense mixture. The next example explains
how exactly one is supposed to behave if one ingredient is not available:

> かえふのほう。かんぞ。二分。ぢん。三両二分。白だん。一分。かい。一両
> 一分。しゆくこん。一分。是はなくはざかう丁子。一両一分。
>
> GO-FUSHIMI TENNŌ 1971, 19/568

Kayō formula: Chinese liquorice: 2 *bu*; aloeswood 3 *ryō* 2 *bu*; sandalwood: 1 *bu*; shell: 1 *ryō* 1 *bu*; *shukukon*: 1 *bu* (if you do not have *shukukon*,
take musk); clove: 1 *ryō* 1 *bu*.

Here the added sentence is written in *warichū* 割注 (small script) in the original. This script type is used for notes and additions. The reasons for such additional information are manifold and may even lie in personal preferences, but
in the context of an author believing these recipes to be the apex of incense-
making and of long trading routes constantly influencing the availability of
exotic ingredients, the insertions here rather give advice in how to react to
shortages in times of need. Also, they nicely illustrate the authors' virtuosity in
handling incense material.

The next example follows after the "mixing procedure," emphasizing the fact that the right way to mix the ingredients is to be followed without any deviation:

是秘伝也。急ぐ時かならずしもしからずといへり。

GO-FUSHIMI TENNŌ 1971, 19/564

This is a secret message. Under no circumstances undertake this if busy.

This sentence expresses concern regarding the time-consuming process of incense-making and warns the reader to attach enough importance to the impending tasks. Time is characterized as "busy," chronotypological information on the personal experience of time. Usually someone interested in such an exotic endeavor like incense contests must have been wealthy as well as powerful, with duties and responsibilities different than those of commoners. In this context, the warning makes much sense and displays awareness of the leaders' full schedules.

8   Conclusion

Inherent in a clock's function is a specific time concept, making it redundant or useful in certain areas of life. By examining three different types of clocks used in medieval Japan, I have differentiated several time concepts, not exclusive in their morphologies but still showing plural temporalities. The incense clock unites different time morphologies, only visible from the perspective of different symbolic forms: incense as an economic factor (economy) and incense as a temporal factor (religion). While the religious function mainly connects myth with history and the present, and cyclic time is prevalent, incense as a commodity in a trading network primarily necessitates a linear temporal organization. Time in economic systems is treated like a commodity itself, qualitatively homogenous and being at odds with other time concepts prevalent in society. The analysis of the incense instruction written or commissioned by Go-Fushimi Tennō repeatedly shows variance in time morphologies simultaneously existing in medieval Japan. Each of the three text types chosen for the partitioning of the text (commentary, formulae, instruction) not only displays content differences but also distinctive differences in their temporal structure. A temporal analysis therefore may be beneficial for the formulation of a not-yet-described text genre or type like the incense instruction, and it may be a subject of further research.

Acknowledgements

This project has received funding from the European Research Council (ERC) under the European Union's Horizon 2020 research and innovation program (grant agreement No. 741166). I would like to thank my colleagues of the ERC grant project Time in Medieval Japan – Raji C. Steineck, Simone Müller, Daniela Tan, Kōhei Kataoka, Georg Blind, Alexandra Ciorciaro, and Etienne Stähelin, as well as Sebastian Balmes and Karl L. Zehnder – for their invaluable help towards the realization of this essay. Nonetheless, all mistakes and flaws are certainly mine. I also would like to thank the participants of the 17th Triennial Conference of the ISST in 2019 for their insightful questions and input on my presentation, which encouraged me to submit this article in the first place. As it is not possible to enter the inner sanctum of Nigatsudō as a female researcher, I was grateful for the accommodating solution of the incense clock being brought outside to the south gate of the hall in between the ceremonies of the busy time of shunie, where I was able to observe and photograph the unique treasure for myself. My heartfelt thanks therefore go to the former abbot Kitakawara Kokei, as well as the present head of the hall, Ueno Shūshin, and his considerate staff.

References

Amino, Yoshihiko. 2012. *Rethinking Japanese History*. Michigan Monograph Series in Japanese Studies, vol. 74. Ann Arbor: University of Michigan Center for Japanese Studies.

Ariga Kōen 有賀要延. (1990) 1995. *Kō to bukkyō* 香と仏教. Second edition. Tokyo: Kokusho kankōkai.

Bedini, Silvio A. 1994. The Trail of Time: Time Measurement with Incense in East Asia: Shih-chien ti tsu-chi. Cambridge: Cambridge University Press.

Canales, Jimena. 2016. "Clock / Lived." In *Time: A Vocabulary of the Present*, edited by Joel Burges and Amy J. Elias, 113–128. New York: New York University Press.

Cassirer, Ernst. (1923) 2010. *Philososphie der symbolischen Formen*. 3 volumes. Hamburg: Felix Meiner Verlag.

Dux, Günter. 1989. Die Zeit in der Geschichte: Ihre Entwicklungslogik vom Mythos zur Weltzeit: mit kulturvergleichenden Untersuchungen in Brasilien (J. Mensing), Indien (G. Dux/K. Kälbe/J. Messmer) und Deutschland (B. Kiesel). Frankfurt am Main: Suhrkamp.

Farris, William Wayne. 1998. "Trade, money, and merchants in Nara Japan." *Monumenta Nipponica* 53(3): 303–334.

Fujiwara Norikane 藤原範兼. (1894) 1971. Kunjū ruishō 薫集類抄. In *Gunsho ruijū* 群書
類従, vol. 19, edited by Hanawa, Hokiichi 塙保己一, 525–561. Third edition. Tokyo:
Zoku Gunsho ruijū Kanseikai.

Go-Fushimi Tennō 後伏見天皇. (1894) 1971. Gofushimiin shinkan takimonohō 後伏
見院宸翰薫物. In: *Gunsho ruijū* 群書類従, vol. 19, edited by Hanawa, Hokiichi
塙保己一, 562–574. Third edition. Tokyo: Zoku Gunsho ruijū Kanseikai.

Griffiths, Jay. 2005. "Living time." In: *About Time: Speed, Society People, and the
Environment*, edited by Tim Aldrich, Forum for the Future, 53–65. United Kingdom:
Greenleaf Publishing Ltd.

Harweg, Roland. 2009. Zeit in Mythos und Geschichte: weltweite Untersuchungen zu
mythographischer und historiographischer Chronographie vom Altertum bis zur
Gegenwart. Berlin: Lit Verlag.

Hashimoto Manpei 橋本万平. 1978. *Nihon no jikoku seido* 日本の時刻制度. Tokyo:
Hanawa Shobō.

Hayakawa Jinzō 早川甚三. 2007. *Bungaku to kōdō* 文学と香道. Nagoya: Arumu.

Horiike Shunpō 堀池春峰. (1996) 2008. *Tōdaiji Omizutori: nigatsudō shunie no kiroku
to kenkyū* 東大寺お水取り:二月堂修二会の記録と研究. Second edition. Tokyo:
Shōgakkan.

Iwasaki Kae 岩﨑佳枝, Takahashi Kiichi, and Shiomura Kō. 1993. *Shichijū ichiban
shokunin utaawase. Shinsen kyōkashū. Kokon ikyokushū* 七十一番職人歌合. 新撰狂
歌集. 古今夷曲集. Shin Nihon koten bungaku taikei 新日本古典文学大系, vol. 61.
Tokyo: Iwanami Shoten.

Jinbo Hiroyuki 神保博行. (2003) 2004. *Kōdō no rekishi jiten* 香道の歴史事典. Second
edition. Tokyo: Kashiwashobō.

Koch, Angelika. 2020, "Diplomatic Devices: The Social Lives of Foreign Timepieces in
Late Sixteenth- and Early Seventeenth-Century Japan." *KronoScope* 20(1): 64–101.

Kōdō bunka kenkyūkai 香道文化研究会. (1989) 2015. *Kō to kōdō* 香と香道. Fifth edition.
Tokyo: Yuzankaku.

Maki, Yūsuke 真木悠介. (2003) 2017. *Jikan no hikaku shakaigaku* 時間の比較社会学.
Twelfth edition. Tokyo: Iwanami Shoten.

Matsubara Mutsumi 松原睦. 2012. *Kō to bunkashi: nihon ni okeru jinkōjuyō no rekishi*
香と文化史:日本における沈香需要の歴史. Seikatsu bunkashi sensho 生活文化史
選書. Tokyo: Yuzankaku.

Minagawa Masaki 皆川雅樹. 2014. *Kodai ōken to tōbutsu kōeki* 古代王権と唐物交易.
Tokyo: Yoshikawa Kōbunkan.

Miyake Kazuo 三宅和朗. 2010. *Jikan no kodaishi: reiki no yoru, chitsujo no hiru* 時間の古
代史:霊鬼の夜、秩序の昼. Tokyo: Yoshikawa Kōbunkan.

Nara National Museum 奈良国立博物館. 2016. *Omizutori* お水取り: *feature exhibition,
treasures of Tōdaiji's Omizutori ritual*. Special line-up 特別陣列. Nara: Tenrijihōsha.

Satō, Michiko. 2009. *Tōdaiji Omizutori: haru wo matsu inori to sange no hōe* 東大寺お水取り:春を待つ祈りと懺悔の法会. Asahi sensho 朝日選書 852. Tokyo: Asahi Shinbun Shuppan.

Segal, Ethan Isaac. 2011. *Coins, Trade, and the State: Economic Growth in Early Medieval Japan.* Harvard East Asian Monographs, vol. 334. Cambridge (USA): Harvard University Asia Center.

Steger, Brigit and Steineck, Raji C. 2017. "Introduction from the Guest Editors to the Special Issue 'Time in Historic Japan.'" *Kronoscope* 17(1): 7–15.

Steineck, Raji C. 2017. "Time in Old Japan: In Search of a Paradigm." *Kronoscope* 17(1): 16–36.

Steineck, Raji C. 2018. "Chronographical Analysis: An Essay in Methodology." *Kronoscope*, 18(2): 171–198.

Tsutsui Kanshō 筒井寛昭; Kajitani Ryōji 梶谷亮治; Bandō Toshihiko 坂東俊彦. (2010) 2018. *Tōdaiji no rekishi* 東大寺の歴史. Motto shiritai もっと知りたい. Fifth edition. Tokyo: Kabushikigaisha Tokyo Bijutsu.

Zerubavel, Eviatar. 1976. "Timetables and Scheduling: On the Social Organization of Time." *Sociological Inquiry* 46(2): 87–94. Wiley Online Library. https://onlinelibrary.wiley.com/doi/abs/10.1111/j.1475-682X.1976.tb00753.x. Accessed Oct. 23, 2016, 14:50.

CHAPTER 12

Understanding Computation Time: A Critical Discussion of Time as a Computational Performance Metric

David Harris-Birtill and Rose Harris-Birtill

Abstract

Computation time is an important performance metric that scientists and software engineers use to determine whether an algorithm is capable of running within a reasonable time frame. We provide an accessible critical review of the factors that influence computation time, highlighting problems in its reporting in current research and the negative practical impact that this has on developers, recommending best practice for its measurement and reporting. Discussing how computers and coders measure time, a discrepancy is exposed between best practice in the primarily theoretical field of computational complexity, and the difficulty for non-specialists in applying such theoretical findings. We therefore recommend establishing a better reporting practice, highlighting future work needed to expose the effects of poor reporting. Freely shareable templates are provided to help developers and researchers report this information more accurately, helping others to build upon their work, and thereby reducing the needless global duplication of computational and human effort.

Keywords

computation time – computation time template – execution time – measuring time – performance metric – reporting time – response time – time template

1   Introduction: Why Computation Time Matters

This paper seeks to provide a non-specialist introduction to computation time – the time that a computational process or program takes to run – and to review the ongoing problems in the way that computation time is reported in current research, demonstrating the negative impact that this lack of reporting can have on developers as they seek to optimize their code. After a review of current problems in the field and their practical implications, recommendations of best practice when measuring and reporting computation time – including freely shareable templates – are also provided in order to avoid perpetuating these critical issues and make these recommendations easier to implement.

The time that it takes for a computer algorithm to finish its task (computation time) is a key performance metric that scientists and software engineers use to determine whether an algorithm is capable of completing its task in a reasonable time frame. However, hardware capability is rapidly increasing, and new techniques are being developed that combine software and hardware to quickly solve problems of increasing complexity. This situation means that, in practice, the supposedly fixed amount of time that it takes one researcher's computer to complete a task will often be vastly different for another. Software developers are also frequently unable to build on the work of others. Poor reporting practices mean that there is currently a substantial duplication of effort required for developers to evaluate if an algorithm will be suitable for an intended task. It is to this end that this paper offers a practical suggestion to help combat this needless duplication of effort, and its human, financial, and environmental costs.

When writing or using any computer program, it is important to first understand how long a process or program would take to run; this time frame is known as the computation time of the task. This temporal aspect to computing is an inherent factor that is often taken for granted by the end user; even the fastest programs take time to calculate across several steps, running on the central processing unit (CPU) or graphical processing unit (GPU), along with the other hardware components, each of which has its own individual frequency at which it is able to update variables and calculate the steps needed to complete the process. For example, when a satellite navigation mapping system, or Sat-Nav, calculates the fastest route from one location to another, the software has to calculate multiple different routes, taking a number of computational steps. The time that the program takes to provide a final mapped route back to the user may be imperceptible (i.e., a few milliseconds), or a noticeable amount of time (i.e., several seconds), depending on the complexity of the route (see Delling et al. 2009 for an example of this in practice in route calculation). Critically, for a system to be useful to a user, not only must it perform the required task successfully, but the computation time from initiating the task to receiving the end result (also known as the response time) needs to be within a reasonable time frame for a busy software user.

This reasonable time frame can vary depending on the task, and the environment in which that task is being conducted. For simplicity, we can arbitrarily split these types of tasks into three groups: "real-time," "quick decision," and "worth-the-wait." For a real-time task the computation time needs to be faster than we can notice; for example, for a video feed this would be less than the frame rate of a camera or screen (approximately less than ⅓0th second, as noted in Wu, Houben, and Marquardt 2017), and this speed is particularly important in gaming and real-time video analysis. A "quick decision" (approximately less than a few seconds) occurs when the user may be prepared to wait a short time to get an answer to a question, for example, in an initial GPS route calculation, when a short delay is not a problem for the user (see Delling et al. 2009 and Delling et al. 2015 for discussions on calculating route complexity). Finally, there is the "worth the wait" timescale, which can be a few minutes (as demonstrated in Kong et al. 2017; Huynh, Li, and Giger 2016), hours (as demonstrated by Wahab, Khan, and Lee 2017; Wang et al. 2014), days (as demonstrated by Melinščak, Prentašić, and Lončarić 2015; Cheng et al. 2017; Mok and Chung 2018) or even months (as demonstrated by Vo, Jacobs, and Hays 2017; Luong and Manning 2016).[1] This timescale occurs when the system may be solving a very complex problem, such as when training a machine learning algorithm on a very large dataset (for example, see Vo, Jacobs, and Hays 2017, and Luong and Manning 2016). Accurate knowledge of the computation time – both predicted and measured – and the ability for the system to adhere to this "reasonable time frame" are both important; if a system is unable to perform within what the user considers to be a reasonable amount of time, it will not be fit for purpose. For example, if a real-time video processing system takes several seconds to process each frame, the system will not be fit for purpose and a more optimal algorithm (or hardware) must be found to enable the system to work.

---

1   To give a few examples from the field of cancer research that demonstrate the vast differences in computation time: see Kong et al. (2017) which uses convolutional neural networks (CNNs) and two dimensional Long-Short-Term-Memory (2D-LSTM) to detect breast cancer metastasis, a process that takes 1 minute and 45 seconds to detect metastasis in a whole slide image (a type of microscopy image with many pixels, e.g. $100,000 \times 200,000$ pixels); see Wahab, Khan, and Lee 2017, which discusses training a deep learning system to detect breast cancer in histopathology images, a computational process that took approximately 15 hours; and see Mok and Chung 2018, which discusses the segmentation of brain tumours in medical images using generative adversarial networks, recording a time frame of approximately four days to train the Coarse-to-fine Boundary-aware Generative Adversarial Networks (CB-GANs). In other areas, computational tasks have been shown to take even longer. As an example, Luong and Manning 2016 explain that in machine translation (e.g. from English to Czech) it took three months to run a program to train a deep learning character-based model using a dataset of 15.8 million sentences and 1.269 billion character tokens.

It is worth noting here that energy consumption has also been proposed as a method of monitoring computational resource usage (for a proposal combining time and energy see Martin 2001, and for examples of energy consumption as a performance metric see Wirtz and Ge 2011; Yang, Chen, and Sze 2017; and Jiao et al. 2010). While this subject is out of the scope of this paper, such methods are likely to be particularly useful in scrutinizing the huge energy cost (both financial and environmental) of some high-performance computing demands. For example, Strubell, Ganesh, and McCallum (2019) highlight the power demands and hence the carbon dioxide output of modern natural language processing systems, while Schwartz et al. (2019) critiques this further. Other examples can be seen in the computational process of mining cryptocurrencies such as Bitcoin, where the energy demands can be colossal: O'Dwyer and Malone (2014) estimate that in 2014 the power consumption for mining Bitcoin was comparable to that of the power consumption of the entire country of Ireland, and Li et al. (2019) estimate that in 2018, Monero, an alternative cryptocurrency, had similar power consumption demands, releasing approximately 21,000 tons of carbon dioxide into the earth's atmosphere through the mining of the currency in China alone. Therefore, in order to determine the efficiency and environmental impact of a program, developers should increasingly measure power consumption as an additional metric alongside their measurement of computation time.

2   How Do Computers Measure Time?

Before we proceed further, it is particularly useful to understand how computers actually *measure* time. Computers can use several different ways to measure the passing of time, each of which has different levels of precision. The most conventional way that computers measure time is using the inbuilt "real time clock" (RTC) which essentially uses the same type of clock that most wristwatches use. With this method, the computer uses a quartz crystal electronic oscillator, which essentially "vibrates" at 32.768 kHz, and is therefore able to provide a precision of 30.5 microseconds – the maximum temporal accuracy of this method, as noted by Horan (2013). This quartz crystal time chip sits on the computer's motherboard (the main circuit board) and provides the time for the operating system to use. This chip can also provide programmers with the ability to measure the computation time of their process, up to the accuracy of 30.5 thousandths of a second.

A programmable interval timer (PIT) can be used in order to provide higher precision. A PIT is another type of microchip that sits on the motherboard and has a clock cycle of 1.193182 MHz, enabling an 0.838 microsecond precision in which the computer can interrupt the CPU and start or stop processes, as noted by Crandall et al. (2006). A time stamp counter (TSC) aligned to the CPU's cycle frequency can be used in order to get greater precision (an example use of TSC is in Skopko and Orosz 2011, and an explanation of TSC by Intel can be found in Intel Corporation 2016). Previously, these CPU clocks ran on a constant clock speed. However, to reach higher speeds, these CPU frequencies are now variable within their use – i.e., the computer can effectively increase or decrease the speed of its own clock (within a defined range) for better performance – which effectively reduces the consistency of the temporal precision, and therefore the reliability of temporal measurement. Finally, there are also high precision event timers (HPET), which, again, are microchips that sit on the motherboard and are able to probe the time of a process when prompted. These operate at a minimum of 10 MHz and typically at 18 MHz, and they therefore have a time precision of less than 0.1 microseconds, or ten thousandths of a second, as noted by VMware (2008).

The measurement of computation time is made even more complicated by the precision of the programmatic function calls (when a programmer requests the amount of time that has elapsed) which easily enables programmers to measure the time that a process has taken. The precision of these function calls can depend on the timing function (or the method of requesting the time used within a program) and which operating system (OS) is being used. Taking an example from the popular programming language Python (Python Software Foundation 2018), this language also allows its users to inspect the commonly-used time.time function (Python Software Foundation 2019).[2] This function provides a temporal "resolution" of approximately 15 milliseconds for a Windows operating system, as tested on Windows 8.1 (see Hudek, Sherer, and Schonning 2017; and Stinner 2017 for more information), but only 1 microsecond for a Linux based operating system (tested on Fedora 26 [kernel 4.12] in Stinner 2017) – an in-practice temporal difference in which one system is effectively 15,000 times more accurate for such short computational processes. When a programmer uses the nanosecond timer function in Python (time.time_ns()) this precision changes again, with a resolution of approximately 84 nanoseconds on Linux, and 318 microseconds on Windows – a temporal difference with a factor of 3,750. When a programmer is documenting

---

2 Python is used by an estimated 8.1 million coders worldwide, according to a 2019 report by
   SlashData (SlashData 2019). A survey of 90,000 coders by StackOverflow also shows that
   41.7% of code developers use Python (Stack Overflow 2019).

computation time using such short processes that will be run many times, it is therefore useful to document the programmatic timing function call that has been used (where possible), along with the operating system that it was used on, *and* the hardware that was used to record it.[3] However, the timing function name and temporal precision are not commonly documented in current practice, leading to a range of issues (see section 5 below).

When measuring and documenting computation time, it is also important to know what type of computation time is being referred to. Computation time normally either refers to response time or execution time (see Figure 12.1). The *response* time is the total time elapsed from starting to completing a process, including the time that any other processes, such as operating system processes, might have taken while the program was running. By contrast, the *execution* time is the total time spent computing the process of interest, not including any other processes (see Audsley et al. 1993; and Baruah and Burns 2006). Therefore, when documenting how computation time is measured, it is important to note which of these is being referred to, so that the reader can understand if other processes are also being included, and, if so, the types of other processes being run in the background should also be documented to provide a complete picture of the situation. A useful early example of a discussion of calculating the execution time of a real-time program by Puschner and Koza (1989) highlights some of the issues around calculating the maximum execution time that persist to this day; this paper also provides an interesting example of a real-time program conducting image analysis on a video camera feed in order to control a robotic arm.

When writing algorithms, it is possible, in controlled environments, to theoretically determine how many computational steps are required in order to complete the process. This calculation does not immediately correspond to the time that it will take to run, but it does provide the best indication of when, given an input of X, then Y number of steps will need to be computed. The careful art (or sub-field of computer science) for determining this temporal scaling is also known as computational complexity.

---

3 Note that the temporal resolution of the programmatic function calls are not normally variable on different hardware across most modern devices, but the hardware needs to be recorded as the time to process is greatly affected by the hardware used, as is discussed in more detail in section 4.

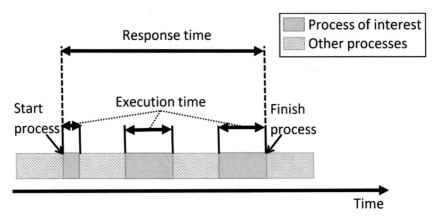

FIGURE 12.1   A block diagram illustrating the difference between response time (total time
elapsed from starting and finishing a process) and execution time (total time
spent actually computing that process, represented by the sum of the green
blocks in the diagram). Other processes might be operating system processes
or other (possibly high priority) processes which interrupt the process of
interest

3   Computational Complexity: Too Complex in Practice?

Computational complexity, a well-studied area within computer science,
investigates how an algorithm scales with various inputs – or how mathematically optimized an algorithm is for its intended purpose. A useful text on computational complexity by Arora and Barak provides a thorough overview of the
field (Arora and Barak 2009). The mathematical representation of how many
steps a program takes to compute, and the time implication due to this complexity, is sometimes referred to as time complexity (for example, see Tomita,
Tanaka, and Takahashi 2006; and Avoine, Dysli, and Oechslin 2005). However,
while an understanding of the following mathematical concepts is required in
order to grasp computational complexity, in practice, many developers do not
have the mathematical training to enable them to optimize their code accordingly. Put simply, the gap between thesis and praxis is effectively too great at
present for the careful research of computational complexity's talented theo-
reticians to be widely applied in a broad range of practice.

3.1   Computational Complexity: A Brief Overview
What follows is a short general overview of the relevant mathematical concepts involved in computational complexity: normally, the exact time to
compute is not of particular interest in this mainly theoretical field. What is

of interest here is determining how a function (or algorithm) scales up with increasing the numbers of a variable n, mostly in order to calculate the worst case scenario (i.e., the most number of steps required to complete a process), and identifying if a method scales well with increasing this variable or not (for example, increasing the number of data samples to be processed). Scaling well in this context means asking whether increasing this variable also increases the complexity (the number of computations that need to be performed), for example, exponentially, in a linear way, or not at all.

The notation for determining the complexity is sometimes called "Big O" (or "Big Oh") notation, as the "Order" of the complexity function is abbreviated to "O" followed by, in brackets, how the method scales with an increase in the number of the variable "n" (for a full discussion of this matter, see Arora and Barak 2009). For example, we may want to calculate whether the upper bound (maximum) of the number of steps scales linearly with the number of the variable "n" referred to as $O(n)$ (also known as "Order n"), or does the maximum number of computational steps scale with the square of the number of the variable "n" referred to as $O(n^2)$ (also known as order n^2). To give a more detailed explanation, a function that takes a constant time to calculate would be an order of 1 ($O(1)$) as it does not change with the value of a variable (n) (for example, the number of data samples). A function that scales linearly with the number of samples (n) would be ($O(n)$), and a function that scales quadratically with the number of samples (n) would be ($O(n^2)$), while a function that scales exponentially with the number of samples (n) with constant c would be ($O(c^n)$). To give a real-world example of this issue in practice, to improve the efficiency of a specific machine learning neural network algorithm, the order of the method may be calculated. A lower order that more effectively scales with increasing numbers of neurons would greatly reduce the time taken to train and predict a result using this model; one example of where this is calculated is in Schmidhuber 1992, which explains an improved method which scales with $O(n^3)$) instead of the compared inferior method, which scales with $O(n^4)$), where n is the number of neurons used in the network. Decreasing this worst-case run time can dramatically improve the efficiency of the program, and can enable the system to be run in a "reasonable time frame."[4]

Calculating the time complexity or "Big O" of an algorithm is reasonably simple for a small function, or a short amount of code, where the coder is able to understand all of the processes involved in the code (i.e., what the code does, and therefore how many mathematical steps that it will take). However, for larger code projects, these often include multiple functions and libraries,

---

4   Discussion of what constitutes a "reasonable time frame" is given in section 1 above.

which are other code functions that the coder can draw on and use as a form of shorthand, but which may not have been written by the coder, or may not even be inspectable by the coder (perhaps to obfuscate the code and maintain the original coder's intellectual property).

For larger real-world projects, it then becomes incredibly hard to calculate the theoretical time complexity of a program or process as there are simply too many unknowns, and even if all of the code is known, it could be far too big a task to inspect what may be many thousands of lines of code that make up the full program. Although there are some efforts to educate non-computer scientists in field of computational complexity, such as Mertens' paper "Computational Complexity for Physicists" (2002), another practical barrier is that the mathematical terminology described in the highly theoretical research papers common to the sub-discipline of computational complexity theory can be difficult for the vast majority of non-specialized programmers to understand, making the existing body of knowledge unusable in practice for many coders (examples of such papers are: Martínez et al. 2007; He and Yao 2001; Ko and Friedman 1982; Chen et al. 2010; Cooper 1990; Ladner 1977; Shanbehzadeh and Ogunbona 1997; Cancela and Petingi 2004; and Peled and Ruiz 1980).[5]

This all means that, in practice, calculating the theoretical time complexity is useful for understanding some important short functions that your code might make use of, in order to determine where your code might take a lot of time to run, or where your code might be improved to make it run faster. However, the nuances of computational complexity are, at present, too complex to give professional non-specialized coders the understanding needed to optimize many real-world programs in practice.

3.2   *Timing-Based Attacks*

It is worth briefly noting that sometimes it's advantageous for a computational task to take a long time. For example, this time delay can be the case when protecting passwords and in encryption processes. Password and encryption protection works because performing a "brute-force" attack – trying all possible variations of a password or encryption key until the correct one is revealed – should take an extremely long time, deterring potential attackers.

---

5   There is a marked lack of studies to evidence the effect or scale of this comprehensibility barrier; however, from many years of practical experience in research and teaching in the field, it is apparent that this is a significant barrier for professional programmers needing to optimize their code. A study which is able to quantitatively or qualitatively assess the extent of this issue would be very welcome to reveal the true scale of this problem.

A timing-based attack on protected information essentially checks to see how long a request process takes to execute, and breaks the encryption in this way; for example the slower the result is returned the closer to the solution the user is, and so those trying to gain access to encrypted information can iterate through different options checking how long they take to return an answer, and then use this information to infer what the correct result is. Timing-based attacks, a type of side-channel attack that takes advantage of security flaws in the implementation or hardware (rather than in the algorithms themselves), have enabled hackers to view private web-content (Felten and Schneider 2000) and to access protected memory on computers (Lipp et al. 2018). As such, computation time can be manipulated for a variety of legal and illegal purposes, demonstrating the real-world importance of computational complexity.

4 The Evolution of Hardware

The time that a computational process takes to run is also dependent on the hardware upon which it runs; for computational processes, given the huge variety of rapidly-developing hardware available, computation time is very much in variance. This point may sound obvious – but as this section will discuss, this variance poses a far greater problem in practice than might initially be assumed. As hardware capabilities change rapidly over time, it is important to take into account that the time taken to run a program on one machine on a specific date is going to be very different for a similarly priced machine just a few years later. Recent technological developments mean that the types of processors that are commonly produced today are quite different from those used before 2004, when most CPUs contained a processor with only one core (see Figure 12.2 for a plot of the evolution of CPU clock frequency and number of cores from 1971 to 2014). Previously, when all processes were sequential (i.e., could only run one at a time on a single core), if the programmer knew that an algorithm process took t seconds to execute and was running on a Y MHz processor, then, moving the same algorithm to a different CPU with a Z MHz processor, the programmer could reasonably expect the process to execute in approximately $(\frac{Y}{Z})t$ seconds.[6] This expectation is no longer the case, as is

---

6   This is a necessary slight oversimplification of the situation as there are also other factors involved here, such as any latency time (i.e. any time delays: for example, it takes time for data to move across the computer circuit boards and initiate a process). However, the basic premise remains that there is in essence a linear scaling between the time to process and the clock speed on a single core system.

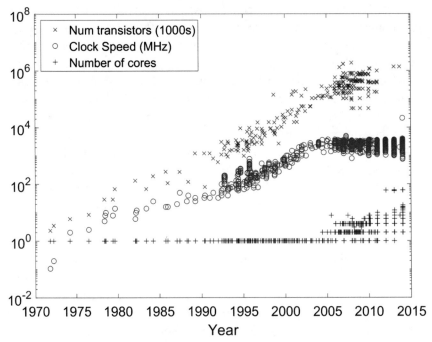

FIGURE 12.2   The evolution of CPUs. This plot shows, on a log scale, the number of
transistors (crosses), the clock speed (circles), and the number of physical
cores (plus signs) in different processors over time. The figure has been
made using data from the Stanford CPU Database, using the details of 1,388
processors in which information of the release dates was published

described in more detail below, and it has now become quite difficult to predict how long an algorithm will take to run across different machines, particularly due to the non-sequential way in which many processes can now be
executed. In order for us to understand these different hardware options and
how this affects the execution time, we must first understand the evolution of
hardware over time.

The database was originally discussed in Danowitz et al. 2012, and the
database has since been updated and can be downloaded from http://cpudb
.stanford.edu/download (accessed 28th October 2019). Note that sometimes
not all three parts (clock speed, number of transistors, and number of cores)
are in the database for some processors, and therefore only the known parts
are plotted here. The data used and code written to produce this plot have
been made freely available as open source at: https://github.com/dcchb/
ComputationTime.

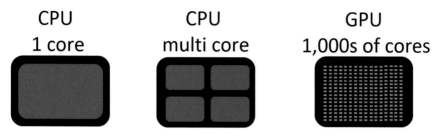

FIGURE 12.3   An illustration of the number of processing units on different processors

In just a few decades, hardware has evolved from single-core central processing units (CPUs) to multi-core CPUs, to graphical processing units (GPUs) (see Figure 12.3 below), and more recently, tensor processing units (TPUs) (see Jouppi et al. 2017 and Jouppi et al. 2018 for more on TPUs). Over the last 50 years the number of transistors in a CPU has grown exponentially, adhering to Moore's law, which states that the number of transistors in a processor doubles every two years (see Moore 1965 and Moore 1975), and the clock speed of CPUs has also been increasing (see Figure 12.2). However, as the visible levelling off of the circles in Figure 12.2 shows, in the last 15 years the increase in clock speed over time has slowed dramatically, and the preferred way to improve performance in practice has been to simply add more cores (as represented with "+" signs in the figure), effectively compensating for this lack of increase in clock speed.

4.1   Single Core CPUs
A single core CPU can process one calculation at a time, for example, calculating 2 × 3 = 6. When there are many parts to a calculation, these are calculated sequentially (in series). CPUs are widely recognised to be appropriate for general purpose computing and are currently the quickest method available for sequential tasks.

4.2   Time Sharing
Throughout the history of computing, there have been multiple attempts to improve the performance that can be squeezed out of a computer processor.[7] One of the earliest such attempts took place in 1959, when John McCarthy – the researcher who coined the name for the field of "Artificial Intelligence" four

---

7   For the purpose of this paper, performance is defined as the tasks or processes that are
     completed per second or per minute.

years earlier – wrote a proposal for "time sharing" on a computer to improve the performance of a computer system when it was used by many people at once (McCarthy 1959). A useful video, filmed at MIT in 1963, describes this time-sharing project, and it has been archived by the Computer History Museum; it is also viewable on YouTube (see Fitch 1963).[8] Time sharing was devised to enable many people to share the same operating system, enabling a more opti-mized use of a computer – a very expensive resource at the time. Using this method, the user's processes are effectively executed in a rolling way, which means that the response time for each user is faster than each user having to wait for the previous user to finish their program (waiting for a program to fin-ish in its entirety is referred to as batch processing). On modern computers, a similar idea is used in multi- or hyper-threading; here, instead of having mul-tiple users, there are multiple processes (normally from the same user) that can share a processor.

4.3   Cores and Threads

If a user has to wait a long time for the computer to give the expected response, this may be due to the CPU not being fully utilized. When the CPU core is wait-ing for some other information in order to progress in its task, it can be put into an idle state. This idle state is effectively wasted time on the processor, and so processors commonly use multi- or hyper-threading to reduce this potentially wasted time by queuing multiple calculations, enabling one calculation to be computed while another is waiting for further information required in order to complete its task (see Marr et al. 2002 for more on hyperthreading).

A CPU may also have a number of physical cores, which means that there are multiple processors built into the same microchip. The use of multiple pro-cessors means that multiple processes can run at the same time (in parallel), which reduces the overall time to run many processes. Programmers can there-fore use a thread (or task) per core, which means that each core has a different job to do, enabling a program (or multiple programs) to run in a shorter time frame.

In addition to this, it's also possible to have more than one thread per core for a multi-core system, which makes it possible to have multiple jobs queued for each core. This thread queuing means that while a process is waiting for completion, another thread can take its place. Having more physical cores

---

8   This video clearly describes the rationale behind this design and what such a design is capa-ble of. As an aside, it is quite remarkable how many fundamental computer science concepts were known and discussed at such an early stage, even when computers did not yet have a computer screen interface (at the time the design was implemented, the computer output display was merely an automated typewriter, typing the output of the machine on to sheets or rolls of paper).

enables genuine parallel processing (where multiple calculations are being computed at the same time). However, having more threads than the number of cores does not allow for any extra parallelization – it only provides less downtime. This difference between real parallelism and reduced downtime is particularly important to understand, as the number of threads distributed across the physical cores is sometimes referred to as the number of logical or virtual cores, which in practice can lead to some confusion in the number of processes that can be calculated at the same time.

4.4   Multi-Core Scalability Issues

The move to multi-core processors means that the time to process a calculation can be reduced when the code is made to work in parallel (i.e., using multiple cores at once). However, it is not always possible to do this, and so improvements in processing in series eventually limit this scalability. This scalability limit is first described by Amdahl (see Amdahl 1967); although not in Amdahl's original paper, the equation that describes this scaling process is called Amdahl's law (Hill and Marty 2008). What this law states is that as long as there is some part of the code that does not run in parallel, the increase in computing speed that can be achieved by running code over many cores (or processors) is limited by the part of the code that runs in series (i.e., the part of the code that can't be split into parallel processes).

A plot of this mathematical law is shown in Figure 12.4, which illustrates the maximum processing "speedup" factor that can be achieved when different percentages of the code are parallelized over different numbers of cores. This figure shows that the increase in speed achieved does not scale linearly with the number of cores, as one might expect; put another way, the speedup benefit does not keep increasing at the same rate when you continue to add more cores.

To provide an extreme example, even when 95% of the execution time is spent using code that can run in parallel, the speedup factor from running the code on 100 cores (or processors) compared with running it on just one core is only 16.8 times faster, and not anywhere near 100 times as fast, as might be expected. What is important here is that, when describing an algorithm, it is not often made clear which parts of the code are possible to run in parallel and which parts of the code have to be run in series (sequentially), and therefore the potential speedup from parallelism is not at all clear in practice. However, steps can be taken to solve this issue. If developers documented the relative parts of their code and how long each part of the code takes to run, and whether these should be run in parallel or series, this would form an incredibly useful collective knowledge base for the developer community, who might use or want to build on this work.

4.5   Ultra Parallel Processing with GPUs

Processing in parallel has become increasingly sophisticated through the use of graphical processing units (GPUs). These were originally built to enable the processing of computer graphics, in order to allow graphics to be displayed on a computer screen. However, these are now widely used in game physics and other general purpose computing tasks (see Owens et al. 2008 for more on GPUs). These have grown in computation power in recent years, primarily due to rapid growth of the computer games industry. In GPUs, the base clock speed is still less than in modern CPUs, but GPUs have many more cores. While most CPUs in recent years have a handful of cores (in the range of 4 to 12 physical cores), GPUs take parallel processing to a whole new level. For example, the laptop on which this paper was written (a Surface Book 2) has an NVidia GTX 1050 graphics card which has 640 GPU cores, with a base clock frequency of 1.354 GHz,[9] while the most high-end GPU "Deep Learning System" at the time of writing, the NVidia DGX-2, has 81,920 GPU cores.[10] GPUs are capable of super-fast processing of images and videos; as images and videos are essentially matrices (tables) of information, any dataset which can be composed in matrix form is suitable for GPU acceleration. In the last few years, processing using GPUs in machine learning research – a sub-field of Artificial Intelligence research – has become commonplace as the datasets to process are very large, and require a lot of processing that can be run in parallel. With this large number of processors, it is possible to achieve an acceleration in performance when parts of the code are made to run in parallel. Nickolls and Dally (2010) found that, when taking Amdahl's law into account, a GPU-only system was fastest when most of the code could be run in parallel; however, when some of the code has to run in series, a CPU and GPU combination was found to be the best option. It is therefore important to provide the details of the GPU hardware specification. When processing on GPUs, as these are so hyper-parallel when compared to CPUs, it becomes even more important to know which parts of the code can be run in parallel and which parts can't in order to find what can be "accelerated" with this parallel architecture.

---

9   For reference or comparison, the specification of the CPU on this Surface Book 2 is an Intel core i7-8650U CPU with a clock frequency of 1.9 GHz, with 4 physical cores and 8 logical processors.

10   Deep learning is a much-used term which benefits from some demystifying here: it actually refers to machine learning, a type of data analysis where many layers of statistics are performed and mathematical models may be learnt automatically from the data provided in order to make either class predictions (e.g., cat or dog) or numerical value predictions (e.g., $10 or $1,000).

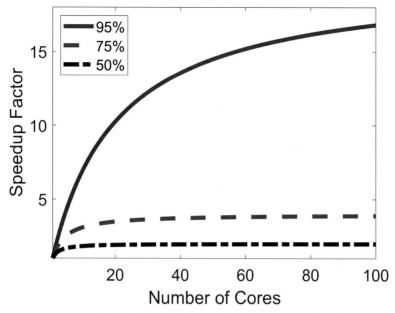

FIGURE 12.4   A plot of Amdahl's law, showing the increase in speed as a function of
the number of cores in three different scenarios, where the execution
time for the part of the code that can be run in parallel is 50%, 75%,
and 95% of the total time to run. The code written for this paper to
produce this plot has been made freely available as open source at:
https://github.com/dcchb/ComputationTime

5   The Problem with Current Practice

In areas of computer science where computation time is a particular focus
of the research, such as the sub-field of scheduling for example, computation time is generally very well reported, with a great level of detail frequently
provided in such papers. Examples of such papers where computation time
is reported in enough detail to help others are as follows: Jiang et al. (2010);
Zhang and Chatha (2007); Camelo, Donoso, and Castro (2011); Radulescu and
Van Gemund (1999); and Wang, Gong, and Kastner (2005). However, in other
application domains in which computation time is important but perhaps
not the central focus of the research, such as medical imaging and machine
learning, important details about the computation time are often neglected in
research papers. For example, if we consider the topic of using deep learning
for the detection of cancer, where computation time is clearly an important

factor in ensuring the result is clinically usable in practice (such calculations often use huge datasets that can take a very long time to process), there
is no mention of the computation time taken in many papers. Such papers
in which this information would be highly useful but is unfortunately missing include, but are certainly not limited to: Esteva et al. (2017); Rakhlin et al.
(2018); Sun, Zheng, and Qian (2016); Abdel-Zaher and Eldeib (2016); Xu et al.
(2016); Cruz-Roa et al. (2013); Fakoor et al. (2013); Tsehay et al. (2017); Dhungel,
Carneiro, and Bradley (2017); Liu et al. (2017); and Bychkov et al. (2018). In
practice, what such an omission may mean is that considerable resources –
financial and human – may be dedicated to implementing such promising
medical research without considering whether a cancer-detecting algorithm
runs in a short enough time to be clinically usable – or whether it takes so
long to complete that it simply cannot be implemented in an already over-
burdened healthcare system. Furthermore, when time is a central focus of the
research, the algorithms or the techniques used to evaluate them are often
too niche to be usable by researchers in other domains. These issues lead to a
problem of inconsistent reporting and inadequate information being provided
to researchers, which hampers efforts to determine whether a particular algorithm or method is sufficient for the requirements of the system.

A major problem here is the lack of information provided in the literature –
including in research papers and code documentation – about the many
important factors that influence computation time, as mentioned above. To
briefly recap, these include the type of computation time being measured or
reported (e.g., the execution time or response time), details of the timing function or method used, details about which parts of the code were run in parallel
and how many processors were used, details of the hardware, and so on. A
lack of this information leads to the significant issue of developers having to
undertake many iterations of trial and error experimentation in their coding
practice, which greatly slows development time. When a researcher or developer doesn't know if an algorithm will perform as required (e.g., if a real-time
video analysis algorithm will calculate the required task before the next frame
in the video), they will often have to manually code and/or test it to see if it will
work, rather than being able to build on the work of others who have already
gone through this process.

To give a basic example of this issue in practice, let's say a developer has been
asked to build a program that can solve a potentially lengthy computational
task, such as using deep learning to detect cancer in medical images, within
two weeks. Without the developer knowing how long the algorithm might take
to execute on a similar dataset on a given machine specification, it's very difficult to estimate if the algorithm will complete within that time without using

trial and error. Without this information the developer could spend the entire two weeks simply working out that they do not have the resources to deliver the task using the initial algorithm and resources available. The task may have been allocated with the expectation that it would be solvable based on the work of others. However, current computational research is broadly driven by reporting results, rather than documenting processes, and so the fact that the task has been possible for another is of limited practical use if the hardware, software and computation time have not been adequately recorded. In this example, the developer would have effectively wasted two weeks to find that the task is not possible with their time frame or resources. However, had they had access to the right information, they would have been able to see how long the algorithms of others had taken to run and know what their hardware and software resources were. A quick look at these specifications could mean discovering early on that a similar process took two weeks to run on a much faster, newer machine – and so a different algorithm, or a longer time frame, would be needed.

A better global development process is required in order to reduce this sizable – and often needless – duplication of effort and waste of developer time. According to a 2019 global developer population and demographic study by software market research company Evans Data Corporation, there are 23.9 million developers worldwide (Evans Data Corporation 2019). This large developer population means that even if only a small percentage of developers are affected by this poor computation time reporting, leading to needless duplication of effort by others, then it is a significant worldwide issue that has potentially large-scale economic implications. There is therefore a need for further research to determine the true economic costs of this issue, revealing the extent of the human, computational, and environmental resources wasted in the needless duplication of effort due to poor reporting.

5.1 The Solution: Computation Time Templates and Automated Reporting

Part of the reason why current practice remains largely inadequate here is that it simply takes time and effort to report computation time. Many developers and researchers may not know how to find and report this information or will be reluctant to do so due to the presumed effort involved.

To reduce the human effort needed in reporting the required computation time details, below are two templates – a simple and an advanced version – for reporting computation time, which can be used in either code documentation or in research publications. To make this process even easier to implement, we have also included an open source program to gather most of this information automatically, populating the template with the CPU processor type, clock speed, number of physical cores, number of logical cores, RAM and operating system. To perform this process, developers simply run the "Computation_time_reporter" program that is included as part of the code repository that goes with this paper at https://github.com/dcchb/ComputationTime, which provides both a Windows executable program, and the corresponding Python script that can be run on any operating system. It is therefore hoped that developers across the world may benefit from these easy-to-use and freely accessible tools and templates to automatically report computation time for the benefit of all.

6   Recommendations

There are several recommendations that we would like to provide for any researchers or software developers for whom the time taken for their programs to run is important (i.e., those working with large datasets, complex algorithms or real-time processing – or indeed any developers who are trying to ensure their code runs as quickly and efficiently as possible), to encourage what we consider to be best practice. As this "best practice" does require effort, and can sometimes appear to get in the way of progressing with the task in hand, we have broken this down into two options: the "Computation Time: Lite" and "Computation Time: Pro."

Documenting the "Lite" version would mean providing the lowest level of basic documentation on the computation time as a form of shorthand, which would still be useful as a reference point for future developers. By contrast, the "Pro" version builds on the information documented as part of the "Lite" version and adds some useful further documentation and recommended best practice. Of course, this added detail may not always be possible, depending on the situation. However, where available, adding these two forms of recorded documentation would enable the developed algorithm or software to be much more easily translatable into other future software or algorithms, and save the colossal duplication of effort that plagues the field today.

With all of the above information in mind we recommend that future researchers explain the following details when reporting on the computation time:

6.1   *Computation Time: Lite (Minimum Details)*
As a minimum, documentation of computation time should list as many of the following as possible:

1. Report the computation time and the type reported (e.g., execution time or response time).
2. Report the hardware used, including the number of cores and type of CPU/GPU.
3. Specify the operating system used.

6.2   *Computation Time: Pro (Additional Details)*

Further details that would be helpful are listed below, and can be documented in addition to the above to create a full "Pro" version of the documentation:

4. Provide code for others to use (preferably open source).
5. Detail the programmatic timing function that has been used to record the computation time (e.g., Python's time.time).
6. Break down the processes and list the time each takes.
7. Show or explain which processes can run in parallel and which have to be run in series.

6.3   *Computation Time: Templates*

Documenting this information does not need to take up a significant portion of a research paper or code documentation; such documentation could be easily inserted in just a few sentences. To make this process even easier, templates are provided below and a program to auto-populate many of the fields is provided in the code that goes with the paper, which can be found here: https://github.com/dcchb/ComputationTime. We would encourage anyone to feel free to reuse and share these templates as needed, citing this paper.

6.3.1   Computation Time: Lite Template

> This program was tested on [insert most important hardware specifications, such as CPU name and clock frequency, number of physical cores, and GPU used], with the Operating System [insert OS, such as Windows 10 Pro or Ubuntu 18.04.3 LTS]. With this hardware and software combination the [execution time (time CPU has taken to execute program only) or response time (time taken from start to end of process, including all background processes)] for this program took [insert time with units] to complete.

6.3.2   Computation Time: Pro Template (Add-On)

Including additional detail need not take up too much space either. A template example of the Pro version that could be added on to the above minimum details is as follows:

This program can be downloaded with a [insert licence type: preferably an easy to share open source licence[11]] license from [insert link to where code can be downloaded from, e.g. GitHub[12]]. The computation time was measured using function [insert name of timing function] with the programming language [insert programming language]. The majority of the computation time is spent in [insert name of the part of the code which is computationally expensive[13]], which on testing took [insert percentage of the computation time spent in that part of the code]% of the time to run. Methods [insert names of parallizable methods] can [completely/mostly/partially] be run using parallel processing, while methods [insert series methods] need to be run in series. The current implementation uses [CPU hyperthreading/GPU acceleration/other type of acceleration] for [insert methods]. It is anticipated that further optimization could be achieved in [insert names of methods], possibly using [insert names of techniques/algorithms to further optimize]; however, this has not yet been investigated.

6.4   *Recommendations for Future Research*

In addition to the above recommendations for software developers, it is also worth highlighting some areas in which further scholarly research is needed, which would greatly benefit both the research and software development communities. The following future studies would greatly help in addressing these identified issues:

– A quantitative review of the current practice of computation time documentation within scholarly research, documenting the number of papers that do and don't fulfil each of the above minimum and additional details, and highlighting areas in which documentation is particularly poor.
– A quantitative study on the current practice of computation time documentation within software documentation, documenting the number of software packages which do and don't fulfil each of the above minimum and additional details.

---

11   A list of open source licenses can be found at: https://opensource.org/licenses, accessed 3rd December 2019. A detailed discussion of open source licences can be found in St. Laurent 2004.

12   GitHub (https://github.com/) is one of the most widely used code repositories and hosts many open source programs, with over 44 million repositories contributed to between 1st October 2018 and 30th September 2019 (GitHub 2019).

13   Note that "computationally expensive" just means that it requires a lot of computation power, which takes time to process.

- A thorough and detailed economic costing of the effects of global duplication of effort that occurs due to insufficient or non-existent documentation of computation time, with practical recommendations of how to improve this in practice.
- Detailed discussion of the global cost of computation time in light of computation power – including computation power templates, as used here for computation time – that raises awareness of ecological costs, and is specifically aimed at reducing computational power consumption where the direct relationship between computation time and power used is unclear (i.e., which types of processes consume more – rather than less – power when execution time is shortened by using more power-hungry hardware, for example). As large-scale computing becomes more prevalent in the twenty-first century (e.g., cloud computing, super-computing, and large-scale distributed computation such as cryptocurrency mining), such computational challenges must now attract greater scrutiny in light of the increasingly urgent problems of climate change.

7   Conclusion

As this paper has demonstrated, computation time is a vitally important measurement which can help us to determine whether a program is fit for purpose and completes within a reasonable time frame, acting as a useful signpost for where further optimization may be needed.

However, without changes to current practice, the practical use value of this knowledge remains limited. To allow coders to more readily speed up the development and optimization of computer programs in practice, then, it would be particularly useful to measure computation time using programmatic timing function calls, as detailed above, and document this information accordingly. Because the precision of these timing functions can vary by operating system, this also needs to be documented, along with the time to process and a detailed breakdown of the software and which parts of it can be run in parallel.

By being able to access these documented measures of computation time when undertaking other similar tasks, the recording developer can track and review the limitations of their own implementation, and other developers and researchers can find the key information needed to determine the viability of such an approach for their own purposes. Documenting and sharing such information on what has already been achieved, alongside sharing the code itself, will help others to optimize their own code, allowing more rapid and cost-effective software development, and enabling coders, researchers and development communities to build on the efforts of those who have gone before. The above computation time templates and the corresponding scripts to automate the population of the required fields are an attempt to reduce the difficulty for researchers and developers in reporting computation time, providing the practical means to do this in a way which is easy to document, and easy to understand.

By helping to reveal the extent to which computation time is a crucial factor in optimizing global software development, the suggested future research studies are likely to add a real incentive for academies, organizations and individuals to take action and put these recommendations into practice for the future benefit of us all. What is required now is a global effort to think beyond the immediate use of a single program, in an ethical move to help future coders who would greatly benefit from sharing these details. As Linus Torvold, the original creator of the Linux Operating System, notes of open source projects: "to really do something well, you have to get a lot of people involved" (quoted in Diamond 2003). It is to this end that we hope that this paper helps to raise awareness about the importance of computation time documentation, and we hope that you will help to spread the message of how to do it well – whether you come to this research as a developer, researcher, or time scholar.

Acknowledgements

Thanks to the International Society for the Study of Time (ISST) for accepting an early version of this paper, which was presented at the 17th Triennial ISST Conference on Time's Urgency, held at Loyola Marymount University in Los Angeles, California, in June 2019, and to its members for their warm feedback at the conference. Thanks also to Patrick Schrempf for directing us to the alternative metric of measuring power consumption as an indication of resource usage.

References

Abdel-Zaher, Ahmed M., and Ayman M. Eldeib. 2016. "Breast Cancer Classification Using Deep Belief Networks." *Expert Systems with Applications* 46: 139–144.

Amdahl, Gene M. 1967. "Validity of the Single Processor Approach to Achieving Large Scale Computing Capabilities." In *Proceedings of the April 18–20, 1967, spring joint computer conference*, 483–485. ACM.

Arora, Sanjeev, and Boaz Barak. 2009. *Computational Complexity: A Modern Approach.* Cambridge University Press.

Audsley, Neil, Alan Burns, Mike Richardson, Ken Tindell, and Andy J. Wellings. 1993. "Applying New Scheduling Theory to Static Priority Pre-Emptive Scheduling." *Software Engineering Journal* 8(5): 284–292.

Avoine, Gildas, Etienne Dysli, and Philippe Oechslin. 2005. "Reducing Time Complexity in RFID Systems." In *International Workshop on Selected Areas in Cryptography*, 291–306. Springer.

Baruah, Sanjoy, and Alan Burns. 2006. "Sustainable Scheduling Analysis." In *2006 27th IEEE International Real-Time Systems Symposium (RTSS'06)*, 159–168. IEEE.

Bychkov, Dmitrii, Nina Linder, Riku Turkki, Stig Nordling, Panu E. Kovanen, Clare Verrill, Margarita Walliander, Mikael Lundin, Caj Haglund, and Johan Lundin. 2018. "Deep Learning Based Tissue Analysis Predicts Outcome in Colorectal Cancer." *Scientific Reports* 8(1): 3395.

Camelo, Miguel, Yezid Donoso, and Harold Castro. 2011. "MAGS – An Approach Using Multi-Objective Evolutionary Algorithms for Grid Task Scheduling." *International Journal of Applied Mathematics and Informatics* 5(2).

Cancela, Héctor, and Louis Petingi. 2004. "Reliability of Communication Networks with Delay Constraints: Computational Complexity And Complete Topologies." *International Journal of Mathematics and Mathematical Sciences* 2004(29): 1551–1562.

Chen, Tianshi, Ke Tang, Guoliang Chen, and Xin Yao. 2010. "Analysis of Computational Time of Simple Estimation of Distribution Algorithms." *IEEE Transactions on Evolutionary Computation* 14(1): 1–22.

Cheng, Ruida, Holger R. Roth, Nathan S. Lay, Le Lu, Baris Turkbey, William Gandler, Evan S. McCreedy, Thomas J. Pohida, Peter A. Pinto, Peter L. Choyke, et al. 2017. "Automatic Magnetic Resonance Prostate Segmentation by Deep Learning with Holistically Nested Networks." *Journal of Medical Imaging* 4(4): 041302.

Cooper, Gregory F. 1990. "The Computational Complexity of Probabilistic Inference Using Bayesian Belief Networks." *Artificial Intelligence* 42(2–3): 393–405.

Crandall, Jedidiah R., Gary Wassermann, Daniela A. S. de Oliveira, Zhendong Su, S. Felix Wu, and Frederic T. Chong. 2006. "Temporal Search: Detecting Hidden Malware Timebombs with Virtual Machines." In *ACM SIGARCH Computer Architecture News*, 34:25–36. 5. ACM.

Cruz-Roa, Angel Alfonso, John Edison Arevalo Ovalle, Anant Madabhushi, and Fabio Augusto González Osorio. 2013. "A Deep Learning Architecture for Image Representation, Visual Interpretability and Automated Basal-Cell Carcinoma Cancer Detection." In *International Conference on Medical Image Computing and Computer-Assisted Intervention*, 403–410. Springer.

Danowitz, Andrew, Kyle Kelley, James Mao, John P. Stevenson, and Mark Horowitz. 2012. "CPU DB: Recording Microprocessor History." *Communications of the ACM* 55(4): 55–63.

Delling, Daniel, Andrew V. Goldberg, Thomas Pajor, and Renato F. Werneck. 2015. "Customizable Route Planning in Road Networks." *Transportation Science* 51(2): 566–591.

Delling, Daniel, Peter Sanders, Dominik Schultes, and Dorothea Wagner. 2009. "Engineering Route Planning Algorithms." In *Algorithmics of Large and Complex Networks*, 117–139. Springer.

Dhungel, Neeraj, Gustavo Carneiro, and Andrew P. Bradley. 2017. "Fully Automated Classification of Mammograms Using Deep Residual Neural Networks." In *2017 IEEE 14th International Symposium on Biomedical Imaging (ISBI 2017)*, 310–314. IEEE.

Diamond, David. 2003. "The Way We Live Now: 9-28-03: Questions for Linus Torvolds; The Sharer." *The New York Times Magazine* (September 28): 23.

Esteva, Andre, Brett Kuprel, Roberto A. Novoa, Justin Ko, Susan M. Swetter, Helen M. Blau, and Sebastian Thrun. 2017. "Dermatologist-Level Classification of Skin Cancer with Deep Neural Networks." *Nature* 542(7639): 115.

Evans Data Corporation. 2019. "Global Developer Population and Demographic Study 2019 Vol. 1." Accessed September 20, 2019. https://evansdata.com/reports/view Release.php?reportID=9.

Fakoor, Rasool, Faisal Ladhak, Azade Nazi, and Manfred Huber. 2013. "Using Deep Learning to Enhance Cancer Diagnosis and Classification." In *Proceedings of the International Conference on Machine Learning*, vol. 28. ACM New York, USA.

Felten, Edward W, and Michael A. Schneider. 2000. "Timing Attacks on Web Privacy." In *Proceedings of the 7th ACM Conference on Computer and Communications Security*, 25–32.

Fitch, John. 1963. "1963 Timesharing: A Solution to Computer Bottlenecks." YouTube. Accessed November 4, 2019. https://youtu.be/Qo7PhW5sCEk.

GitHub. 2019. "The State of the Octoverse – 2019." Accessed December 3, 2019. https://octoverse.github.com/.

He, Jun, and Xin Yao. 2001. "Drift Analysis and Average Time Complexity of Evolutionary Algorithms." *Artificial Intelligence* 127(1): 57–85.

Hill, Mark D., and Michael R. Marty. 2008. "Amdahl's Law in the Multicore Era." *Computer* 41(7): 33–38.

Horan, Brendan. 2013. "Adding an RTC." In *Practical Raspberry Pi*, 145–162. Springer.

Hudek, Ted, Tim Sherer, and Nick Schonning. 2017. "Windows Driver Docs: High -Resolution Timers." Accessed November 28, 2019. https://docs.microsoft.com/en -us/windows-hardware/drivers/kernel/high-resolution-timers.

Huynh, Benjamin Q., Hui Li, and Maryellen L. Giger. 2016. "Digital Mammographic Tumor Classification Using Transfer Learning from Deep Convolutional Neural Networks." *Journal of Medical Imaging* 3(3): 034501.

Intel Corporation. 2016. *Intel R 64 and ia-32 Architectures Software Developer's Manual: Volume 3B: System Programming Guide, Part 2, section 17.15 time-stamp counter.* Technical report.

Jiang, Yunlian, Kai Tian, Xipeng Shen, Jinghe Zhang, Jie Chen, and Rahul Tripathi. 2010. "The Complexity of Optimal Job Co-Scheduling on Chip Multiprocessors and Heuristics-Based Solutions." *IEEE Transactions on Parallel and Distributed Systems* 22(7): 1192–1205.

Jiao, Yang, Heshan Lin, Pavan Balaji, and Wu-chun Feng. 2010. "Power and Performance Characterization of Computational Kernels on the GPU." In *Proceedings of the 2010 IEEE/ACM Int'l Conference on Green Computing and Communications & Int'l Conference on Cyber, Physical and Social Computing*, 221–228. IEEE Computer Society.

Jouppi, Norman P., Cliff Young, Nishant Patil, David Patterson, Gaurav Agrawal, Raminder Bajwa, Sarah Bates, Suresh Bhatia, Nan Boden, Al Borchers, et al. 2017. "In-Datacenter Performance Analysis of A Tensor Processing Unit." In *2017 ACM/IEEE 44th Annual International Symposium on Computer Architecture (ISCA)*, 1–12. IEEE.

Jouppi, Norman, Cliff Young, Nishant Patil, and David Patterson. 2018. "Motivation for and Evaluation of the First Tensor Processing Unit." *IEEE Micro* 38(3): 10–19.

Ko, Ker-I, and Harvey Friedman. 1982. "Computational Complexity of Real Functions." *Theoretical Computer Science* 20(3): 323–352.

Kong, Bin, Xin Wang, Zhongyu Li, Qi Song, and Shaoting Zhang. 2017. "Cancer Metastasis Detection Via Spatially Structured Deep Network." In *International Conference on Information Processing in Medical Imaging*, 236–248. Springer.

Ladner, Richard E. 1977. "The Computational Complexity of Provability in Systems of Modal Propositional Logic." *SIAM Journal on Computing* 6(3): 467–480.

Li, Jingming, Nianping Li, Jinqing Peng, Haijiao Cui, and Zhibin Wu. 2019. "Energy Consumption of Cryptocurrency Mining: A Study of Electricity Consumption in Mining Cryptocurrencies." *Energy* 168: 160–168.

Lipp, Moritz, Michael Schwarz, Daniel Gruss, Thomas Prescher, Werner Haas, Anders Fogh, Jann Horn, Stefan Mangard, Paul Kocher, Daniel Genkin, et al. 2018. "Meltdown: Reading Kernel Memory from User Space." In *27th {USENIX} Security Symposium ({USENIX} Security 18)*, 973–990.

Liu, Saifeng, Huaixiu Zheng, Yesu Feng, and Wei Li. 2017. "Prostate Cancer Diagnosis Using Deep Learning with 3D Multiparametric MRI." In *Medical Imaging 2017: Computer-Aided Diagnosis*, 10134: 1013428. International Society for Optics and Photonics.

Luong, Minh-Thang, and Christopher D. Manning. 2016. "Achieving Open Vocabulary Neural Machine Translation with Hybrid Word-Character Models." In *Association for Computational Linguistics (ACL)*. Berlin, Germany. https://nlp.stanford.edu/pubs/luong2016acl_hybrid.pdf.

Marr, Deborah T., Frank Binns, David L. Hill, Glenn Hinton, David A. Koufaty, J. Alan Miller, and Michael Upton. 2002. "Hyper-Threading Technology Architecture and Microarchitecture." *Intel Technology Journal* 6(1).

Martin, Alain J. 2001. "Towards an Energy Complexity of Computation." *Information Processing Letters* 77(2–4): 181–187.

Martínez, Sonia, Francesco Bullo, Jorge Cortés, and Emilio Frazzoli. 2007. "On Synchronous Robotic Networks – Part II: Time Complexity of Rendezvous and Deployment Algorithms." *IEEE Transactions on Automatic Control* 52(12): 2214–2226.

McCarthy, John. 1959. *Memorandum to P. M. Morse Proposing Time Sharing.*

Melinščak, Martina, Pavle Prentašić, and Sven Lončarić. 2015. "Retinal Vessel Segmentation Using Deep Neural Networks." In *10th International Conference on Computer Vision Theory and Applications (VISAPP 2015).*

Mertens, Stephan. 2002. "Computational Complexity for Physicists." *Computing in Science & Engineering* 4(3): 31.

Mok, Tony C. W., and Albert C. S. Chung. 2018. "Learning Data Augmentation for Brain Tumor Segmentation with Coarse-To-Fine Generative Adversarial Networks." In *International MICCAI Brainlesion Workshop*, 70–80. Springer.

Moore, Gordon E. 1965. "Cramming More Components onto Integrated Circuits." *Electronics* 38(8): 114.

Moore, Gordon E. 1975. "Progress in Digital Integrated Electronics." In *Electron Devices Meeting* 21: 11–13.

Nickolls, John, and William J. Dally. 2010. "The GPU Computing Era." *IEEE Micro* 30(2): 56–69.

O'Dwyer, K. J., and D. Malone. 2014. "Bitcoin Mining and Its Energy Footprint." In *25th IET Irish Signals Systems Conference 2014 and 2014 China-Ireland International Conference on Information and Communications Technologies (ISSC 2014/CIICT 2014)*, 280–285.

Owens, John D., Mike Houston, David Luebke, Simon Green, John E. Stone, and James C. Phillips. 2008. "GPU Computing." *Proceedings of the IEEE* 96(5): 879–899. doi:10.1109/JPROC.2008.9177571.

Peled, Abraham, and Antonio Ruiz. 1980. "Frequency Domain Data Transmission Using Reduced Computational Complexity Algorithms." In *ICASSP'80. IEEE International Conference on Acoustics, Speech, and Signal Processing* 5: 964–967. IEEE.

Puschner, Peter, and Ch. Koza. 1989. "Calculating the Maximum Execution Time of Real-Time Programs." *Real-Time Systems* 1(2): 159–176.

Python Software Foundation. 2018. *Python.* Version 3.7, June 27. Accessed November 28, 2019. https://www.python.org/.

Python Software Foundation. 2019. "Time – Time Access and Conversions." Accessed October 8, 2019. https://docs.python.org/3/library/time.html.

Radulescu, Andrei, and Arjan J. C. Van Gemund. 1999. "On the Complexity of List Scheduling Algorithms for Distributed-Memory Systems." In *International Conference on Supercomputing*, 68–75. Citeseer.

Rakhlin, Alexander, Alexey Shvets, Vladimir Iglovikov, and Alexandr A. Kalinin. 2018. "Deep Convolutional Neural Networks for Breast Cancer Histology Image Analysis." In *International Conference Image Analysis and Recognition*, 737–744. Springer.

Schmidhuber, Jürgen. 1992. "A Fixed Size Storage O (n 3) Time Complexity Learning Algorithm for Fully Recurrent Continually Running Networks." *Neural Computation* 4(2): 243–248.

Schwartz, Roy, Jesse Dodge, Noah A. Smith, and Oren Etzioni. 2019. *Green AI*. arXiv: 1907.10597.

Shanbehzadeh, Jamshid, and Philip O. Ogunbona. 1997. "On the Computational Complexity of The LBG and PNN Algorithms." *IEEE Transactions on Image Processing* 6(4): 614–616.

Skopko, Tamas, and Peter Orosz. 2011. "Investigating the Precision of the TSC-based Packet Timestamping." *Carpathian Journal of Electronic & Computer Engineering* 4(1).

SlashData. 2019. "Global Developer Population Report – Community Edition." Accessed November 29, 2019. https://slashdata-website-cms.s3.amazonaws.com/sample_reports/EiWEyM5bfZe1Kug_.pdf.

St. Laurent, Andrew M. 2004. *Understanding Open Source and Free Software Licensing: Guide to Navigating Licensing Issues in Existing & New Software*. O'Reilly Media, Inc.

Stack Overflow. 2019. "Stack Overflow's Developer Survey Results – 2019." Accessed November 29, 2019. https://insights.stackoverflow.com/survey/2019#most-popular-technologies.

Stinner, Victor. 2017. "Python Developer's Guide: PEP 564 – Add New Time Functions with Nanosecond Resolution: Annex: Clocks Resolution in Python." Accessed November 28, 2019. https://www.python.org/dev/peps/pep-0564/#annex-clocks-resolution-in-python.

Strubell, Emma, Ananya Ganesh, and Andrew McCallum. 2019. "Energy and Policy Considerations for Deep Learning in NLP." In *Proceedings of the 57th Annual Meeting of the Association for Computational Linguistics*, 3645–3650. Florence, Italy: Association for Computational Linguistics, July. doi:10.18653/v1/P19-1355.

Sun, Wenqing, Bin Zheng, and Wei Qian. 2016. "Computer Aided Lung Cancer Diagnosis with Deep Learning Algorithms." In *Medical Imaging 2016: Computer-Aided Diagnosis*, vol. 9785, 97850Z. International Society for Optics and Photonics.

Tomita, Etsuji, Akira Tanaka, and Haruhisa Takahashi. 2006. "The Worst-Case Time Complexity for Generating All Maximal Cliques and Computational Experiments." *Theoretical Computer Science* 363(1): 28–42.

Tsehay, Yohannes K., Nathan S. Lay, Holger R. Roth, Xiaosong Wang, Jin Tae Kwak, Baris I. Turkbey, Peter A. Pinto, Brad J. Wood, and Ronald M. Summers. 2017. "Convolutional Neural Network Based Deep-Learning Architecture for Prostate Cancer Detection on Multiparametric Magnetic Resonance Images." In *Medical Imaging 2017: Computer-Aided Diagnosis*, 10134: 1013405. International Society for Optics and Photonics.

VMware. 2008. "Timekeeping in VMware Virtual Machines." Accessed October 8, 2019. https://www.vmware.com/content/dam/digitalmarketing/vmware/en/pdf/techpaper/Timekeeping-InVirtualMachines.pdf.

Vo, Nam, Nathan Jacobs, and James Hays. 2017. "Revisiting im2gps in the Deep Learning Era." In *Proceedings of the IEEE International Conference on Computer Vision*, 2621–2630.

Wahab, Noorul, Asifullah Khan, and Yeon Soo Lee. 2017. "Two-Phase Deep Convolutional Neural Network for Reducing Class Skewness in Histopathological Images Based Breast Cancer Detection." *Computers in Biology and Medicine* 85: 86–97.

Wang, Gang, Wenrui Gong, and Ryan Kastner. 2005. "Instruction Scheduling Using MAX-MIN Ant System Optimization." In *Proceedings of the 15th ACM Great Lakes Symposium on VLSI*, 44–49. ACM.

Wang, Haibo, Angel Cruz Roa, Ajay N. Basavanhally, Hannah L. Gilmore, Natalie Shih, Mike Feldman, John Tomaszewski, Fabio Gonzalez, and Anant Madabhushi. 2014. "Mitosis Detection in Breast Cancer Pathology Images by Combining Handcrafted and Convolutional Neural Network Features." *Journal of Medical Imaging* 1(3): 034003.

Wirtz, Thomas, and Rong Ge. 2011. "Improving Mapreduce Energy Efficiency for Computation Intensive Workloads." In *2011 International Green Computing Conference and Workshops*, 1–8. IEEE.

Wu, Chi-Jui, Steven Houben, and Nicolai Marquardt. 2017. "Eaglesense: Tracking People and Devices in Interactive Spaces Using Real-Time Top-View Depth-Sensing." In *Proceedings of the 2017 CHI Conference on Human Factors in Computing Systems*, 3929–3942. ACM.

Xu, Tao, Han Zhang, Xiaolei Huang, Shaoting Zhang, and Dimitris N. Metaxas. 2016. "Multimodal Deep Learning for Cervical Dysplasia Diagnosis." In *International Conference on Medical Image Computing and Computer-Assisted Intervention*, 115–123. Springer.

Yang, Tien-Ju, Yu-Hsin Chen, and Vivienne Sze. 2017. "Designing Energy-Efficient Convolutional Neural Networks Using Energy-Aware Pruning." In *Proceedings of the IEEE Conference on Computer Vision and Pattern Recognition*, 5687–5695.

Zhang, Sushu, and Karam S. Chatha. 2007. "Approximation Algorithm for the Temperature-Aware Scheduling Problem." In *2007 IEEE/ACM International Conference on Computer-Aided Design*, 281–288. IEEE.

CHAPTER 13

Variations of Narrative Temporalities in John Farrow's 1948 Film *The Big Clock*

Raphaëlle Costa de Beauregard

Abstract

This paper first shows that, in John Farrow's 1948 film *The Big Clock*, time, though by nature silent, is present whenever audible. Secondly, time being also invisible, this paper shows how time can be made visible in the film by the piecemeal reconstitution of events on a blackboard. Thirdly, the focus is on the manipulation of screen time by characteristic cinematographic devices – the flashback in the initial sequence of the film and the careful space-time continuity in the editing of sequences of different temporalities. If our attention is thus compelled to concentrate on the logical condensation of an action of 36 hours into 96 minutes, our cultural expectations are also necessarily addressed. Among these, Western scientific theories about time's contingency, irreversibility, and entropy clearly become specific representations of time in *The Big Clock*. The *topos* of time used in classical culture to express a human tragedy, is re-appropriated by the film's visuals and becomes properly cinematographic.

Keywords

contingency – *The Big Clock* – clocks – entropy – John Farrow – film editing – flashbacks – Hollywood cinema – Industrial Revolution – irreversibility – modernity – temporal visual icons

∙ ∙ ∙

The clock strikes. Faustus says: "O lente, lente currite, noctis equi!"
CHRISTOPHER MARLOWE *Dr Faustus* (1589)

∙ ∙
∙

The explicit references to industrial civilization that document the cultural background of John Farrow's 1948 film *The Big Clock* (based on Kenneth Fearing's 1946 eponymous novel), the numerous clocks in the film, give us an aural, as well as visual, representation of Newtonian time (Canales 2016, 114–115).[1] Perhaps more significant regarding the concept of variations in temporalities with which we are concerned in this paper is the opening sequence of the film [02:16–02:59], surprisingly introducing the action as it is drawing to its end. We discover Manhattan's skyline at night thanks to a slow high-angle panning from left to right that leads us to a building's clock tower and a mysterious prisoner. We hear a voice-over beginning with an injunction: "Think fast, George," which introduces the ensuing narrative in a flashback (Bordwell 2017, 80–84) and thus sets a deadline to the plot when this initial, but also ultimate, crisis is resumed. We are made to share the speaker's point of view, that of a man who is currently trying to escape a false accusation for a crime; in order to answer his questions, such as "How did I get into this? What happened?" and, more important still, "When did it all start?," he begins to recollect events in chronological order [02:59 to 01:23:00]. Jumping back into past events, we thus enter the world of George Stroud and the other characters working at Janoth Publications, and we have a special interest in the search for the cause of this crisis, which, we discover, has been triggered by the murder of Pauline, Janoth's mistress.

A brief summary of the chronological action in *The Big Clock* may prove helpful before our discussion of the film's representation of time as both "invariant" and "in variance." Earl Janoth and his close adviser Steve Hagen run Janoth Publications and its several magazines, among which is *Crimeways*, directed by George Stroud. About to go on vacation, Stroud misses a train appointment with his wife and child, and unwisely spends the evening with Pauline York, Earl Janoth's mistress. He comes across Janoth on Pauline's landing as he is leaving her flat in the morning, but the latter fails to identify him. While Stroud goes to the station to catch an early train, Janoth and Pauline quarrel over her

---

1 John Villiers Farrow (1904–1963) was an Australian-born American film director, producer, and screenwriter, who was married to Maureen O'Sullivan and was the father of Mia Farrow. He began directing with Paramount in 1928 and he later worked with United Artists, RKO, MGM and Warner Bros B-pictures. The film information is as follows: *The Big Clock* 1948, US, 96 min, b/w. Paramount. Director: John Farrow; producer: R. Maibaum; adapted from *The Big Clock* (1946) by Kenneth Fearing (1902–1961); script: J. Latimer; editing: Eda Warren; music composer: Victor Young. Cast: Ray Milland (George Stroud), Charles Laughton (Earl Janoth), Maureen O'Sullivan (Georgette, Stroud's wife), George Macready (Hagen, Janoth's secretary), Rita Johnson (Pauline, Janoth's mistress), Elsa Lanchester (the painter Louise Patterson). The video used for references to film time in this paper is DVD video BD592; ISBN 7-88588-046-X.

visitor, Pauline throws a clock at Janoth, and her lover then kills her. The pair Janoth/Hagen hastily launch a speedy manhunt for a substitute murderer. They entrust this assignment to George Stroud and call him back. He realizes that they have planned to charge Pauline's visitor, whom they call "Randolph" (himself, but unknown to them), with her murder. Stroud has his own plan to find proof that Janoth killed Pauline out of jealousy, and he tries to delay the enquiries about the unknown Randolph as much as he can until he has enough proof to accuse Janoth of the murder. Witnesses to Stroud's and Pauline's evening activities, among whom is the artist Louise Patterson, come to Janoth Publications to identify "Randolph," and they recognize Stroud as Pauline's companion that night. Stroud hides in the clock tower, then escapes, and, drawing on the evidence that he has been collecting with his wife Georgette, accuses Hagen of the murder. Janoth seizes this alibi and shoots Hagen but accidentally falls to his death in the elevator shaft.

From the opening clock tower scene, paradoxically, we viewers soon get to know more than Stroud does, when we see the murderer of Pauline, Janoth, set up his own plot with Hagen – the false accusation of an as yet unidentified murderer – in order to escape the police [35:42–37:07]. The two plots (Stroud's attempt to escape being charged with murder and Janoth's parallel attempt) rely on the manhunt narrative pattern, but another plot is added, the supposed existence of the third party, Randolph, which gives us viewers an omniscient point of view above all the characters, including Stroud himself. The three plotlines eventually lead us to the final issue, when a single temporality in the present time [02:16–02:59] is restored [01:23:00 – end of film 01:36:00], which merges Randolph's disappearance with Stroud's final victory over Janoth.

In this paper I first examine the way silent time is identifiable as "clock-time" (Canales 2016, 125) in that it is made audible – and visible – by clocks and their avatars in the film. Secondly, I examine the uncanny way in which invisible time is also made visible by other devices, such as frames within frames. Thirdly, and more specific to cinema, I discuss "continuity editing" – an achievement that gives cinema its specificity as a unique form of art. Moreover, equally specific to the linearity of film time, I explain how we viewers are compelled to follow the condensation of an action of 36 hours (movie time) into 96 minutes (film time) (Maltby 1995, 420). Finally, in light of Western theories of time in parallel with the progress of their industrialization, I look at three modern time models – contingency, irreversibility, and the threat of entropy – as aspects of film time in *The Big Clock*. As I argue in more detail below, what makes these time models relevant to the subject of this paper is the fact that they were theorized during the same period as, or slightly prior to, the emergence of cinema and thus used for mapping film time (Doane 2002, 1–4, 22–25).

1   Clocks, Phones and Audible Time

Just like the clock striking for Dr Faustus in Marlowe's play, clocks striking in
The Big Clock call our attention to the present moment as a step in the march
of time, giving coherence to the overall plot and the plots-within-the-plot. As
a result, hearing clocks striking the hour focuses our attention on a particular scene, and we recall the immediate past event and start anticipating the
immediate future one, while we are struck by either their strident or their deep
sound as well as their threatening mechanical regularity.[2] In other words, we
inscribe the scenes within a linear time structure.

Moreover, whenever the clocks are heard in this film, their sound is synchronous with the image of a clock, thus associating the aural and the visual effects.
By recalling the conventions of the invention of an invariant "objective" clock-
time, they are the vehicle of an inaudible but forever present power presiding
over the characters' destiny.[3] In addition, they also represent the simultaneity
between the film's screen continuum and our own time continuum as we step
inside the film world with our imagined body – that is, the body that we believe
is ours in "lived time" but have only seen in mirrors (Schefer 1997, 107–108).

Several inserts (that is, shots addressing only us viewers) frame clocks in
close-up, but, instead of providing interruptions in the flow of cinematic continuity, these shots punctuate the action by reminding us that time never stops.
They appear in the film either as the huge clock that towers over the name of
the firm in the impressive entrance hall or clocks on walls in rooms. The scene
in the meeting room where the board assembles (described below) is a case in
point, giving both an aural and a visual image of the presence of time as almost
another character in the film's plot.

The camerawork here deserves a detailed analysis. A crane dolly moves
from a close-up of Hagen to a symmetrical high-angle long shot of the conference room, having at its center Janoth's empty chair and, behind it, the elevator
and the clock above, reading 10:59. The composition of the frame concentrates
on the clock and the elevator, while the staff seated around the circular table
are busy with their files and only focus on this mise-en-scene of power when

---

2   These four levels of observers' psychological reactions to experience were transferred as early
    as 1916 to analyses of the film viewer's reactions by William James's German colleague Hugo
    Munsterberg in *The Photoplay*.

3   In the eighteenth century, the earlier association between clocks and a proportioned, law-
    ful, and God-governed universe "had been given support by Newton's classical mechanics"
    (Canales 2016, 115). In the nineteenth century, clock time becomes an emblem of capitalism
    for Karl Marx. Whatever authority it is made to endorse, clock time is regarded as the enemy
    of human lived time.

the clock strikes eleven. The elevator doors open, and Janoth enters exactly when the clock strikes [00:08:23]. He then slowly moves around the table while expressing in a soft voice and slow phrasing his business worries. He twitches his moustache in the middle of his fleshy features with a weird regularity, which seems to echo the clock ticking the seconds. When he reaches the section of the table in the foreground, an attendant standing at attention jumps forward like an automaton to pour some coffee in a cup for him; when he reaches the other end, Hagen, equally automatically, rises to collect the empty cup, but Bill, Janoth's threatening mute bodyguard, is faster and collects it himself. The robot-like servility of these men introduces the theme of enslavement caused by the necessity of keeping time or acting just in time or being on time. The scene characterizes business magnate Janoth as if he stood for the economic law of "time is money."

This idea is reinforced as the scene continues. Janoth then sits down slowly at the head of the table and asks his directors to explain in one single minute the project they intend to develop in order to redress the dwindling number of sales of Janoth Publications, thus making them compete with one another within an equal time span. The circumference of the circular table now mirrors the working of a clock, and the clock's hand relentlessly moving around comes to mind, though Janoth does reverse the clockwork mechanism by stepping left to begin circling round, as if he were the ruler of his firm's time.[4] He actually spends more time watching his wristwatch in order to reckon the allotted minute than paying attention to the speaker's words. When each director's time is over, he rudely interrupts him and declares that his proposal is quite rotten. Only George Stroud is able to resist Janoth's impact on the group of directors: having joined the meeting late [09:36], he does not rise to present his project [10:36], contrary to all his other colleagues, who rise when summoned to speak [10:01]. Stroud is eventually rewarded by Janoth's compliments [11:02], having proved to his boss that he is neither a robot nor a slave, and, moreover, that he too can control his own time.

Throughout the film, Stroud and Janoth compete in an increasingly futile attempt to control time's flight, attempting to impede the speed of clock-time. Ironically, it is their gradual loss of control over time that becomes an issue once the murder has taken place [35:40]. This loss of control is transposed into visual elements, such as witnesses testifying about what they have seen, portraits of individuals, or other visual proofs. Janoth is aware that he has been

---

4   A particular clock whose hands can be moved by the boss in order to control the speed of the workers' production comes to mind as a possible intertextual reference in Farrow's film – the clock in Fritz Lang's 1926 *Metropolis*. UFA, Germany, b/w, silent.

seen by an eyewitness, and he attempts to bring to life a perfectly convincing reconstitution of the murderer before Pauline's dead body is discovered by the police. Stroud has strewn his evening spree with objects that may prove he is Pauline's visitor, such as a painting and a sundial. In addition, he has accidentally left a handkerchief with Pauline [58:14], which Hagen collects upon inspecting the victim's room and which Georgette later discovers and exhibits as a proof of Hagen's guilt [1:32:00]. While each of these objects eventually provide clues to reconstruct the time map, Janoth wishes to accelerate the hounding of the mysterious Randolph, while, on the contrary, Stroud devises means of slowing the process. Time has become an antagonist that both men attempt to control, personified by the invisible Randolph's increasing potential "visibility."

The two characters' attempt to control time's flow underscores the human quality of Newtonian time (Canales 2016, 115). After the crime, because the clock Pauline threw at her lover has stopped as it crashed on the wall right before he killed her, it gives the hour and minute of the crime. The next day, when Hagen and Stroud separately inspect Pauline's flat, where her dead body is seen lying, each man eventually picks the broken clock from the ground and changes its time: Hagen moves the clock back by 30 minutes, from 12.55 to 12.23, and Stroud moves the clock's hand forward by one hour, from 12.23 to 1.20. While both are using the clock's hands to create fake clues for the police, who will inevitably arrive, Stroud stands a better chance of succeeding because he changed the clock after Hagen. Moreover, 12.55 is the time reference that enables Stroud, with the help of a taxi driver's office secretary, to reconstruct Janoth's flight from the scene in a taxi-cab, [1:15:47].

Ironically, however, the futility of controlling clock time is highlighted in the scene when Stroud hides in the clock tower. This episode takes place as the camera cuts directly back to the opening scene in the present, which takes place in the clock tower [0:01:21–02:59], matching the action in the same scene so that it appears as the logical climax of the manhunt [1:25:01]. Suddenly, Janoth realizes the big clock has stopped [1:25:01], which also stops all clocks in the building. After a short moment of immobility when time's progress is apparently suspended throughout the whole building, the camera cuts to Stroud [1:25:20]. He realizes, that while he was examining the control board, he interfered with it by mistake, and he quickly gets the mechanism in working order again, though he has also inadvertently betrayed his presence inside the clock tower. Janoth thinks it is "Randolph" and immediately mandates Bill, his mute bodyguard, to go and shoot him dead [1:25:20]. Cross-cutting shows the two men hiding from each other, but Stroud is the faster of the two, and, to our relief, knocks Bill down the tower steps with his gun [1:25:40]. We then hear a phone ringing and see a close shot of Bill, who has recovered from the blow, moving to the private elevator door. But this allows Stroud swiftly to cut the power of Janoth's elevator [1:28:05], locking Bill helplessly inside. Slow music is used during the suspense crosscutting, as if Stroud was about to faint with fear. The episode is symbolically important, and ironical as well, as if stopping clocks could stop the march of time.

The phone is bizarrely heard ringing when we do not expect it, and it causes Bill to use the elevator and get eliminated from the final confrontation between Stroud and Janoth. It is one of the many phones we hear ringing throughout the film, their sound echoing the ticking of clocks and contributing to the general atmosphere of tension in the film. The phones create an uncanny multiplication of voices and temporalities, with the phone connected to the intercom, in particular, allowing different people to participate in remote conversations, as onscreen characters are actually following what is going on in another room. For example, early in the film, while Stroud is having an argument with Hagen regarding his vacation time [00:11:59], the camera cuts to the dark silhouette of Pauline's back in a nearby office while the sound of the men's argument is heard going on. Here the continuity of the sound adds a sense of threat – she is clearly eavesdropping – to the simultaneity of the temporalities expressed by the crosscutting.

Other equally erratic time punctuations are heard in the phone calls from the distant employees whom Stroud has sent on various more or less impossible missions in order to delay the fatal issue of his being discovered. The sharp sound of their ringing comes as a surprise, interrupting ongoing dialogues with a ghostly voice in a more distant place outside the building. For example, while we have seen Stroud in conversation with someone in the street outside, either getting information about his official enquiries or leading his own secret enquiry about the taxi-driver who picked Janoth up after the murder [1:23:59], when he returns inside he reaches Janoth's office just in time to pick up the phone and hear his own associate Cordette shouting in a loudly triumphant voice: "We have our man. He has just been seen entering the building" [1:05:29]. The suspense that was created by Stroud's private parallel enquiry outdoors is suddenly sharply increased by this phone call, as action is speeding up. In this instance, two aspects of action make time visible – movement and change – and clearly prove the working of the "'engine' of time [...] in the sense that it continuously renews the present" (Klein 2016, 190). The danger for Stroud is now so great that, at this point, he takes the decisive step of hiding in the clock tower [1:25:01], and the camera smoothly picks up the continuity editing of the present moment, resuming the action where it had been interrupted by the initial flashback.

Thus do clocks create a background of reliable, invariant Newtonian time, as the fatal issue, the discovery of the murderer, draws nearer. However, in a very different manner, the shrill, ubiquitous intrusions upon conversations without warning, brought about by calls from the intercom and the phones, reinforce our sense of a time that appears quite unstable and even unreliable. We now understand that the concept of time can be given symbolical meanings, in addition to its commonly agreed feature of a succession of moments divided between past and future by our awareness of the present. When clocks are literally used as lethal weapons in Pauline's fight with Janoth [33:51–35:42], the couple's quarrel over her visitor involves two "time machines." Pauline sends an actual clock flying at Janoth's face but misses him. In return, Janoth grabs an ornamental sundial from a nearby shelf and throws it back at her, killing her – offscreen, the Hollywood Code still prevailing [00:35:08]. We return to the present and realize it is now irreversible, caught between an action in the near past and its consequences in the near future: the "deed is done" to quote Shakespeare's Macbeth.

The ornamental sundial is now rife with connotations related to the representation of time as an inescapable reality. In this scene, the use of the sundial shows in a literal manner Janoth's loss of control over time, as well as Stroud's. During their evening spree, Stroud has been entertaining Pauline by window-shopping and pretending to look for a "green clock," as some kind of comically fanciful non-existent object. When they enter the shop of an antiquary and Stroud challenges him with finding such a very unlikely object, the antiquary tells them about a sundial and makes it a "green clock" by decorating it with a green ribbon. Later, while they are in her flat, Pauline unexpectedly hears Janoth's car outside and urges Stroud to leave at once. She hands him the sundial, but he tells her to keep it as a "souvenir," which is indeed what it comes to signify as a testimonial of her murder. Though the sundial, unlike the clocks, is quite silent, it nevertheless signifies the passing of time by translating it visually. In the film, this "green clock" sits on a shelf near the entrance and immediately catches Janoth's eye as an anomalous object. This alternative time prop, the so-called green clock, comes to signify Stroud's and Janoth's loss of control over their destiny, as illustrated by the short sequence devoted to Stroud's brief intoxication on a sofa in Pauline's flat [00:32:21–00:32:29].[5] Furthermore, the

---

5   Green is, of course a color, that has numerous connotations in Western culture. It is reputedly an unstable, changing color that can only be obtained by combining other colors, and it is therefore difficult to use and easily erased. The fact that it is here discussed here in a black-and-white film dialogue highlights its cultural connotations. For example, the O.E.D. specifies "sickly," "jealous," and "fresh," "young," "immature" – that is, "unreliable." On color

superimposition of an image of a Patterson painting that he bought at an art dealer's shop with images of bottles, nightclub signs, and a whirl of unsteady frames dancing like waves in an ocean of bubbles all cast an ominous light over his destiny.

The Patterson painting plays a key role in the film. In an earlier scene set during Stroud's and Pauline's spree, we had seen Stroud and an unidentified woman fighting over the painting, which they both earnestly want to buy. They are both under the influence of a wild desire to get hold of the picture. George Stroud allows himself to be lured by greed, gazing at collections of purely useless objects, grabbing two of them, and becoming intoxicated with excess of drink.

2   Blackboards and Canvases in the Film

Apart from the faces of clocks, which appear when they are heard ticking, other stage props express time's presence by visual means, such as screens embedded within the screen, which draw the characters' attention as well as ours. For example, Farrow's film shows how the progress of the manhunt depends on *Crimeways*'s central blackboard, upon which the members of the staff register the slow construction of the puzzle as information pours in. This large blackboard, literally a black board, is not unlike a black film screen within the screen, upon which there gradually appears the figure of the unknown man – a Wellsian invisible man – Randolph. The *Crimeways* staff use it for their investigations, and they believe that an accurate picture will eventually come into shape and make the missing individual identifiable to the crowd. This is the catastrophe that Stroud realizes is threatening him as soon as Janoth calls him on the phone, telling him of someone who was "in Van Barth's with a girl" [41:29–41:50], someone whom they need to find. From this moment on, each piece of information added to the blackboard provides visual proof that Stroud is the man they are looking for. When all the available witnesses are finally invited inside the building, one of them, the art dealer, eventually catches a glimpse of "Randolph," actually Stroud, going by. We viewers hear the art dealer shouting, and we see him pointing in Stroud's direction as the latter vanishes through a doorway [1:03:28], an event which is repeated later with another witness, an employee of the Van Barth restaurant [1:19:50]. Until then, while being in charge of locating the invisible man, Stroud himself had remained quite

---

symbolism, see Gage (1999), and, for a history of color and film, see Dalle Vacche and Price (2006).

invisible as Randolph to all the members of the staff, including Janoth, Hagen, and even Stroud's secretary, Cordette. They are screened talking to Stroud as a familiar colleague whom they know well, reporting to him about the search, while they cannot see that they are gradually describing him in detail on the blackboard [1:06:40].

An additional witness, the painter Louise Patterson plays a major role in the visible – only to us – onscreen progress of the plot because of her paintings [1:10: 41]. She is actually the women in the art dealer's shop who had argued with Stroud about the paining he wished to buy. Their confrontation is reported to Janoth and Hagen. As a result, Patterson's painting, bought by "the blonde's companion" provides Janoth/Hagen with the hope of getting hold of solid proof. They keep repeating that finding "the man with the painting" will mean that they have found Randolph [51:40], seemingly unaware that the blonde's identity, having been established by Patterson, is now known to Klausmeyer, the police, and Stroud himself [57:23], who at once gets rid of Klausmeyer by sending him on a trip to some distant place. Patterson also tells Klausmeyer that she agrees to help with the investigations at *Crimeways* by making a sketch from memory of "the murderer" [56:10], and we viewers anticipate that she will inevitably identify Stroud as this murderer when she goes to his office.

The mise-en-abyme of Patterson's canvases and the exciting play upon the visibility/invisibility dichotomy is reinforced when we see that another painting by Patterson is actually hanging in Stroud's office. Despite the fact it might well betray him as the buyer of a Patterson painting in the art dealer's shop, he decides not to take it down because he believes people around him have identified it already. But such is not the case, and, the first time we see the painting on Stroud's office wall, Janoth goes near it and sits beneath it, quite unaware of its significance [00:20:59]. It is entirely invisible to him, a fact that is emphasized by the camera framing frontally the painting on the wall and the armchair in which he sits immediately underneath.

This focus on the use of screens within the screen underscores the "surprise versus suspense" dialectic at work here, which rules the spectator's empathy because we anticipate a crisis when the two circumstances concerning Patterson's paintings actually collide.[6] This climactic moment occurs when Patterson enters Stroud's office in order to help him identify the murderer, and she is confronted with both the man and her work on the wall [1:10:06]. Under the shock of the discovery, she utters a scream of surprise, but we understand

---

6 Hitchcock illustrates this dialectic in his well-known distinction between the viewer's two attitudes to time by a bomb under a table. Either we have not seen the bomb and are surprised when it explodes, or we have seen it, unlike the characters, and expect it to explode at any moment but do not know when this will happen (Truffaut 2017, 15).

her astonishment is not only caused by finding herself in front of the "murderer"; it is also caused by her discovery of another of her works [1:10:41]. When Stroud tells her blandly: "I've always been an admirer of your work.", we share with him the double-meaning of her scream, and expect them to reach an agreement: she promises not to betray him in order to get back her painting. Later, having complied with Janoth's above-mentioned request to make a sketch from memory of the customer in the art dealer's shop, she produces an avant-garde non-figurative portrait, which she comments upon with a smirk and a catch in her shrill voice: "I think I've sketched his mood rather successfully, don't you?" [1:18:00].

Patterson ironically produces an alternative to the unknown figure whose image is tentatively outlined on the blackboard by undoing the image and thus protecting Stroud's invisible status. The use of such screens within the screen relates to the issue of the visibility of time in several ways. The blackboard recovers fragments of time and reconstructs piecemeal a coherence between them, some of which appear as causes and others of which appear as consequences. On the one hand, for example, on the blackboard, once Randolph's entrance in the art dealer's shop is established, its consequence is the belief he is indeed Randolph. On the other hand, collecting witnesses is part of the framework necessary to make the succession of events legible, and only witnesses can draw the portrait of Randolph and make his identification certain. While Janoth and Hagen plan to subvert the identification process in order to erase the fake murderer's image by shooting him before the police catches him, their own conception of linear time is merely wishful thinking. In terms of cinematography's representation of time, the addition of fragments, the elaboration of a logical link of causality between them as clues, or the manipulation of witnesses and images featuring the role of time in the film are actually entirely dependent on editing (Bordwell 1997, 270–314), a fact which requires a study of the editing of the film's sequences.

3 Editing and the Illusion of Linear Time

Time is the raw material that is sculpted by editing, so to speak. Fragmenting a person's life into a succession of bits and pieces of time is what cinema is about. The screen shows us linear film time as if onscreen action were really happening as in the linear lived time we know, but, of course, the real nature of cinematic time is far from being a continuous flow (Canales 2016, 118).[7] Marey's

---

7 Henri Bergson describes our scientific thinking as "cinematographic" in *Creative Evolution*, by which he means fragmenting reality into shots, turning real things into abstract

work is rife with reflections about the necessity of creating an illusion of movement by an optical effect of continuity with an apparatus that is discontinuous, that is, "the very technology of the photographic apparatus" (Doane 2002, 46–60), with 24 frames per second being the speed necessary for an illusion of continuity in movement. To this initial fragmentation of time is added another fragmented form, the composition of shots into sequences – or, more precisely, scenes – that allow us to move from one moment to the next (Albera 2002, 31–46), what David Bordwell calls the classical Hollywood "continuity system" (Bordwell 2002, 194–213). In *the Big Clock*, the sequencing of the 36 hours of Stroud's life is gradually reconstructed chronologically thanks to the editing of sequences in the flashback, but some of its fragments also reach the blackboard as pieces from another "film" that more and more matches the events. The artifice of continuity achieved by editing is revealed to us by this blackboard, since it shows the juxtaposing of fragments and their assemblage, offering a hypothetical succession of events from the Randolph character's actions.

Let us turn to a detailed discussion of editing as the means of achieving the fluidity of film time in *The Big Clock* by the "continuity system" (Bordwell 2002, 184–213). A first characteristic of classical screen narrative is the belief that the practice of an invisible camera is necessary to the illusion of reality, no gazing at the camera by the actor being allowed, which creates what is usually called the fourth wall of representation. The camerawork is carefully providing us with the expected depiction of spatial and temporal coherence, a world in which the characters are unaware of the presence of the camera or of an observer in the movie theater. We enter the mind of George Stroud thanks to the voice-over monologue: he believes he has become part of a machine that seems to him, as he says to himself, to be "independent of human labor" [00:02:16]. When the camera cuts to the beginning of the embedded flashback narrative, in a well-lit scene, it pans on Stroud and his self-assured steps towards the elevator on the morning of the first day, Thursday April 24 10:48 AM [00:02:60]; we then fall, as it were, 'into step' with Stroud. From then on, his 'becoming', i.e. his experienced present time, is ours. The camera frames George turning his back to us inside the elevator cabin, which confirms that we must identify with him as he watches and overhears other members of the staff, and then it cuts to him stepping out from the elevator on his way

---

discontinuous objects. His own work is quite the opposite – an attempt to coin concepts appropriate for the "becoming" continuum of reality, also called the "stream of consciousness" (*Creative Evolution* 1998, 306–310). The effect of the time continuum sought by film editing remains unaddressed in his work, which caused him to be later criticized. But, of course, his research started before 1895.

to his office. He soon disappears behind a secretary hurrying past a desk into another room, then reappears as he lifts a phone receiver, jumps as he hears a piece of good news, and shouts instructions around him, while his secretary, Cordette, keeps calling for his attention. There is no editing here, the *tempo* of employees as busy as bees being closely sustained by the continuous tracking of the camera. The smooth continuum makes the haste of the characters a sign of their seemingly being wired by some invisible source of energy which they share with Stroud. The words he utters for himself are in the present tense, such as "Think fast ...", and "How did it all start?" [02:16–02:59], and the tracking establishes this present as his memory at work. We viewers never question the division of temporalities of the ensuing narrative flow, on the one hand the present 'becoming' of his reminiscing, and on the other hand the past of the events unraveling under our eyes in their own "present" chronology. The continuity which is established for us viewers relies on the visual continuity of camera tracking which picks up on his voice-over monologue, wiping away, so to speak, the time hiatus achieved by the flashback. Once our belief in the narrative present of the unraveling past events is established by the camera tracking which follows Stroud entering the lift, watching people inside with him, and stepping out briskly, then moving to his office and literally identifying himself with the other employees, the present time is one we share with Stroud's memory which has now become implicit, and the events, which he shared in the past, become shared in the present of the storytelling.

The editing of sequences which structures the story-telling emphasizes the double level of time, present memory and past events. Once the manhunt for Randolph has started, the use of match cuts in continuity editing ensures that Stroud's constant motion from one place to another remains in smooth harmony with the other characters' movements (Hayward 2002, 94–97). Stroud remains hidden in the crowd, being carefully followed by the camera in longshots or medium-shots so that he is filmed apparently casually moving into and out from doorways, acting as if intent upon his purpose but also mindful of matching the global fluidity of everyone's movements. One might add here that the past chronology is itself complex since Stroud decides that he is best hidden, ironically, when being most visible to Janoth and Hagen, which means behaving in order to seem to them actively looking for Randolph, until this strategy fails him, as a first witness spots him entering the building [1:03:28].

Moreover, the film's editing is also a means of creating tension by the increased shortening of the time of each shot in the last act of the drama, which starts with Janoth's loud voice shouting: "Now!" [1:05:29] as he is told that Randolph has been seen entering the building and orders all doors to be shut. As a result, the members of the staff also accelerate the *tempo* as they make a thorough search of each room one after another, and Stroud's own steps get caught in a tightening web [1:20:44]. Contributing to this pulsing rhythm, suggestive of someone's heartbeat, is Janoth's voice shouting orders until the film returns to the opening sequence in the present, which, we now understand, shows the end of the manhunt [1:23:05]. Editing, therefore, not only shows Stroud's carefully hidden emotions but also controls our own. As Eisenstein has argued, acceleration and deceleration of editing by shortening or lengthening the projection time of shots is an important source of our emotion and empathy with the character's feelings ([1947] 1975, 95–96).

The film also uses parallel editing, and screens the simultaneous timelines of embedded spectators in the story: Janoth and Hagen watch the progress of the manhunt for Randolph while the camera intercuts scenes framing Stroud, as the antagonistic embedded film viewer, and his assistant Cordette, who is ubiquitous but blind to what is really going on under his eyes. For example, the intercutting of short scenes in which the blackboard of *Crimeways* is shown [00:52:40] with scenes devoted to the hunt for the picture that Randolph bought [00:52:06–00:52:25] addresses parallel observers whose viewpoint is edited for us in succession: Hagen/Janoth, who create the fictitious Randolph in order to hide Janoth, and ignorant Cordette. One should add Bill to this group: whenever he is framed by the camera, being deaf and dumb, he seems to be keeping an even closer watch over everyone including Stroud, increasing the suspense of impending discovery for us viewers. These observers are thus only gradually kept informed of the growing visibility of the "invisible man." Stroud is the only character who actually knows he is the much sought for Randolph, a situation which reinforces our identification with the character initiated in the opening sequence telling us about his present nightmarish plight.

It might be argued that the creation of a fictional character within the film who gradually grows visible – Randolph – not only literally gives shape to the progress of the investigation but also works as a visual metaphor of the invisible/visible essence of time. The presence of these embedded spectators within the film and the discrepancy between how much truth they are gradually, or suddenly, discovering, protracts and stretches to seemingly infinite limits the present time of the story. While this generates tension as time goes by and clocks strike, it also causes our own gaze to become fractured because we know that they are watching each other in a dissymmetrical manner. Watching and time continuum eventually become welded in an infinite maze. Stroud is aware of the Randolph mystification but lacks proof to unmask Janoth, while Janoth is hiding behind this mystification. Moreover, the intercutting of the blackboard "invisible man" and shots of Stroud standing next to it adds a particularly ironical significance to such parallels for us viewers. Stroud is plainly visible to us, but he remains invisible to his interlocutors as they are reporting to him about the gradually improved definition of the image. To all characters in the plot, whoever they are, he remains actually invisible until he becomes visible to Louise Patterson and the other two witnesses, the art Dealer and the waiter from Barth's restaurant. Janoth and Hagen are also quite invisible to the same crowd although visible, of course, to Stroud himself. Finally, we see Georgette and McKinley as an inspector [01:28:25]: they are the eye witnesses of the final truth. Stroud is in Janoth's office stoutly declaring to the two men that Hagen is the murderer and lists his proofs: Pauline's handkerchief was found by Georgette in Hagen's cigar box, the maid being away could not testify anything, the taxi driver was bribed to go on vacation. Stroud then calls the inspector in, who enters along with Georgette [1:29:40]. We know Janoth has a gun in his pocket [01:32:30] and when Cordette enters still looking for Randolph, Janoth says: "call it off ... everybody gets home." He then shoots Hagen and races to the elevator to his death. In these very short sequences which create an acceleration of movement, the moment when Janoth and Hagen are unmasked (that is, become visible for two new witnesses, Stroud's wife Georgette and the lawyer/ policeman McKinley), the fracture of the gaze into several observers, including ourselves, recovers the unity necessary for a final coherence.

These examples of devices aiming at the creation of linear time show how far an illusion of continuity is essential to the production of an onscreen coherent narrative. As Maltby reminds us, the illusion of continuity which is created by editing as well as tracking or panning in different sequences is also the result of an overall pattern[8] in which the opening scene establishes the plot situation, the present time in the opening clock tower scene in Farrow's film, and sets a deadline which occurs when we resume this present and establish a continuity between present and past (Maltby 2013, 416). The fragmentation of the narrative in film-time is a logical process that makes the necessary juxtaposing of fragments intelligible and pleasant.[9] There is a continuity between classical drama and film time which shows how time as an agent in

---

8   The problem is to adjust film time and action time (Maltby 420). In Farrow's 36 hours (movie time) into 96 minutes (film time). "The rule that one page of screenplay equals one minute is a calculation of film time, and the iron laws of the three -act structure [...] are devices by which movie time must be fitted to the commercial aesthetic of film time, which determines that a movie is 'about two hours long'" (Maltby 425).

9   This is found early in film history with screenwriting manuals. For instance Syd Field *Screenplay: The Foundations of Screenwriting*, New York: Dell, 1979, Michael Hauge, *Writing Screenplays that Sell*, New York: McGraw Hill, 1988 (Maltby 416). The earlier works by Eisenstein, Pudovkin, Koulechov and French writers such as Dulac, Epstein, or Marcel L'Herbier and Louis Delluc, are a matter of common knowledge. See Dudley Andrews, *The Major Film Theories- An Introduction*, Oxford: Oxford University Press, 1976.

the unraveling of the action is an essential component of Western culture, as Maltby underlines.[10] The characters' plight shows them struggling against Fate in attempts to control the flight of time, while events happen that provide necessary information about causality and parallel plots, and overlap with coincidences caused by chance. Just as they fail to control the progress of time and destiny in tragedies, so do we recognize Janoth's desperate struggle against the inevitable discovery of Pauline's murderer as a struggle against time. What remains to be seen, however, is how cinema re-appropriates these classical models and re-writes them in terms of a modernist understanding of time as action.

4   Cinematic Time

I now examine Mary Ann Doane's problematics of "cinematic time," which are grounded on a study of science and technology in the Industrial Era, in light of my previous remarks. The Hollywood continuity system discussed above has been seen to combine timelines in which time is at variance within the invariant film time structure representing the lived time of both characters and audience. Deeply related to the history of the origins of cinema, this invariance lies at the heart of the modern definitions of three qualities of time's action: irreversibility, contingency and entropy. These ruled in the world of the Industrial Revolution, a period contemporary to the birth of cinema, and, according to Doane, are the backbone of cinematic time. Hence the proposal to examine Farrow's film in the light of these three characteristics in a film whose title explicitly refers to the workings of time.

The choice of introducing the end of the action to begin the narrative, which then proceeds in a flashback, expresses time's irreversibility by creating a deadline. The opening scene of Stroud as a prisoner in the clock tower, is like Hitchcock's "bomb under the table" (Truffaut 2017, 15); we expect an explosion, but we do not know when this will happen. The suspense created by the jump back in time gives us a sense that time is a reality that we can apprehend by giving the irreversibility of time the logical pattern of cause and consequence. The opening scene is the consequence of causes which the narrative will be hunting out in order to answer Stroud's question "How?" and "Why?" But what

---

10   Just like classical drama, a film narrative should be divided into three parts: a "movie screenplay should have three acts. Act 1, the set-up, establishes the plot situation [...] Act 2, the confrontation, builds it during the next half. Act 3, the resolution, brings the story to its conclusion" (Maltby 416).

is properly cinematic is to create a visual trick which actually reverses the order of linear time by showing us a moment of time that is normally impossible to see until one reaches it. Conversely, from the beginning of Stroud's story, time is being rewound so to speak, when we are shown what took place before this moment.[11]

Janoth's accidental use of a sundial as a weapon has several connotations related to our understanding of the action of time, as suggested above. To start with, this sundial, a "bow sundial" specifically, should reckon hours by capturing sunrays passing through a narrow slot in order to form a uniformly rotating sheet of light that falls on the circular bow. But its designer added an additional arrow as a sort of crest, which is quite unusual. If the "arrow of time" commonly means that a phenomenon is irreversible (Klein 2016), then it is ironic that neither Stroud nor Pauline are able to decipher the implicit message lying at the core of the sundial's ornamental "arrow of time" as they handle it, while Janoth unconsciously uses it very aptly.

While time in *The Big Clock* appears irreversible as far as Pauline's death is concerned, the film also makes time's contingency equally perceptible by the coincidences (forms of classical digressions) that allow for Hitchcockian surprise effects. For example, the invisibility/visibility dichotomy that rules Stroud's survival during the manhunt is also an opportunity for contingency, or chance, to play its part in the plot. In the scene of Stroud's first chance encounter with the as yet unidentified Louise Patterson [00:27:26–00:28:40], the painter's work, as well as her presence, actually become personifications of time's secret rule of contingency in all its potent influence upon individual destinies. The second encounter between Patterson and Stroud in his office repeats the initial surprise effect. Introduced as a witness who can identify "the man with the painting," Patterson quite unexpectedly turns out to be more interested in making a deal with a man who collects her work (and thus recover her painting) rather than unmasking this man as a criminal. She subsequently outwits the art dealer's offer and produces the requested portrait from memory of the man she saw, but the painting is a non-figurative abstract design. Louise Patterson thus appears to be shamelessly capitalizing on the cliché of artists who are supposed to be unreliable in a world governed by clocks and consumerism – a characteristic that she is careful to emphasize by her shrill voice and erratic fits of laughter.

---

11   The ability to project a film backwards and forwards, and thus reverse the action or develop it, was part of the initial tricks which delighted viewers as a specifically cinematic visual trick. Louis Lumière, *Démolition d'un mur*, early 1896.

Chance also rules Stroud's choice of pastimes in Pauline's company. This episode is essential to the characters' destiny but from their point of view, they believe that such moments of vacation when ordinary rules are suspended are of no consequence. And yet, ironically, Stroud's jokes are all about time and as he jokes loudly about clocks, and "green clocks" in particular, in the art dealer's fateful shop. The object with a green ribbon that Stroud buys seems at first more like a Dadaist joke, which makes it into something of a quibble. But it soon becomes a vehicle for the invisible contingency of time as it has the power to smash our hopes and dreams, metaphorically, but also literally when Janoth picks it up by mistake and kills Pauline more or less unintentionally.

Moreover, it might also be suggested that "the green clock," silently sitting next to the dead body with malevolent indifference, is an ominous sign of the general entropy threatening the world to which it belongs – that is, Janoth's as well as Stroud's world, which is ruled by the motto "time is money." Invisible entropy grows gradually, but unmistakably, visible, thanks to the digressions from the manhunt plot, such as Stroud's spending spree and final intoxication, or Janoth's fit of jealousy, that all actually keep us informed about the progress of the final disaster, as in, for example, the gradual loss of control over time that the two main characters suffer. In Janoth's case, his ambition to improve his magazines' commercial attractiveness – and control the flight of time – is shattered to pieces by his untimely mistake. As for Stroud, he buys "the green clock" in a moment that he literally steals from his allotted time that evening, with dire consequences for his immediate vacation plans. To sum up, among its several possible connotations, the bow sundial can also be seen as a metaphor of the general entropy so characteristic of time in modernist films.

5   Conclusion

The Big Clock remains in our memory, among other things, as drawing a clever parallel between two closely linked observers who are seeking clues of each other's involvement with a murder as they race against time. They stand for our own presence in the world of the film as observers and hearers of clocks, who are also being manipulated into an exploration of the essentially elusive reality of time. During the time we spend in the movie theatre, we are unaware of time's passage in the world outside. But the cinematic time that rules our free time uses the deadline of the opening sequence and the flashback which follows to express irreversibility, as well as coincidences such as Stroud's encounters with Patterson to express contingency. But among the several digressions (different temporalities) such as Stroud's intoxication or Janoth's fit of jealousy, some clearly express a threat of general entropy. Like Stroud, we get trapped into a present nightmare and look for answers to questions about "when" and "how." What we remember as well is the sundial's totally silent but "howling" visibility, underscoring its erratic inscription within the chain of irreversible events. Its very presence is wholly due to chance, and, under its malevolent green 'shining' like a signal, the transformation from its initial destiny as an ornament in a middle-class sitting room into a deadly weapon causes disaster to the three characters. Ironically, lack of causality in its original appearance within reach of Janoth's hand, transforms it into the cause of Pauline's death and Janoth's own.

It might thus well be argued that our pleasant obliviousness of invariant time in the movie theater is clearly upset by time in variance in *The Big Clock*. The tension between regular, quantitative clockwork time and the irregular intrusions, collisions, and even erratic situations in which the characters get caught makes invisible time visible on the screen. As to the suggestion of a form of correspondence between scientific concepts of irreversibility, contingency and entropy, as characteristics of time in the Industrial Era, and story-telling devices such as deadline, coincidence and digression, the film shows that time is indeed, as suggested above, the raw material of which cinema is made.

References

Albéra, François, Marta Braun, André Gaudreault, eds. 2002 *Arrêt sur image, fragmentation du temps- Aux sources de la culture visuelle moderne* [*Stop Motion, Fragmentation of Time: Exploring the Roots of Modern Visual Culture*]. Lausanne: Editions Payot.

Andrews, Dudley. 1976. *The Major Film Theories: An Introduction.* Oxford: Oxford University Press.

Bergson, Henri. (1911) 1998. *Creative Evolution.* Translated by Arthur Mitchell. New York: Dover Publications.

Bordwell, David. (1985) 1993. *Narration in the Fiction Film.* London: Routledge.

Bordwell, David. 2017. *Reinventing Hollywood: How 1940s Filmmakers Changed Movie Storytelling.* Chicago: Chicago University Press.

Bordwell, David, Janet Staiger and Kristin Thompson. (1985) 2002. *The Classical Hollywood Cinema: Film Style and Mode of Production to 1960.* London: Routledge.

Bordwell, David and Kristin Thompson. (1979) 1997. *Film Art: An Introduction.* New York: McGraw Hill.

Burges, Joel and Amy J. Elias, eds. 2016. *Time: A Vocabulary of the Present.* New York: New York University Press.

Canales, Jimena. 2016. "Clock/Lived." In *Time – A Vocabulary of the Present*, edited by Joel Burges and Amy J. Elias, 113–128. New York: New York University Press.

Dalle Vacche, Angela, and Brian Price. 2006. *Color: The Film Reader*, London: Routledge.

Doane, Mary Ann. 2002. *The Emergence of Cinematic Time: Modernity, Contingency, the Archive*. Cambridge: Harvard University Press.

Eisenstein, Sergei. (1947) 1975. *The Film Sense*. Translated by Jay Leyda, London: Harcourt Brace and Co.

Gage, John. 1999. *Colour and Meaning: Art, Science and Symbolism*. London: Thames and Hudson.

Hayward, Susan. (2001) 2002. *Cinema Studies: The Key Concepts*. London: Routledge.

Klein, Étienne. 2016. "What Does the 'Arrow of Time' Mean?" *KronoScope* 16(2): 187–198.

Maltby, Richard. (1995) 2013. *Hollywood Cinema*. London: Blackwell.

Munsterberg, Hugo. (1916) 2007. *The Photoplay: A Psychological Study*. New York: Bibliobazaar.

Schefer, Jean Louis. 1997. *L'Homme ordinaire du cinema*. Paris: Cahiers du cinéma.

Truffaut, François. 1993. *Alfred Hitchcock, Entretiens*. Paris: Gallimard. English translation: (1983) 2017. *Hitchcock, The Definitive Study of Alfred Hitchcock by François Truffaut*. London: Faber & Faber.

CHAPTER 14

Transcending Temporal Variance: Time-Specificity, Long Distance Performance and the Intersubjective Site

Emily DiCarlo

Abstract

The *Daylight Saving Time series* offers four visual art projects that engage in interventionist strategies to push against the infrastructure of clock time. In *I Need To Be Closer To You (DST IV)* (2018), the most recent iteration in the series, artists Emily DiCarlo and Evan Tyler collaborate on a long-distance performance-for-video, which examines the history and inconsistency of Canadian Daylight Saving Time through an intimate lens. With consideration to the lineage of Romantic Conceptualist thought, this article traces site-specificity's evolution, establishes "time-specificity" as a key component in the work, and explores telepresence as a means for remote collaboration, considering the phenomenological and empathetic potential of being "both here and there."

Keywords

Canadian time zones – collaborative video performance – contemporary art – Daylight Saving Time – *I Need To Be Closer To You* – intersubjectivity – long-distance synchronization – Romantic Conceptualism – site-specificity – telepresence – temporal variance – time-specificity

1 Introduction: *I Need To Be Closer to You (DST IV)*

I Need To Be Closer to You (DST IV) (2018) is a collaborative, long-distance project between myself and visual artist Evan Tyler, which focuses on the shifting temporal proximity of two long-distance lovers during spring Daylight Saving Time. At the time of its conception, we lived two time zones apart, separated by more than 2500 kilometers: Tyler residing in the flat prairie plains of Regina, Saskatchewan, while I called the metropolis of Toronto, Ontario, my home. Significantly, this spatiotemporal distance impacted both our artistic collaboration and long-term relationship dynamic.

Scheduled annually at 02:00:00 EST on the second Sunday in March, clocks across Canada turn ahead to lose an hour.[1] However, Saskatchewan is an anomaly to this rule, and upon time change, the one-hour time difference between us extends into two.[2] In anticipation of this exception, we devise a digitally-mediated performance using video chat technology and a geolocational, timestamped smartphone application. We know that only I will 'time travel' forward, an hour of clock time evaporating. For Tyler, time will remain uninterrupted. Approaching the anticipated variance and waiting for our temporal distance to dilate, we record ourselves, simply existing for the camera but also for each other while our placements in space and time are documented on-screen by a GPS locator and real-time counter.

My time zone dictates that my performance happens first. Tyler, existing in the universally coordinated future, watches live and during the minute leading up to my temporal leap, I look directly into the camera's lens and say the words "I need to be closer to you" at the 30-second mark. Soon after, the on-screen counter displays my 'time travel' as the clock's seconds jump from 1:59:59 EST to 3:00:00 EDT. Acting as registered evidence, the minute-long footage serves as a blueprint for the hour-long final work. In the premeditated decision to loop my minute-long video-performance 60 times, Tyler now has an event score to follow for his own performance.[3] Through the act of looping, the missing Daylight Saving hour takes form.

---

1   Signed into law in 2005 but enacted in 2007, the Energy Policy Act in The United States extended the yearly Daylight Saving Time period by a total of five weeks, shifting the DST spring period from the first Sunday of April to the second Sunday of March and the fall period from the last Sunday of October to the first Sunday in November. Following the United States' lead, Canada followed suit and also extended their DST period to coordinate with their American neighbours, a practice done since the late 1960's to maintain consistency in travel and trade. 109th U.S. Congress. "Sec. 110. Daylight Savings." An act to ensure jobs for our future with secure, affordable, and reliable energy. United States Government Printing Office. Pub.L. 109–158, 119 Stat. 594, enacted August 8, 2005. Accessed March 8, 2019. https://www .congress.gov/109/plaws/publ58/PLAW-109publ58.pdf.

2   Since the Time Act in 1966, all of Saskatchewan observes Central Standard Time year-round with the one exception of Lloydminster, a city that straddles the provincial border between Alberta and Saskatchewan. The Government of Saskatchewan. "Tools, Guides, and Resources for Municipalities: Saskatchewan Time System." Accessed on March 10, 2019. https://www .saskatchewan.ca/government/municipal-administration/tools-guides-and-resources/ saskatchewan-time-system.

3   The term "event score" refers to a short script or set of instructions needed to carry out a performance work. It is derived from Fluxus, an interdisciplinary art movement active throughout the 1960s and 70s, whose work took interest in the immaterial, the experiential, and the mundanity of daily life to counter the values of the commercial art market and academic

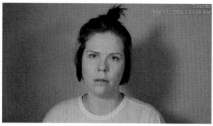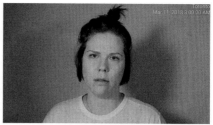

FIGURE 14.1   *I Need To Be Closer To You (DST IV)*, 2018. Two-channel video, Daylight Saving Time performance on March 11, 2018. Performance video stills
PHOTO © EMILY DICARLO

Tyler waits his turn to perform knowing that he will not experience the same 'time travel' I did and begins to imagine my earnest words of longing, repeating over and over. With synchronization serving as a substitute for physical togetherness, he prepares to perform in tandem to the final looped performance score. He begins recording at 01:59:00 CST and 30 seconds later, he imitates my cadence and vocalization from earlier, saying "I need to be closer to you." The on-screen counter ticks forward ... 01:59:57, 01:59:58, 01:59:59... 02:00:00 CST. In the absence of a Daylight Saving Time change in Saskatchewan, Tyler performs for the next hour, each minute on the 30-second mark, echoing the same sentence, "I need to be closer to you" into the camera as I watch on. Although his commitment pushes him forward, his endurance waivers, and occasionally he falls behind on his punctual cue or stumbles with the phrase's wording. But he persists through his exhaustion, performs to the hour's end at 03:00:00 CST, and turns off his camera to complete the performance.

Months later in a darkened gallery, adjacent monitors play the two-channel video, each depicting our framed faces.[4] We stare in silence, patiently waiting, until finally we say in unison our incantation of desire: "I need to be closer to you." We are coordinated again, back in sync, the shared Daylight Saving system glitch amended through our collaborative efforts.

---

institution discourses of the time. Originating with the movement's co-founder Dick Higgins, a contemporary of experimental composer John Cage, an event score manifests as "short, terse descriptions of performable work" that act as an execution guide for specific actions or ideas.

    Lisa Wainwright, "Fluxus." *Britannica Academic* (*Encyclopedia Britannica Online*), 2016. Accessed April 8, 2020. https://www.britannica.com/art/Fluxus.

4 The first viewing of *I Need To Be Closer to You (DST IV)* took place in E.E.L Gallery at the University of Toronto, Canada.

2   *The Daylight Saving Time Series* (2007–2018)

As the fourth iteration in my ongoing *Daylight Saving Time series*, *I Need To Be Closer to You* (*DST IV*) pushes the limits of "time-specific," long-distance collaboration. Each installment in the *DST series* offers an intervention that pushes against the infrastructure of clock time. These gestures are intended as poetic acts of resistance, asserting that the personal is always political.

When North American Daylight Saving Time regulations first shifted on November 4, 2007 to extend the DST period by five weeks, *Taking Back The Night* (*DST I*) materialized as a two-hour long, performance-for-video. With camera recording, I silently sat for the hour leading up to the mandated time change at 02:00:00 EDT, only to watch the clocks turn back to 01:00:00 EST again, prompting me to "relive" the hour's recent past. As I endured the second hour's passing while observing the clock's minutes tick by just as it had, waves of déjà vu and delirium washed over me. This act of waiting, an exercise in slowness, seemed to dilate into an ever-extending moment.

Holding affinities with Faith Wilding's performance art piece *Waiting* (1972), Kate Brettkelly-Chalmers discusses in her chapter on "Duration and Endurance: Minimalism and Performance" the temporal dimension in this germinal feminist work: a 15-minute monologue in where the artist slowly explains how a woman's life is structured by sociopolitical systems as an experience of waiting and submission. Between Wilding's drawn-out pauses, she utters phrases such as "Waiting to have a good figure," "Waiting for him to come home, to fill my time," "Waiting for some time to myself," and "Waiting for sleep." As Brettkelly-Chalmers poignantly notes, "The act of waiting is the distention of time in experience: the inescapable duration that swells with conscious awareness. In Wilding's *Waiting* the audience is asked to contemplate a mode of passive anticipation – to sit with an individual experience of duration." Similarly,

FIGURE 14.2   *I Need To Be Closer To You* (*DST IV*), 2018. Two-channel video, Daylight Saving Time performance on March 11, 2018. Performance video stills
PHOTO © EMILY DICARLO

Taking Back The Night (*DST I*) provokes a suspended experience, forcing the viewer to also slow down and wait.[5]

In a world where immediacy is the temporal standard and perpetual growth, productivity and efficiency are the tenets of late-capitalist pursuits, waiting can be viewed as resistance to the demands of an accelerated culture. Notable countertrends such as the Slow Movement have sprung up as a way to reclaim both personal and collective time. Instead of doing everything at snail's pace, the movement works to reclaim situational tempos that strike a balance of varying paces, fighting against the reigning expectation for speed at all times. Since its humble beginnings in 1989, the relatively nascent trend continues to gain traction and has now organized enough to host an annual conference under the Society for the Deceleration of Time's guidance. They leverage the German word *eigenzeit* to ground their views. With *eigen* meaning "own" and *zeit* meaning "time," they believe that every living being, event, process or object has its own inherent time or pace.[6]

Similar to *I Need To Be Closer To You* (*DST IV*)'s post-production process, *Taking Back The Night* (*DST I*)'s two hours of performance footage are split, installed parallel from one another, and played simultaneously as a two-channel video installation. If closely investigated, the timestamps of the two video sequences read identically. The project identifies a time-specific loophole in the Daylight Saving Time regime and exposes the conceptual impossibility of one individual existing twice within the "'exact'" same time frame – a triumph of the uncanny.

While *Taking Back The Night* (*DST I*) concentrates on slowness as a resistant act, *Composing Absence* (*DST II*) looks at lost time and the impulse to fill its void. During the first minute of the hour-long work, the single-channel video shows a computer clock's seconds ticking towards the automatic time skip, jumping from 01:59:59 EST to 03:00:00 EDT, only to loop again 59 more times in an attempt to "fill" the missing hour. The anxious reality of not having enough hours in a day directly results from the demands of our always-on, 24/7 networked world. With little distinction between work-time and leisure-time anymore, multi-tasked activities often overlap and occur simultaneously. While our natural biological rhythms require pause and rest, we have been conditioned for constant engagement. *Composing Absence* (*DST II*) reconstructs the

---

5   See Kate Brettkelly-Chalmers, *Time, Duration and Change in Contemporary Art: Beyond the Clock* (Bistol and Chicago, IL: Intellect, 2019), 82.
6   See Carl Honoré, *In praise of slow: how a worldwide movement is challenging the cult of speed* (Toronto: A. A. Knopf Canada, 2004), 38.

FIGURE 14.3   *Taking Back The Night (DST I)*, 2007. Two-channel video, Daylight Saving Time
performance on November 2, 2008. Duration: 120 mins. performance, 60 mins.
video. First projection: "Origins and Futures," The International Society for the
Study of Time Twelfth Triennial Conference, Monteverde, Costa Rica
PHOTO © EMILY DICARLO

vanished time but resists a model of productivity. Instead, the project creates
a meditative space through its visual repetition and encourages the viewer to
reflect on the modernist myth of endless progress.

A decade later, *Streaming My Stolen Hour (DST III)* presents the discrepancy felt between the autonomy of an analogue quartz clock and the flawless
precision of a digitally connected computer.[7] Broadcast as a public livestream
on November 5, 2017, a laptop webcam records the clock's face pacing towards
the regulated autumn time change. In contrast to my filmed presence in *Taking*

---

7   Still used today to power consumer home clocks and wristwatches, quartz crystals vibrate
at 32,768 Hz, which is equal to a one-second beat. Used as the national standard in most
countries between the 1930s to 1960s, quartz's accuracy drifts only one second in three years.
Today, a network of national meteorology labs maintains our universal time standard. The
universal standard is the nanosecond, an accuracy maintained by using cesium-driven
atomic clocks. Cesium's much higher resonant frequency (9,192,631,770 Hz) will only err by
one second in 1.4 million years, making it the superior choice for time-keeping today. Justin
Rowlatt, "Caesium: A brief history of timekeeping," *BBC News*, October 4, 2014. Accessed
December 18, 2019. https://www.bbc.com/news/magazine-29476893.

FIGURE 14.4   *Composing Absence (DST II)*, 2008. Single-channel video, Daylight Saving
Time performance on March 9, 2008. Duration: 60 mins. First projection:
"Origins and Futures," International Society for the Study of Time Twelfth
Triennial Conference, Monteverde, Costa Rica
PHOTO © EMILY DICARLO

Back The Night (DST I), the ticking clockface stands in for my performing body.
At the moment of "falling back an hour," the wall clock also falls out of sync
with the automatically networked time change on the laptop counter, prompting a manual reset. A hand sweeps into the frame, removing the clock on the
wall and turns back the hands to 01:00:00 EST, and, during this brief moment
halfway through the two-hour webcast, we are reminded of the invisible labor
needed to keep time running. But the time change is imprecise and a lagged
variance now persists between the recording device and the broadcasted content. Perfect synchronization is impossible when the current universal time
standard runs on nanosecond accuracy. As a simple gesture, *Streaming My
Stolen Hour (DST III)* illuminates the discrepancy between human ability and
networked precision.

In all four iterations of the *Daylight Saving Time series*, each work's aim
remains the same: to identify the absurd qualities of artificially-made temporal structures and, through working within their set parameters, reveal an
experiential entanglement with these systems. With each iteration acting as
an expanded foundation to the last, the projects gain complexity despite the
time structure's parameters remaining fixed. Through visual discourse, the
added variables of site-location, performer identity and quantity, and leveraged technology illustrate a range of strategies for creating alternative temporal realities.

FIGURE 14.5   Emily DiCarlo, *Streaming My Stolen Hour (DST III)*, 2017. Single-channel,
               livestream video, Daylight Saving Time performance on November 5, 2017.
               Duration: 120 mins. livestream
               PHOTO © EMILY DICARLO

3   Daylight Saving Time: Temporal Variance in an Arbitrary System

The non-stop push of 24/7 culture has resulted in a flattening of temporal
tenses, leaving little distinction between past, present and future, both in a
geographic and experiential sense. As the art critic and essayist Jonathan Crary
describes, "24/7 announces a time without a time, a time extracted from any
material or identifiable demarcation, a time without sequence or recurrence."[8]
In our always-on world where ceaseless production operates and supply chain
movement never sleeps, time-based legislations such as Daylight Saving Time
are rendered obsolete. More now as an archaic imposition, DST has been cited
to cause more harm than good with reported increases in traffic accidents and
heart attacks the day after the time change.[9]

---

8   See Jonathan Crary, *24/7: Late Capitalism and the Ends of Sleep* (London; New York: Verso,
    2013), 29.
9   See Stanley Coren. "Daylight Savings Time and Traffic Accidents," *The New England Journal
    of Medicine* 334, no. 14 (1996): 924–925; Rickard Ljung and Imre Janszky, "Shifts to and from
    Daylight Saving Time and Incidence of Myocardial Infarction." *The New England Journal of
    Medicine* 359, no. 18 (2008): 1966–1968.

Time-shifting in North America was first referenced by Benjamin Franklin in a letter to *Journal de Paris* in 1784 when he pointed out the discrepancy between allocated working hours and the available sunlight in a day. In response, he proposed an alternative working schedule by shifting sleep schedules earlier, citing how the change would save on all the candles and lamp oil that were being unnecessarily used during the evening hours.[10] A practical solution proposed amid the First Industrial Revolution, Franklin's austere approach reflected the manufacturing priorities of the time, which aimed to increase factory production while saving on energy costs.

With rapid economic growth and technological advancement well underway at the turn of the twentieth century, the world grew smaller as capitalism's production accelerated. Temporal coordination of the now not-so-distant geographies became necessary. Up until 1840, towns and cities functioned on their own local mean time derived from solar observance. Running local time this way had worked for centuries, but, with trains speeding across the terrain from destination to destination, stations that were only one degree of longitude apart had clock times with an average four minutes difference. In more extreme examples, although New Orleans, USA, was only 130 kilometers west of Baton Rouge, a twenty-three-minute difference pertained.[11] While most of Britain accepted time transmitted by telegraph from the Royal Observatory Greenwich by 1855, North America struggled to coordinate.

Wolfgang Schivelbusch, author of *The Railway Journey: The Industrialization of Time and Space in the Nineteenth Century*, describes how "[i]n stations used by several different lines there were clocks showing different times: three of these in Buffalo, six in Pittsburgh."[12] Finally, after many policy revisions, the Association of American Railroads resolved to create four coordinated time zones spanning the country. On November 18, 1883, which became known as "The Day of Two Noons," each mandated railroad station reset its local clock to standard noon. This coordinated gesture erased any temporal gradients and brought the sharpened precision needed to keep the flow of capitalism moving. The following year marked the birth of global standard time when twenty-seven nations agreed to recognize Greenwich as the prime meridian, eventually leading to the creation of Coordinated Universal Time. Finally, in

---

10   See Benjamin Franklin, "Essay on Daylight Saving: Letter to the Editor of the Journal of Paris 1784." *WebExhibits online museum*. Accessed on February 18, 2019. http://www .webexhibits.org/daylightsaving/franklin3.html.

11   See Carl Honoré, *In praise of slow: how a worldwide movement is challenging the cult of speed* (Toronto: A.A. Knopf Canada, 2004), 26.

12   See Wolfgang Schivelbusch, *The Railway Journey: The Industrialization of Time and Space in the Nineteenth Century* (Oakland: University of California Press, 2014), 57.

a 1895 paper to the Wellington Philosophical Society, George Vernon Hudson, a British-born New Zealand astronomer and entomologist, officially proposed a legislative model for Daylight Saving Time, which became the foundation of Daylight Saving Time today.

Although Daylight Saving Time's aim was to optimize labor productivity and standardize, the most baffling aspect of its contemporary usage is its inconsistency across the globe. Naturally, countries further from the equator are subject to more drastic experiences with summer-winter light patterns and therefore, still make appropriate venues for the practice. However, DST's sporadic implementation reveals an absurdity worth noting. In their collaborative essay "Time Zone Politics and Challenges of Globalisation," geographers and cultural theorists Karl Benediktsson and Stanley D. Brunn provide an in-depth survey on the conflicting interplay between biological temporalities and chronopolitically motivated time structuring. In their section devoted to the complexity and irregularity of Daylight Saving Time usage, they point out that "there are entire countries on DST; some parts not on DST (Australia, Canada and Mexico). Other countries have implemented DST for a year or two. There are also some non-DST 'islands' within countries where DST is otherwise used: Eastern British Columbia and Western Saskatchewan and the Navajo reservation in Northeast Arizona."[13]

Unlike the rest of Canada, the anomaly province of Saskatchewan does not participate in Daylight Saving Time. The province has a rebellious history when it comes to agreeing on temporal unification. Before the Time Act of 1966, which sought to establish a uniform provincial time zone, small Saskatchewan municipalities remained defiantly divided, and with just cause. Geographically, the 105-degree standard meridian for Mountain Standard Time runs right down the centre of the province. In a 1958 provincial election on the matter, the results concluded that urban populations favoured Central Standard Time and rural communities preferred Mountain Standard Time,

---

13   The term "chronopolitics" refers to the relationship between the political behavior of an individual or group and their time-perspectives, which often produce the formation and dissolution of institutions, resulting in a shifting social order. An example of chronopolitically motivated time structuring can be found in China, when in 1949 the Communist Party mandated the entire country to amalgamate their five time zones into one, enforcing an ideology of "enhancing national unity." Now running solely on Beijing Standard Time, a time standard rooted in the far west of China, the discrepancy of solar time and clock time increases the further east one travels. Karl Benedicktsson and Stanley D. Brunn. "Time Zone Politics and Challenges of Globalisation." Royal Dutch Geographical Society KNAG, vol. 106 (2015): 278–280.

with an equal split on whether or not to follow Daylight Saving Time.[14] The polarizing debate led the government to establish a Time Committee comprising members from the Saskatchewan School Trustees Association, the Association of Rural Municipalities and the Urban Municipalities Association. Frustratingly, the committee concluded that "the observance of one uniform time throughout the province all year, for instance, would work except that the people of the province will never agree on whether it should be Mountain or Central Time."[15] Finally, the Time Act passed the uncommon legislation that Saskatchewan should not follow Daylight Saving Time, and instead conform to one uniform time year-round, abiding by Central Standard Time. Even with this Canadian exception, further time variances exist within the province, as shown with Lloydminster, a small prairie city that straddles the border of Manitoba and observes Mountain Standard Time during the winter.

The irregularity of Daylight Saving Time legislation and malleability of time standards in general not only highlights the power of geopolitical interests at play but also presents an opportunity for imaginative intervention. In response, *I Need To Be Closer To You (DST IV)* leverages the system's arbitrary construction and exploits its temporal variance. Because Saskatchewan disregards the observance of DST the temporal distance between Tyler and me dilates at Daylight Saving Time's juncture, producing an oscillating yet consistent format of experiential time variation every year. For six months we remain temporally close, with only one hour between us, and for the remainder of the year during the rest of Canada's Daylight Saving Time shift, we feel the two-hour distance. *I Need To Be Closer To You (DST IV)* proposes that this perceived disruption in time can be collaboratively amended, but as I explain in the following section, it takes a mystic jump of Romantic Conceptualist logic to reach this proposed conclusion.[16]

---

14   See The Government of Saskatchewan, "Tools, Guides, and Resources for Municipalities: Saskatchewan Time System." Accessed on March 10, 2019. https://www.saskatchewan .ca/government/municipal-administration/tools-guides-and-resources/saskatchewan -time-system.

15   Earl R. V. Milton, *A Submission to the Government of Saskatchewan Regarding Time Zoning in Saskatchewan.* 1966. Accessed on March 10, 2019. https://www.saskatchewan .ca/government/municipal-administration/tools-guides-and-resources/saskatchewan -time-system.

16   Sol LeWitt. "Sentences on Conceptual Art." *Art-Language: The Journal of conceptual art* 1, no. 1 (1969).

4   Romantic Conceptualism: A Method of Potentializing

Emerging during the 1960's, Conceptual Art asserted that an idea alone could be a work of art.[17] Rising out of failed modernist ideologies, Conceptual Art reacted against the purity of formalism and rejected the commodification of art that rose with Abstract Expressionist painters. Instead, language became the artists' preferred medium, a vehicle to disseminate art into the world. Dismissing sublime universality altogether, they focused instead on the environmental and sociopolitical injustices of the time, making space for marginalized perspectives of feminist, queer, and visible minorities. Preferring analytic rigor to sentimental pathos, Conceptualism rejected emotionality in favour of cool, critical intellectualism. Sol LeWitt, one of the movement's founders famously said, "It is the objective of the artist who is concerned with Conceptual Art to make his work mentally interesting to the spectator, and therefore usually he would want it to become emotionally dry. [...] The expectation of an emotional kick, to which one conditioned to Expressionist art is accustomed, [...] would deter the viewer from perceiving this art."[18]

However, Romantic Conceptualism, a movement retroactively defined by writer and curator Jörg Heiser in 2007 from his namesake exhibition, recast select Conceptual artists through an emotional lens – artists such as Yoko Ono, Bas Jan Ader, and Felix Gonzales-Torres taking note of their intuitive production styles and their interest in personal transcendence.[19] While not always overtly political, represented artists created work in times of social unrest whether during the Vietnam War or the AIDS crisis. A parallel held with Romanticism, the cultural movement that emerged in Europe during the early 1800s after the French Revolution's failure, art critic Jan Verwoert describes a value shift in certain Conceptual artists that rejected a logos of universality, noting a pendulum swing "from a limiting force of reason to the delimiting power of imagination, from the closed system to the open form of the fragment."[20] He goes on to emphasize that this approach to meaning-making

---

17   There is a long history of proto-conceptual art, such as with Marcel Duchamp's infamous readymade *Fountain* (1917), but for the purpose of this essay, focus will remain on the Conceptual Art movement from the 1960–70's.

18   Sol LeWitt. "Paragraphs on Conceptual Art." *Artforum* 5, no. 10 (1967).

19   See *Romantic Conceptualism*, edited by Jörg Heiser. Exhibitions: Kunsthalle Nürnberg, Nuremberg, May 10–July 15, 2007 and the BAWAG Foundation, Vienna, September 14–Decmber 1, 2007. Exhibition catalogue.

20   Jan Verwoert. "Impulse Concept Concept Impulse: Conceptual Art and its Provocative Potential for the Realization of the Romantic Idea." *Romantic Conceptualism* (Vienna: Kunsthalle Nürnberg, BAWAG Foundation Vienna, 2007), 167.

represents a shift towards trusting "in art's practice potential to convey experiences, set processes in motion, and open up social spaces that expand the horizon of existing conditions."[21]

During a time of struggle for gay rights, Felix Gonzales-Torres created *Untitled (Perfect Lovers)* (1987/1990), a sculptural work consisting of two ordinary wall clocks, synchronized and displayed side-by-side. With the assumption of one falling out of sync or stopping before the other, the intimate work responds to his partner's terminal AIDS diagnosis by not only publicly raising awareness of the pandemic but also suggesting "the dream of victory of the moment over eternity, of love over death."[22] The work served as a poetic gesture with political ambitions: Gonzales-Torres sought to produce an open field of imaginative potential, proving that the layering of romanticism on conceptual methodologies is about more than just sentimentality.

I Need To Be Closer To You (DST IV) works within a Romantic Conceptualist tradition, proposing the possibility of temporal transcendence through a collaborative, conceptual act. By recognizing the difference between each region's temporal rules and configurations, Tyler and I draw a framework, set parameters, and establish a procedure to achieve the project's objective. Our discrepancy is both experiential and represented.

In real-time during our performance, Tyler and I feel the dilated distance firsthand at the moment the clock advances an hour only in my time zone, and through documentation, our shared experience transforms from an insular exchange instead to signify a more broadly understood discrepancy between time zones.

In his short essay on the work, *Notes on Temporal Displacement*, visual artist and writer Lee Henderson elaborates by saying,

> Each needs to be closer to the other, but they occupy separate, antithetical ontologies: one is linear and causal, the other looped and untouchable. Emily's time warp happens again, and again, each minute. Her jump cut is the moment where the effect of DST becomes physical, where the

---

21   Verwoert, "Impulse Concept Concept Impulse," 175.

22   Heiser also recounts the words Gonzales-Torres wrote to his partner Ross Laycock before his passing, saying "Don't be afraid of clocks, they are our time, time has been so generous to us. We imprinted time with the sweet taste of victory. We conquered fate by meeting at a certain TIME in a certain space. We are a product of the time, therefore we give back credit where it is due: time. We are synchronized, now and forever. I love you." Heiser, Jörg. "Felix Gonzales-Torres" *Romantic Conceptualism* (Vienna: BAWAG Foundation and Nuremberg: Kunsthalle Nürnberg: 2007), 80.

FIGURE 14.6   *I Need To Be Closer To You* (*DST IV*), 2018. Two-channel video, Daylight Saving
Time performance on March 11, 2018. Duration: 60 mins. First projection:
"Time in Variance," International Society for the Study of Time Seventeenth
Triennial Conference, Los Angeles, California
PHOTO © EMILY DICARLO

temporal displacement occurs for that half of the pair. She keeps reliving
that minute ... he doesn't.[23]

The Romantic Conceptualist approach offers a field of potential, encouraging
its creator to imagine and act upon those possibilities, as *I Need To Be Closer To
You* (*DST IV*) aspires to do. With Romanticism's "capacity to break open the
narrow horizon of existing conditions," Verwoert insists, "It would be wrong to
dismiss this method of potentializing as a mere mind game [...] To realise the
possibilities of thought in Conceptual works here became a means to create
intellectual, social, emotional and not least practical freedoms for oneself and
others."[24] Holding this power, Tyler and I imagine that our longing will psychi-
cally suture the gap between us. By reconstructing our recorded performances
into a collaged simultaneity in a physical space, we creatively achieve what
we set out to do. We overcome the expanded variance of imposed Daylight
Saving Time regulations and through this exercise, for a moment, we feel closer
to one another.

---

23   Lee Henderson, "Notes of Temporal Displacement." *Emily DiCarlo and Evan Tyler: I Need
To Be Closer To You*, ed. by Robyn York (Toronto: Anchorless Press, 2019) Artist multiple.

24   Verwoert, "Impulse Concept Concept Impulse," 173.

5   Untethered: Site-Specificity as Temporal Process

Having established *I Need To Be Closer To You (DST IV)*'s temporal framework and its Romantic Conceptualist ties, it is necessary to explore the role that site-specificity plays by using both traditional and expanded contemporary definitions. Site-specificity, most notably associated with the material-intensive Land Art movement during the 1960–70s, argues that a site's context is integral to a work's reading and that moving it from its intended environment destroys its meaning and essence.[25] In a rejection of the commodified art object, emphasis was placed on the phenomenological experience of the viewers, their presence in a shifting spatial relationship to the work, and the context of place it inhabited. To fully understand the nuance and intention of an Earthwork, the viewer had to physically be there, a body moving in visual dialogue with the sculpture and its environment. Experiencing Land Art was a durational encounter. In many Earthworks, entropy became a central theme within these phenomenological sites housing minimalist installations. For example, in Robert Smithson's *Partially Buried Woodshed* (1970), the artist covered half of a decaying woodshed with 20 truckloads of soil until its center beam cracked, setting the scene for its inevitable collapse years later. While many Earthworks explored time through entropic processes, works like Nancy Holt's *Sun Tunnels* (1973–76), an installation comprising four large concrete cylinders arranged to align with the sunrise and sunset on summer and winter solstices in Utah's Great Basin Desert, provide an early example of a time-specific work, underlining that to fully experience it, the viewer must witness the piece on a particular day and time.

Concurrently, site-specificity took on new forms in immaterial, performance-based practices, where communal and shared experiences became the focus. Allan Kaprow, a major figure in Fluxus, developed "Happenings" or scripted events for specific locations where performers and audiences followed

---

25   The sanctity of site-specificity was legally tested in 1981 with Richard Serra's famously controversial work, *Tilted Arc*. Commissioned by the General Services Administration (GSA) for the Federal Plaza in lower Manhattan, the 120-foot long, 12-foot high solid steel, curved, wall-like structure caused extensive public debate and led to an eight-year legal battle advocating for its removal and relocation. Serra argued that the site's context became part of the work and integrally completed it, stating that the work could therefore not be moved without destroying its essence. After a drawn-out public dispute, the sculpture was finally removed in 1989 and at Serra's request has never been exhibited since. Harriet Senie. "Richard Serra's 'Tilted Arc': Art and Non-Art Issues." *Issues in Public Art* 48, vol. 4 (1989): 300.

instructions to experience an art piece. In his happening *Eat* (1964), the event took place in a log hut and small cave with a running stream, where participants were invited to drink wine and eat food prepared in real-time by the performers. Created by another prominent member of the New York art movement, Yoko Ono's *Morning Piece* (1965) originated as an instructional text-based work but was later developed as an invitation to a three-day event on her apartment building's rooftop. There, she distributed small shards of glass attached to pieces of paper, each reading a future date with specific period of the morning, such as "until sunrise," "after sunrise," or "all morning," and encouraged participants to look through the glass to the sky at the designated times – always during periods of transition – to contemplate our universal connection through daily rhythms. Susan Hiller, another Conceptual artist dabbling in romantic tendencies, created the site-specific event *Dream Mapping* (1973), where seven participants slept for three consecutive nights in an English meadow marked with "fairy rings" or circles of Scotch bonnet mushrooms, an event that spoke back to British folklore. Each participant's dreams were charted the morning after, and, at the end of the performance, Hiller layered the group's information to produce a single, shared dream map – a portrait of a collective subconscious in a transitory state.

As notable precursors of contemporary art practices today, these works provide an evidentiary shift in the way we understand the limits of site-specificity and its temporal dimension. Eventually, a rejection in the privilege of place occurred. Moving away from corporeal, in-situ encounters instead to abstracted sites of knowledge production, curator Miwon Kwon explains that site-specificity's definition "has been transformed from a physical location – grounded, fixed, actual – to a discursive vector – ungrounded, fluid, virtual" and notes that "the guarantee of a specific relationship between an artwork and its site is not based on a physical permanence of that relationship [...] but rather on the recognition of its unfixed impermanence, to be experienced as an unrepeatable and fleeting situation [...] the 'work' no longer seeks to be a noun/object but a verb/process."[26] The definition's migration from a passive to active status implies that a project's site can now be located in a liminal space of becoming or be experienced through a durational unfolding. Demonstrating a rhizomatic evolution, the untethered concept migrates from a static territory into a dynamic field, open for interpretation and implementation.[27]

---

26   Miwon Kwon, "Genealogy of Site Specificity," *One Place after Another: Site-Specific Art and Locational Identity* (Cambridge: MIT Press, 2002), 24, 29–30.

27   In botany, a rhizome is a continuously growing horizontal underground root system that produces lateral shoots with no distinct core, resulting in expanding interconnected

In the case of *I Need To Be Closer To You* (*DST IV*), the site-specificity of the work shifts focus from the two physical locations to a third shared site of exchange. While the geographic locality of Regina, Saskatchewan, and Toronto, Ontario, still hold importance in regards to the variance found between their DST time changes, the real site is *elsewhere*, positioned in the hour-long inter-subjective unfurling. Within this third alterior space, site-specificity emerges through real-time, remote interfacing. In Kwon's words, sites such as this one manifest as a "field of knowledge, intellectual exchange, or cultural debate [... and] unlike in previous models, this site is not defined as a precondition. Rather, it is generated by the work (often as 'content'), and then verified by its convergence with an existing discursive formation."[28] Meaning is generated through our exchange, producing a shared authorship through collaborative, dialogic action.

This expanded concept of site-specificity originates with James Meyer's notion of the "functional site," whereby he argues that site-specificity is structured "intertextually" rather than spatially, instead acting as "a process, an operation occurring between sites, a mapping of institutional and discursive filiations and the bodies that move between them (the artist's above all)."[29] He goes on to describe it as "an informational site, a palimpsest of text, photographs and video recordings, physical places and things [...] a temporary thing; a movement; a chain of meanings devoid of a particular focus."[30] *I Need To Be Closer To You* (*DST IV*) presents a vectored relationship of literal and functional sites: the instructional text of the event score, the happening of collaboration, the recorded artifact of the performance, the video installation in the

---

pathways. Although now this term is often used in contemporary network theory to describe complicated telecommunication systems, philosophers Gilles Deleuze and Félix Guattari first leveraged this term to describe non-hierarchical forms of theory and research. In the introduction of *A Thousand Plateaus*, the second volume of their Capitalism and Schizophrenia project, they describe contemporary knowledge production by saying, "The rhizome operates by variation, expansion, conquest, capture, off-shoots [...] the rhizome pertains to a map that must be produced, constructed, a map that is always detachable, connectable, reversible, modifiable, and has multiple entry-ways and exists in its own lines of flight ... the rhizome is acentered, non-hierarchical, nonsignifying system without a General and without an organizing memory of central automaton, defined solely by a circulation of states." Gilles Deleuze and Felix Guattari, *A Thousand Plateaus*, translated by Brian Massumi (Minneapolis: University of Minnesota Press, 1987), 21.

28 Miwon Kwon, "Genealogy of Site Specificity," 26.

29 James Meyer, "The Functional Site; or, The Transformation of Site Specificity," *Space, Site, Intervention: Situating Installation Art* (Minneapolis: University of Minnesota Press, 2002), 25.

30 James Meyer, "The Functional Site; or, The Transformation of Site Specificity," 25.

gallery, and finally this published critique. In short, the project's specific site is an interwoven, sum of its parts.

6   Both Here and There: Time-Specificity, Telepresence, and Desire

Following in the performative art practices of the 1960s–70s, a movement that valued relational encounters and the immediacy of happenings, the success of *I Need To Be Closer To You* (*DST IV*) relies on the emphasis it places in real-time presence. Historically, this artistic interest in *liveness* flourished from the proliferation of new communication technologies and the newfound ubiquity of broadcast television in households. With homes connected to events happening on the other side of the world, distance collapsed, and a new temporal standard was established. Conceptual Art's obsession with the immediacy of time echoes the excitement during the nineteenth century with the telegraph's introduction, an invention able to travel great distances almost instantly through electrical pulses over the wire. "This is indeed the annihilation of space," the Philadelphia Ledger said in response to the first long-distance telegram in the United States. The acknowledgement of this space-time compression, one that would only increase with the Internet and other wireless technologies, held awe in its metaphysical overcoming.[31] In tandem with the creation of the railway and subsequentially Coordinated Universal Time zones, the relationship between communication and transcendence is undeniable.[32]

Collaboration through long-distance communication has a rich history in Canadian visual art. While Fluxus and Land Art practices were in full force in the United States, one of the first artworks to leverage technology to facilitate dialogic creative exchange was *Trans V.S.I.* (1969) by Iain Baxter from the Vancouver-based art collective N.E. Thing Co. Connected through telephones and fax machines using telex, a network operating on telegraph-grade circuits, Baxter relayed creative prompts to graduate students at the Nova Scotia College of Art and Design in Halifax. During the course of the three-week experiment, Baxter sent instructions such as "live in Vancouver time," prompting the

---

31   First referenced by geographer David Harvey in 1989, "space-time compression" builds on Karl Marx's theory of the "annihilation of time and space" and refers the felt effects of technological innovations that reduce both spatial and temporal distance. David Harvey, *The Condition of Postmodernity: An Enquiry into the Origins of Cultural Change* (Cambridge: Blackwell, 1990).

32   See Lee Rodney. "Long Distance Performance: Autobiography and Globalization." *A Journal of Performance and Art* 34, no. 2 (2012): 27.

students to time-shift by four hours, document their experiences and report back via fax.

Playing with the conceptual limits of time travel, Fluxus artist Ken Friedman created *In One Year and Out the Other* (1975), a dialogical telephone performance that allowed participants to magically connect two years apart in real-time. Executed on New Year's Eve, the time-specific performance score instructs participants to "make a telephone call from one time zone to another to conduct a conversation between people located in different years."[33] Recently celebrating its 45th anniversary, this event has been performed every year since its first iteration and now includes the use of computers, tablets and cellphones to post and send digital copies of the performance score, along with photos from wherever the participants are at midnight.

While the idea was not articulated until 1991 by artist and theorist Roy Ascott, *telepresence* describes the condition of "being both here and there" when one is engaged with telecommunication technologies, such as video conferencing. Telepresence bridges the dilated temporal difference and lengthens our reach through space and time, producing a quantum effect of being in two places at once. When considering the potential of collaboration through this medium, this phenomenological conundrum of the user's bodily status invites expanded site and time-specific intervention. New media artist Eduardo Kac argues that "Dialogical telepresence events combine self and other in an ongoing interchange, dissolving the rigidity of these positions as projected remote subjects. Telepresence art has the potential to conciliate the metaphysical propensity of cyberspace with the phenomenological condition of physical environments."[34] By redefining shared experiences across distant geographies, telepresence encourages empathy, reciprocity and mutual understanding.[35] Theatre critic Wilmar Sauter developed his "theory of yearning in time-specific performance" as a response to Kista Theater's multi-site production *Lise & Otto* (2012), where actors simultaneously performed in separate venues while being digitally broadcast in real-time to two sets of live audiences.[36] Drawing on telepresence's phenomenological potential and the fundamentals of site-specific performance, he explains that "a time-specific presentation is based on the

---

33 See Ken Friedman, *52 Events* (Edinburgh: Show and Tell Editions), 118.

34 Eduardo Kac, "Negotiating Meaning: The Dialogic Imagination in Electronic Art," *Bakhtinian Perspectives on Language and Culture; Meaning in Language, Art and New Media* (New York: Palgrave Macmillan, 2004), 216.

35 See Roy Ascott, "Connectivity: Art and Interactive Telecommunications," *Leonardo* 24, no. 2 (1991), 116.

36 Wilmar Sauter, "Interference between Present and Absent Performers: Time-Specific Performance as Phenomenal Experience." *Nordic Theatre Studies* 24 (2012): 77.

simultaneity of actions happening in different places. From a phenomenological point of view, time-specificity creates a new kind of experience, in which absence plays as dominant a role as presence. Presence creates a yearning for the absent."[37] As a result of one of the performer's bodily absence, desire and longing become the connective tissue between what is and is not there. Sauter further describes that this "unification of absence and presence finds a theoretical explanation in the phenomenon of time. The split stage is united by the timeframe of the performance [... and] becomes time-specific." Our shared experience redefined through networked lines, Tyler and I perform both in the *here* and *there*, our temporal positions blurring between geographic sites. As our bodies extend, cross, double and interface, *I Need To Be Closer To You (DST IV)* proves that through repetition and will, our coordinated performance can unify and amend our temporal distance.

7   Conclusion

German Conceptual artist Joseph Beuys famously proposed that every human being is an artist, concluding that "only art is capable of dismantling the repressive effects of a senile social system that continues to totter along the deathline."[38] *I Need To Be Closer To You (DST IV)* is driven by much more than a desire for togetherness; it is about autonomy, freedom, and the release from an enforced and burdening system. By recognizing our potential as artists and by considering our capacity to dream and create, we are afforded the power of meaning-making in this world. Believe in the delimiting force of our imagination and its potential to actualize alternative realities. Perhaps, if only for a brief moment, you too will find autonomy and step outside the system's temporal standard.

References

109th U.S. Congress. "Sec. 110. Daylight Savings." *An act to ensure jobs for our future with secure, affordable, and reliable energy.* United States Government Printing Office. Pub. L. 109–158, 119 Stat. 594, enacted August 8, 2005. Accessed March 8, 2019. https://www.congress.gov/109/plaws/publ58/PLAW-109publ58.pdf.

---

37   Wilmar Sauter, "Interference between Present and Absent Performers: Time-Specific Performance as Phenomenal Experience," 83.
38   See Joseph Beuys. "I Am Searching for Field Character." Society into Art, London, 1974. Exhibition catalogue.

Ascott, Roy. "Connectivity: Art and Interactive Telecommunications." *Leonardo* 24, no. 2 (1991): 116.

Benedicktsson, Karl and Stanley D. Brunn. "Time Zone Politics and Challenges of Globalisation." *Royal Dutch Geographical Society KNAG*, vol. 106 (2015): 276–290.

Beuys, Joseph. "I Am Searching for Field Character." *Society into Art*. London, 1974. Exhibition catalogue.

Botting, Gary. "Happenings." *The Theatre of Protest in America*. Edmonton: Harden House, 1972.

Brettkelly-Chalmers, Kate. *Time, Duration and Change in Contemporary Art: Beyond the Clock*. Bristol and Chicago, IL: Intellect, 2019.

Coren, Stanley. "Daylight Savings Time and Traffic Accidents." *The New England Journal of Medicine* 334, no. 14 (1996): 924–925.

Crary, Jonathan. *24/7: Late Capitalism and the Ends of Sleep*. London; New York: Verso, 2013.

Deleuze, Gilles and Felix Guattari. *A Thousand Plateaus*, translated by Brian Massumi. Minneapolis: University of Minnesota Press, 1987.

Franklin, Benjamin. "Essay on Daylight Saving: Letter to the Editor of the Journal of Paris 1784." *WebExhibits online museum*. Accessed on February 18, 2019. http://www.webexhibits.org/daylightsaving/franklin3.html.

Friedman, Ken. *52 Events 2002*. Edinburgh: Show and Tell Editions, 2001.

Government of Saskatchewan. "Tools, Guides, and Resources for Municipalities: Saskatchewan Time System." Accessed on March 10, 2019. https://www.saskatchewan.ca/government/municipal-administration/tools-guides-and-resources/saskatchewan-time-system.

Harvey, David. *The Condition of Postmodernity: An Enquiry into the Origins of Cultural Change*. Cambridge: Blackwell, 1990.

Heiser, Jörg. "A Romantic Measure." In *Romantic Conceptualism*, edited by Ellen Seifermann, 134–149. Vienna: BAWAG Foundation and Nuremberg: Kunsthalle Nürnberg: 2007.

Henderson, Lee. "Notes of Temporal Displacement." In *Emily DiCarlo and Evan Tyler: I Need To Be Closer To You*, edited by Robyn York. Toronto: Anchorless Press, 2019.

Honoré, Carl. *In Praise of Slow: How a Worldwide Movement Is Challenging the Cult of Speed*. Toronto: A. A. Knopf Canada, 2004.

Hudson, G. V. "Art. LVIII – On Seasonal Time." *Transactions and Proceedings of the Royal Society of New Zealand* 31 (1898): 577–583. Accessed on April 4, 2020. http://rsnz.natlib.govt.nz/volume/rsnz_31/rsnz_31_00_008570.pdf.

Kac, Eduardo. "Negotiating Meaning: The Dialogic Imagination in Electronic Art." In *Bakhtinian Perspectives on Language and Culture; Meaning in Language, Art and New Media*, edited by Finn Bostad, Craig Brandist, Lars Evensen Sigfred, and Hege Charlotte Faber, 199–216. New York: Palgrave Macmillan, 2004.

Krauss, Rosalind. "Sculpture in the Expanded Field." *October* 8 (1979): 30–44.

Kwon, Miwon. "Genealogy of Site Specificity." *One Place after Another: Site-Specific Art and Locational Identity.* Cambridge: MIT Press, 2002.

LeWitt, Sol. "Paragraphs on Conceptual Art." *Artforum,* vol. 5, no. 10 (June 1967): 79–83.

LeWitt, Sol. "Sentences on Conceptual Art." *Art-Language: The Journal of Conceptual Art,* vol. 1, no. 1 (May 1969): 11–13.

Ljung, Rickard and Imre Janszky. "Shifts to and from Daylight Saving Time and Incidence of Myocardial Infarction." *The New England Journal of Medicine* 359, no. 18 (2008): 1966–1968.

Meyer, James. "The Functional Site; or, The Transformation of Site Specificity." In *Space, Site, Intervention: Situating Installation Art,* edited by Erika Suderburg, 23–37. Minneapolis: University of Minnesota Press, 2002.

Milton, Earl R. V. *A Submission to the Government of Saskatchewan Regarding Time Zoning in Saskatchewan.* 1966.

Rodney, Lee. "Long Distance Performance: Autobiography and Globalization." *A Journal of Performance and Art* 34, no. 2 (2012): 22–36.

Rowlatt, Justin. "Caesium: A Brief History of Timekeeping." *BBC News,* October 4, 2014. Accessed December 18, 2019. https://www.bbc.com/news/magazine-29476893.

Sauter, Wilmar. "Interference between Present and Absent Performers: Time-Specific Performance as Phenomenal Experience." *Nordic Theatre Studies* 24 (2012).

Schivelbusch, Wolfgang. *The Railway Journey: The Industrialization of Time and Space in the Nineteenth Century.* Oakland: University of California Press, 2014.

Senie, Harriet. "Richard Serra's 'Tilted Arc': Art and Non-Art Issues." *Issues in Public Art,* Art Journal 8, No. 4 (1989): 300.

Solnit, Rebecca. *River of Shadows: Eadweard Muybridge and the Technological Wild West.* New York: Viking, 2004.

Verwoert, Jan. "Impulse Concept Concept Impulse: Conceptual Art and its Provocative Potential for the Realization of the Romantic Idea." In *Romantic Conceptualism,* edited by Ellen Seifermann, 65–175. Vienna: Kunsthalle Nürnberg, BAWAG Foundation Vienna, 2007.

Wainwright, Lisa. "Fluxus." *Britannica Academic (Encyclopedia Britannica Online),* 2016. Accessed April 8, 2020. https://www.britannica.com/art/Fluxus.

Wallis, George "Chronopolitics: The Impact of Time Perspectives on the Dynamics of Change." *Social Forces* 49, no. 1 (1970): 103, 107.

CHAPTER 15

Temporal Experience in George Benjamin's *Sudden Time*

Martin Scheuregger

Abstract

This chapter examines the orchestral composition *Sudden Time*, by British composer George Benjamin (b. 1960) in relation to ideas of temporal perception. In his program note, the composer makes a connection to malleable time in how the work was conceived, and in so doing opens it up for analysis based on these ideas. Starting with context from various approaches to musical time, distinctions are made between how we may experience time, broadly speaking as Newtonian and chronometric, or Einsteinian and psychological. The chapter then applies these ideas to a reading of *Sudden Time* in which perceptual time is understood to be manipulated through a range of compositional devices that manifest in different "timezones" – sometimes heard successively, sometimes concurrently. The work may be seen as grounded in temporal ideas that help us understand its structures and material, whilst these very structures also illuminate temporal ideas in themselves.

Keywords

chronometric time – Einsteinian time – George Benjamin – musical composition – Newtonian time – perceptual time – psychological time – spectralism – stasis – *Sudden Time* – timezones

Sudden Time (1989–93) by British composer Sir George Benjamin (b. 1960) is a musical work that deals explicitly with the perceived flow of time. In his program note, Benjamin describes a number of features that explain why time is important. The ideas from his description of the work's origin form the starting point for this study.

> Some of the concepts behind this piece can be illustrated by a dream I once had in which the sound of a thunderclap seemed to stretch to at

least a minute's duration before suddenly circulating, as if in a spiral, through my head. I then woke, and realised that I was in fact experiencing merely the first second of a real thunderclap. I had perceived it in dream-time, in between and in real time. Although this is but analogy, a sense of elasticity, of things stretching, warping and coming back together, is something what I have tried to capture in this piece.

BENJAMIN 1997, n.p.

A reading of the work is proposed here in which our perception of the flow of time is understood to be manipulated in various ways, echoing Benjamin's experience of elastic temporality and following his noted interest in "the organic evolution of simple elements and their morphology through time" (Lack 2001, 13). Furthermore, structural and formal techniques can be understood through temporal concepts that may not translate directly into perceivable phenomena. The primary means through which time is manipulated is in material of various speeds, but whilst a traditional interpretation would see musical themes changing in relation to time as a constant, this essay suggests that we may interpret different presentations of material as changing types of time from stretched to compressed. These sections of different temporalities can be considered as different "timezones" – a term used throughout this chapter – in which material is presented in different temporal contexts. These different timezones may be present concurrently (in layers of material) or sequentially (in adjoining sections): the concept of timezones is useful for understanding the perceptual possibilities explored below. In this model, time, not speed, becomes the variable that is manipulated to affect the flow of music.

A visual analogy may help to clarify this distinction. In two different videos, a car is shown driving slowly, covering the same distance over the same amount of time. In one, this is achieved by the vehicle travelling at 30 mph; in the other, the car is driven at 60 mph, but the image is played at half speed. We see the car achieving the same thing in both videos, but in the slowed-down film we will notice people and other vehicles moving more slowly in comparison to our apparent 30 mph driver. Now, imagine filming a car ambling at 1 mph then speeding up the playback by 60 times so that it matches a video of a car driving at 60 mph. The car once again moves across the same distance in the same time in both videos, but in one, pedestrians become a blur, the pulsating rhythms of controlled traffic become apparent, and we start to perceive the fading afternoon light. The two videos once again represent the same type of event, but the means of achieving them, and the way we interpret each, is quite different. This study demonstrates that, whilst *Sudden Time* can be seen in terms analogous to a car merely driving at different speeds, it is most illuminating to consider the speed of "playback" being increased or decreased resulting in a changing interrelationship of musical timezones. We are presented with a work full of different speeds, but by considering how *time* is treated, we may come to a fuller understanding of the work.

This chapter considers a variety of perspectives on the perception of musical time before applying this to a listener-focused musical analysis of Benjamin's *Sudden Time*. This process offers a part phenomenological, part music-analytical reading of the work that proposes how a listener may understand it in relation to time. Time and movement are discussed first through reference to the spectral school of composition with which Benjamin's work can be associated. Specifically, spectral approaches to temporality are explored with a focus on perception. Various temporal dualities are explored next: chronometric and psychological time, differentiated and undifferentiated continua, stasis and movement, stretched and compressed time, Newtonian and Einsteinian time, and self-motion and relative motion. These pairs of ideas are brought together to form a basis for interpreting *Sudden Time*. The work forms the focus next, as its structure is explored in relation to types of time and various timezones. Moments of "frozen time" are outlined as structurally important first, then the discussion moves to teleological motion and the manner in which timezones are negotiated both horizontally and vertically. Conclusions are made with reference to time, speed and distance.

1 Time and Movement

There are different ways of considering time, which each have their own bearing on the perception of the work aurally and which may also influence how the ideas with which it is underpinned are considered. By recognizing these conceptual concerns, we can understand the music in ways which may not be immediately accessible from its surface alone. In acknowledging both perceptual and conceptual manifestations of time (those that we can hear and those that are more abstract), we may achieve a nuanced impression of the workings of *Sudden Time*.

The conception and perception of time are crucial to spectral music, and, given Benjamin's formative association with spectralism, framing the use of temporal devices in *Sudden Time* with some of these ideas is useful.[1] The terms

---

1　Graham Lack has provided one of the few attempts to place Benjamin's music within the context of spectralism and concludes that he is influenced by the approach but builds on it with new devices and ultimately does not use it as a method (Lack, 2001).

"spectral" or "spectralism" are rejected by many so-called spectral composers (Anderson 2000, 7) and can be understood "more as an attitude than as a set of practices" (Cross 2018, 6).[2] A simplistic explanation would be that it is an attitude that is focused on the nature of sounds, with composers often using the properties of sound to generate musical material in works that pay particular attention to timbre. Furthermore, transformations of material are important, as is the notion of liminality and the interrelatedness of musical elements – pitch, rhythm, timbre and so on – that are usually thought of as separate.

Spectralism is traditionally seen as emerging in France in the late 1970s and reflects a trend in Western art music across the twentieth century to move beyond means of musical composition commonplace in previous centuries. Earlier musical forms tend to be goal-oriented as tension is set up at the beginning that will ultimately be resolved at the end of a work. This teleological approach differs from some spectral music that instead is focused on the starting point – a single sound, for example – and moves outward from there. Jonathan Harvey – a British composer who was associated with spectralism much more than Benjamin – explains that "[t]he tonal system, unlike spectralism, exists in a temporal context" that comes from the goal orientation set up by the clear "hierarchy of components" (2000, 12). Although he goes on to explain his own hierarchies of spectra, this notion of spectral music sitting somehow outside of time is important. Others see spectralism as temporal in origin (Grisey, 2000), but it is the conscious manipulation of temporal perception through musical devices that sets this mindset apart from more traditional teleological music.

When we examine and magnify the properties of a sound – as spectralists often do – we find even the most basic musical material (an instrument playing one note, for example) is internally complex. Multiple spectra are present – different pitches that combine at specific relative volumes to create the sound we hear – that are themselves the result of vibrations at different speeds. Furthermore, the sound fluctuates in other unpredictable ways and changes from the attack of the note, through its decay, sustain, and release.[3] In short, a single apparently atemporal phenomenon like a note on an instrument can be seen as deeply temporal. In Gérard Grisey's landmark 1975 work *Partiels*, the entire opening is constructed from an analysis of one such note

---

2   The special edition of *Twentieth Century Music* that Cross is introducing reflects the multiple spectralisms that are now understood to exist.

3   These terms may be familiar to some in reference to synthesizers, which model these components of a sound to imitate specific instruments or create entirely new sounds.

(an E on a trombone) and is an ideal work to begin with for a listener wanting to becomes more acquainted with spectral music.

Whilst an affinity with the spectralists can be felt in Benjamin's language today, he differs from them in his "predilection for reassembling discrete acoustic phenomena as opposed to splitting them asunder [... and] as a writer of music based on the fusion of sonic materials, as opposed to their fission" (Lack 2015, 11). For Grisey "spectral music has a temporal origin" (2000, 1) and his thoughts on time provide a useful starting point for addressing Benjamin's approach to temporality. A fundamental distinction made by Grisey is that between chronometric and psychological time: time as measured by beats and seconds versus that experienced by a listener. Grisey believes psychological time may be shaped by musical content, explaining that, for example, by inserting an "unexpected acoustic jolt" into a passage, the "linear unfolding of time" is disrupted, causing us to feel as though *"Time has contracted"* (1987, 259; emphasis in original). By contrast, time will be felt to have expanded given an uninterrupted and predictable section. For Grisey, this perceptual expansion of time is a necessary part of spectral music, allowing a listener fully to experience the composed sound: "The more we expand our auditory acuity to perceive the microphonic world, the more we draw in our temporal acuity, to the point of needing fairly long durations" (1987, 259). Perceptual time can be stretched by compositional means, but disruptions are needed to avoid undifferentiated continuity:

> At its most extreme, if this continuity is maintained throughout the duration of a work, it is virtually impossible to memorize anything. With no prominent event making an impact on our consciousness, the memory slips. It has nothing to latch on to – hence the effect of intense fascination or hypnosis – and all that emerges is a hazy memory of the contours of the sound's evolution. Time past is no longer measurable: I would call this process psychotropic, or better still chronotropic.
>
> GRISEY 1987, 273

As with many issues related to spectral music, these temporal concerns often fuse. Indeed, time and sound become functions of each other: "spectral music no longer integrates time as an external element imposed upon a sonic material considered as being 'outside-time', but instead treats it as a constituent element of sound itself" (Grisey 2000, 2).

Grisey summarizes the temporal consequences of spectral music in five points, the first three of which are directly relevant here. He explains that spectralism has led to:

- More attentive attitude towards the phenomenology of perception;
- Integration of time as the very object of form;
- Exploration of "stretched" time and "contracted" time, separate from that of the rhythms of language.

GRISEY 2000, 2–3

These points all play out in the interpretation of *Sudden Time* presented here. First, perceptual time is essential to understanding how musical material is heard, and indeed the spectral "fascination with the psychology of perception" (Anderson 2000, 8) echoes Benjamin's own admission that he is "obsessed with harmony and the perception of harmony, from very simple fusion to extremely complicated diffraction and simultaneity" (Service 2008, 93). Second, form is articulated by various timezones and "frozen" moments which create the structure explored below and so time does become the object of form. Lastly, the stretching and contracting of time are fundamental to the work's very basis as Benjamin points out in his program note. The totality of *Sudden Time*'s musical argument is reflected in these temporal-spectral concerns, providing a strong basis for interpreting this music in terms of spectral thinking.[4]

The spectral notions of chronometric and psychological time can be enriched with further considerations. David Epstein makes an important distinction between time as a differentiated and an undifferentiated continuum. The first involves demarcation either by "mensural means external to a subject" (1979, 55) (objective), or by the experiences of a listener (subjective). An undifferentiated continuum contains no markers by which to judge the passing of time: such a passage will be perceived according to a listener's own experiential framework. In *Sudden Time*, both types are found. Extensive sections are differentiated by many sonic objects and structures, whilst others sound empty. In the extreme, some passages are so saturated with activity that demarcation of time is just as difficult as when there are no changes at all. Epstein's ideas can be expanded to reflect the correlation between perceived time, level of activity (texture), and perception of motion (Figure 15.1). When there is too much activity, the music may appear to stop. There is therefore a tension between stasis and movement that does not map directly on to textural density. This model is fundamentally linked to Grisey's comments cited above in relation to the perceptual contraction and expansion of time.

---

4   Benjamin distances himself from spectralism in some ways, claiming "He [Tristan Murail] is passionate about processes and I am against; I don't like the obvious or didactic" (Nieminen 1997, 12); however, an analysis based on spectral ideas still remains fruitful.

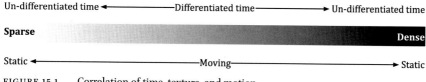

FIGURE 15.1   Correlation of time, texture, and motion

In *Sudden Time*, the context provided by the composer's program note suggests an interpretation that events are shaped by time being stretched and compressed. Such an interpretation reflects Grisey's claim that a composer may manipulate how a work is perceived through the musical material presented: the idea of differentiated time demonstrates how this may be achieved. Taking into account the composer's conceptualization of a work may appear redundant in phenomenological terms but in this case reflects the likely situation that a listener will read Benjamin's program note when hearing a performance (although perhaps this is less likely with a recording) and its specific evocation of temporal experience might affect how they perceive the music.[5]

A further distinction noted by Epstein is that between chronometric and integral time. The first is continuous, marking the regular stream of time as noted above in relation to Grisey; the latter "denotes the unique organizations of time intrinsic to an individual piece – time enriched and qualified by the particular experience within which it is framed" (1979, 57), and it is broadly comparable to psychological time. *Sudden Time* may be seen as either *suggesting* types of time (from stretched to compressed) that we, nonetheless, hear chronometrically (our internal clock keeps ticking at the same speed), or as distorting our very perception of time (where our internal clock itself speeds up and slows down). This can be broadened out to include the idea of Einsteinian versus Newtonian time explored by Michael Rofe (2014, 346–347), whose duality is fundamental to the ways in which time can be understood in Benjamin's music. Newtonian time is absolute and chronometric and can be used to observe structure, but it is the concept of malleable Einsteinian time that is particularly potent in *Sudden Time*. Rofe highlights the parameters that can be manipulated to achieve this plasticity: density of activity; type of activity; patterns of tonal shifts; "clarity and regularity of the metre"; the order of events (2014, 346–347; quote from 347). These acutely map onto Grisey's understanding of psychological time (1987, 258–259). Whilst Rofe shows these

---

5   Eric Clarke highlights the importance of the "available information for a viewer/listener" with the example of a dramatic context, but this approach is equally useful in a concert work such as this (2005, 87–88).

to be empirically true, he also points to Stockhausen's more intuitive observation that periods of low density may be perceived as stretching time whereas dense musical activity can appear to pass at a faster rate (1959, 64). This view echoes the correlations shown in Figure 15.1 and further enhances the understanding derived from spectral thinking.

We have two related concepts to deal with, therefore. *Sudden Time* can be understood as presenting different types of time that we interpret as *representative* of stretched or compressed time, or as inducing within us the sensation of time being malleable through devices that specifically manipulate our temporal experience. Is our internal clock constant or variable?

Eric Clarke considers the perception of movement either in terms of self-motion or the relative motion of objects: the former involves a listener moving with the music as if part of its narrative; the latter, a static observer perceiving objects moving relative to their own position and to each other (2005, 75–76). We are either on the train seeing the trees flash past, or by the track watching the trains rush by. This distinction mirrors that of our understanding of time as Einsteinian or Newtonian. In *Sudden Time*, the constant shifting of objects in relation to each other suggests a stationary listener observing the music moving along (or with a regular internal second hand ticking); however, the compositional warping of time also invites the listener to jump on board and experience its slowing down and speeding up "in-person" (our internal clock fluctuating).[6] Analogizing relative movement, Clarke calls upon the sensation felt when one is aboard a train and believes it has started to move, only to realize it is the adjacent vehicle that is pulling away (2005, 75). And so in *Sudden Time* a listener may be both traveling with the music and hearing it shift against their position. We are comfortable on a train, but the myriad movements of carriages on both sides means we are never sure exactly what is stationary: what the original speed is, where things are drawn out, and where they are sped up is not clear. This ambiguity is present in the temporal considerations of spectral music where the aim is often to achieve self-motion in a listener, to get them aboard the train and moving with the composer. Grisey is, in fact, hostile to the idea of a fixed listener:

> What a spatial view of musical time – but also what anthropocentrism there is in this image of a man at the center of time, a listener fixed at the

---

6   Another of Benjamin's works, *A Mind of Winter* (1981), engages with both perspectives, depending on whether one perceives its sonic objects as static in space (chronometric time continuing to pass) or as static in time (psychological time slows down).

very center of the work to which he is listening! One might say that a truly Copernican revolution remains to be fought in music....

GRISEY 1987, 242–243

We can understand *Sudden Time* as moving freely between different temporal approaches. We are not fixed to seeing *Sudden Time* as suggesting types of time or experiencing the feeling of time being manipulated. It does both, so we may find ourselves regularly alighting and boarding the various perceptual trains, or our perceptual clock may sometimes run regularly and at other times speed up and slow down.

Intrinsically linked to time is tempo. In another context these terms might be mathematically related – speed = distance ÷ time – but, whilst music fills a certain duration, we cannot talk about its speed in such literal terms, whilst the notions of space and distance are even more nebulous.[7] An important observation in this context is that "sounds specify motion by means of change" (Clarke 2005, 75). *Sudden Time* features a continual shifting of parameters, pulse in particular changing regularly and existing in layers of different tempi (or timezones). Motion and time coalesce here, as we can understand *Sudden Time* as distorting time rather than using musical ideas that move at different speeds. The equation can be rearranged to reflect this: time = distance ÷ speed. Distance and speed become variables that can be manipulated by the composer to create the desired length of time, or – in *Sudden Time* – type of time (expanded, contracted or regular, as Benjamin writes in his program note). Although the music may be palpably fast, slow, or both at once, we are invited to perceive these variations in terms of time itself being altered: the change in speed or space (distance) is a result of the composer's active shaping of time. Speed and distance act as functions of time, and it is time which may be most usefully considered as the primary domain of the work's musical argument.

2 Stasis: Frozen Time as Formal Scaffolding in *Sudden Time*

Temporal concerns explored thus far relate to ways of perceiving a musical work and have encompassed a range of perspectives that a listener may take. Many of these have related to the concept of movement, particularly whether a work can be seen as moving against a consistent chronometric grid or through a variable stream contingent on a listener's perception. Stasis is also important

---

7 The idea of motion in music has been usefully questioned: see Adlington 2003 and Clarke 2005, 62–90.

in *Sudden Time*, and, just as movement can be created through different temporal lenses, so stasis can be seen as either a pause in motion as time continues or as stopping the clock. Whichever is applied, static and moving material are both important here. Time and movement act as structure-forming parameters as *Sudden Time* exhibits a form guided by the motion of material and through the structural use of different timezones.

In his program note, Benjamin says that *Sudden Time* "basically divides into two continuous movements, the first (lasting about five minutes) acting as a turbulent introduction to the second, where a subliminal metre is perpetually distorted and then re-assembled" (1997, n.p.). The second movement starts at the second bar of rehearsal mark N: there is a pause marked "long c. 15 seconds" during which the strings hold an eight-note chord at *ppp* so that when the music recommences, a sense of newness is achieved. The tempo also drops from ♪ = 90 to ♪ = 48, and the level of activity reduces. The frozen chord allows this change of movement to be perceived readily and appreciated as significant. It provides one of a number of successive moments of stasis that I refer to as "frozen time." Each moment demarcates one section from the next, providing aural pillars between different types of material. These give a listener the opportunity to absorb the sound and understand what is happening in a manner similar to that described by Grisey (1987, 259), and each presents stasis or frozen time in various ways.[8] The formal frozen moments may be perceived through chronometric or psychological time, but in the reading presented here, considering these moments as freezing or suspending time (our internal clock stopping for a moment) is most fruitful. Figure 15.2 outlines their placement and basic features, Figure 15.3 shows the pitch material of each, whilst Figure 15.4 demonstrates the two-movement form that they delineate.

The second instance of frozen time (henceforth referred to with the roman numeral labels from Figures 15.2 and 15.3) is the most striking, and it clearly marks the start of the second movement: I, on the other hand, maintains a degree of internal movement that provides a sense of continuity with the previous and following music. Subsequent moments have similar levels of impact on the perceived flow of time: III is a short hiatus more than a structural change, although it still marks a pivot from one type of material to the next; IV echoes II in its complete stillness and similar texture, and could very well mark the start of a new movement.

---

8   These moments echo the icicle-like repeated chords which open Benjamin's *A Mind of Winter*. These give a sense of physical freezing through "cold" sonorities – some of which are found in *Sudden Time* – as well as musical material with little or no harmonic movement.

Label	Figure*	Timing	Duration	Description
I	3 < L	2:55	8"	Sparse held notes then low, quiet double bass triplets
II	N	3:41	9"	Held *ppp* chord in strings during long pause
III	T+3	5:15	3"	*ff* bass drum sparks brief G.P.
IV	1 < W	5:47	9"	Held *ppp* chord in strings during pause
V	5 < G¹	8:23	7"	Held chord in flutes and strings with limited internal movement
VI	R¹	10:40	10"	Held unison F♯ in piccolos gradually melting into further notes (6 bars)
VII	V¹+3	11:35	10"	Held unison B with a lot of internal timbral flux
VIII	Z¹	12:24	14"	Held string chord with movement coming from tabla (3 bars)

*) '<' indicates bars before subsequent figure; '+' indicates bar of that figure.

FIGURE 15.2   *Sudden Time*, table of "frozen" moments
ALL DURATIONS ARE TAKEN FROM BENJAMIN 2005, TRACK 6

Frozen moments II and IV act as aural pillars, framing a section that may be seen not only as a transition from the first movement to the second, but as a means of large-scale transformation of material types. II is presented in the strings *ppp*, but, whilst IV maintains this basic profile, it consists of string harmonics, creating an icy sonority more delicate than II. Furthermore, whilst both use chromatic saturation, II includes major seconds and IV further adds minor thirds and sixths. An upwards shift in register occurs between the two, the basic tessitura moving by an octave. These two variations on frozen time embrace the idea of stillness completely: marked non-vibrato, the strings maintain stasis by limiting the micro-variations in pitch that might otherwise be exploited by spectral means to generate momentum.

Towards the end of the work, VI and VII bookend the climax. Here, the frozen time of the earlier developmental section is condensed into single pitches played with ferocious intensity and internal sonic flux (F♯ in VI, and B in VII). The effect of freezing is the same, but the quality of this stillness is markedly different. Whereas in the previous example, every element remains alike, in VI and VII – despite much simpler pitch material – there is a high degree of flux, as instrumental color constantly changes. But, rather than creating momentum, this results in equilibrium, seen in the intricate writing of the three piccolos and underlying celli of VI (Figure 15.5), with similarly dense changes of color used in VII. If the sound is frozen in II and IV, then a thaw has taken place here; with this extra freedom comes movement, but the music is swimming against the tide and remains motionless. The feeling of stretching out a moment is truly achieved in these examples of what we can call "dynamic stasis." Their connection with instrumental synthesis in spectral music is

FIGURE 15.3   *Sudden Time*, frozen time

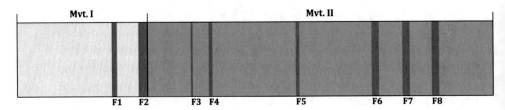

FIGURE 15.4   *Sudden Time*, simple form

FIGURE 15.5   *Sudden Time*, frozen time VI
© COPYRIGHT 1997 BY FABER MUSIC LTD, LONDON. REPRODUCED BY KIND
PERMISSION OF THE PUBLISHERS

unmistakable,[9] and they act as clear examples of a feature of spectral music that Joshua Fineberg describes as "micro-variations [in] instrumental timbres, mimicking this attribute of natural sounds" (2000, 90).

Moment VIII heralds a final sectional change. Coming after a melting away from the climax, this moment of stillness is mediated by the sonority of tablas, so far unheard in the piece. The drums vitalize the otherwise static harmony, giving a layer of movement to a sonority reminiscent of V. In this final section (from VIII), the basic frozen sound is in place throughout, with a solo viola line providing a horizontal, melodic focus. There are two simultaneous timezones here: the chords that slowly transform but remain essentially frozen, and the soloist proceeding through unmediated, "normal" time.[10] This

9   The term "instrumental synthesis" refers to the use of live instruments to recreate a sound. Each instrument might, for example, be given a constituent pitch derived from the spectral analysis of a sound, with microtonal tunings and balancing of relative dynamics used to mimic the original. The opening of Grisey's *Partiels*, discussed above, is a quintessential example of such a technique.

10   Once again, a precursor to this is found in *A Mind of Winter*, where a tension between active horizontal and static vertical material is exploited, especially during the opening section.

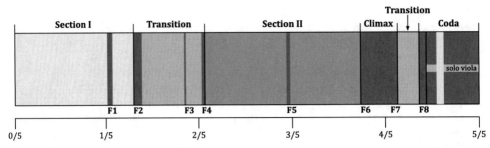

FIGURE 15.6   *Sudden Time*, nuanced form

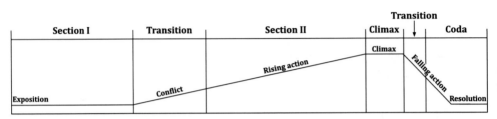

FIGURE 15.7   *Sudden Time*, dramatic form

conclusion to *Sudden Time* summarizes the interactions of different timezones heard throughout; it juxtaposes static and mobile material and subtly transforms the overall texture from frozen to moving. Up until this point, moments of frozen time have acted as pillars to delineate structure; the coda inverts this relationship by swapping the frozen pillar for a brief section of fast activity (three bars before rehearsal mark B2 to the fifth bar of B2) interrupting the otherwise frozen flow. The coda places frozen and flowing time together in an inverted microcosm of the piece.

Figure 15.6 sets out a more nuanced formal scheme which takes into account the impact of each point of frozen time. Although the two-movement model is useful, this reading better reflects the segmentation suggested by the changes within the second movement. Bringing the impressions from both models together, it is possible to observe an evolving, even organic, structure. Reflecting this organicism, a narrative view can be taken. An exposition is followed by a passage of conflict between different states of time and between types of thematic material; next, a section of rising action sees ideas develop, shift, and feed into the climax which follows; in the second transition and coda, tensions begin to resolve as textures melt away through a period of falling action to reveal the transparent texture of the final resolution. This version is outlined in Figure 15.7.

This fluent narrative thread overlays the block-like structure suggested by the moments of frozen time. However, the distinct structure underpins these fluid changes and gives the work its fundamental foundation. On the surface, there is evidence of the guiding shape outlined by the frozen moments, but their scaffolding does not dictate every texture and angle of the exterior. This reading appears to tally with Benjamin's own thoughts on structure and form:

> A really organic form is not constrained by breaks, it flows over them. Structure is the passage of material through time. Material in time can't always bump into partitions, or else the narrative of the music will be constantly broken.
>
> NIEMINEN 1997, 24

Such smooth transitions have been noted in relation to earlier works too: Benjamin's breakthrough work, *Ringed by the Flat Horizon* (1980), is "a piece of many beautiful sounds and very few hard edges" (Schiffer 1988, 78) that "aspire[s] to a Carter-like atonal athematicism which is organically evolutionary rather than statically non-developmental" (Whittall 1983, 138). These observations – noted in quite critical appraisals of the work of a composer only twenty years old – prefigure traits of organic development found across Benjamin's work since this formative composition.

Adding more descriptive detail to the formal schemes already outlined provides an overall representation of the piece (Figure 15.8), which reflects the continuity between sections. This will act as the starting point from which to look to further detail.

3   Teleological Motion

Benjamin's music is energetic and mobile, and even its most tranquil moments can be composed of highly active lines. During many static sections in *Sudden Time*, the level of detailed instrumental writing – the granularity of the texture – suggests fast motion, like the black and white dots of static television noise that move frantically but create an immobile image. Crucially, though, whilst such moments suggest motion, they do not indicate trajectory. In order to create the latter, there must be a sense of directed change: it requires teleological movement. These types of motion may be perceived as distinct, with words such as "busy" being used to describe motion without

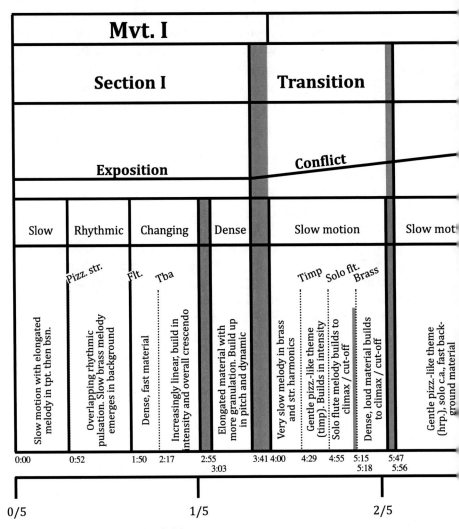

FIGURE 15.8   *Sudden Time*, detailed form

trajectory and "progression" to talk about teleological movement. In *Sudden Time* we can observe both types of movement separately and in combination. There is goal-oriented movement in the outline of the dramatic form of the work explored above, whilst movement without trajectory is seen at points of climax and stasis. Both types of movement are explored next.

From the fifth moment of frozen time (rehearsal mark F1) through to the coda at rehearsal mark Z1, there is a portion of rising action, a climax and the subsequent falling action (outlined in Figure 15.8). The rising and falling parts have a trajectory, whereas the climax – whilst containing significant internal motion – is formally static and non-developmental. These demonstrate the

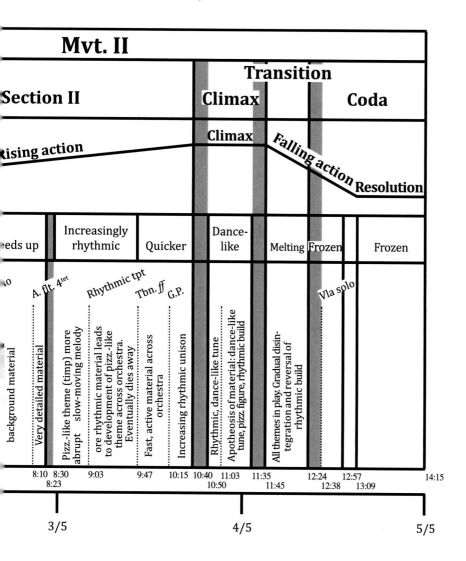

type of organic shifts and the treatment of time and motion used throughout. The segment of rising action nearest the climax (rehearsal marks P1–R1) shows the most palpable transformation from one state to another as the focus shifts from a complex, polyphonic texture to a clear, ascending line. This is a subtly orchestrated change that involves the interaction of two timezones, each of which contain two types of material. These four layers interact to achieve this sophisticated transformation.

Figure 15.9 outlines the pitch material of the melodic line that emerges over this transition, making clear the switch from falling to rising motives. This texture is complicated from rehearsal mark Q1, where the strings maintain

FIGURE 15.9   *Sudden Time*, rising action into climax

a downward trajectory, moving through the sequence of pitches with larger leaps: the upper strings, for example, move down via sixths, in contrast to the overall rising in thirds. This movement becomes less towards rehearsal mark R1, dovetailing with the more obvious rising of the percussion, harp, and piano. This Escher-like construction conflates rising and falling gestures, relating to the idea of a Shepard tone (something explored in various spectral works).[11] These gestures are overlaid, creating a cumulative ascent of pitch and demonstrating that even in this relatively transparent texture, there is internal motion that, in this case, runs counter to the overall trajectory. Figure 15.10 shows a representation of how pitch is perceived, with a Shepard tone for comparison.

Throughout this section, the woodwinds and brass similarly employ upward and downward gestures, but these inhabit a different timezone characterized by

---

11   A Shepard tone is an auditory illusion in which we perceive the pitch of a sine tone as infinitely ascending or descending. It is created through the superimposition of sine tones an octave apart that smoothly increase (or decrease) in pitch and gradually fade out as they reach their highest or lowest point.

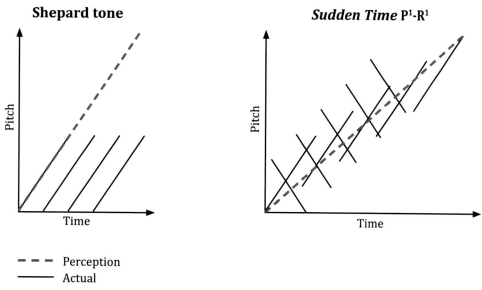

FIGURE 15.10   Shepard tone versus cumulative rising of *Sudden Time*

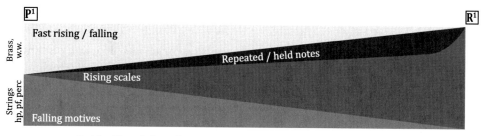

FIGURE 15.11   *Sudden Time*, shifts in focus

faster rhythms (semiquavers and dotted semiquavers). Even within this instrumental layer, a further transformation takes place, as the lines gradually morph into repeated and held pitches. The two instrumental strata (strings plus percussion, and brass with woodwind) can be considered as separate timezones within which changes of focus between material takes place (Figure 15.11). They are initially heard together, but one gradually becomes the main focus before the frozen stillness of rehearsal mark R1 takes over (Figure 15.12).

There are multiple shifts taking place within and between distinct strata of activity, and whilst the result is a complex soundworld and an intricate score, the aural result is one of clarity. Benjamin's use of granulated detail gives a higher resolution to his sound, allowing changes from one state to the next to be gradual. This type of transformative process is found throughout *Sudden Time*, giving an ebb and flow of textures, sonorities, and themes, which can be

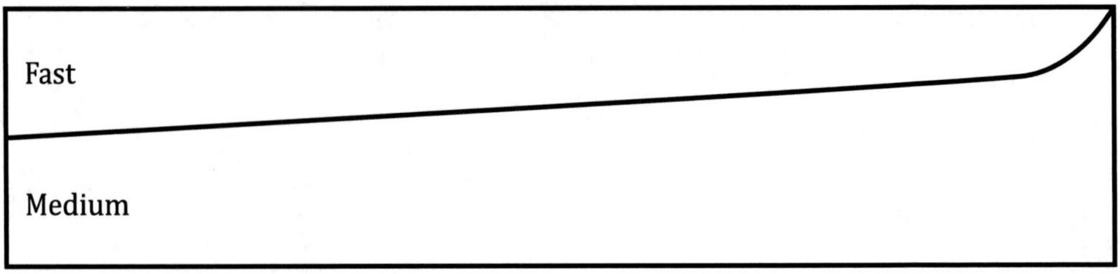

FIGURE 15.12   *Sudden Time*, shifts in timezone

perceived in terms of the multiple approaches to temporality (and the resulting different timezones) explored above. That a piece called *Sudden Time*
exhibits seemingly organic shifts brings into question what is "sudden" about
it. I would suggest that the suddenness is found, if anywhere, in the moments
of frozen time which provide structural jolts that are, to some degree, suggestive of Benjamin's sudden awakening from his dream as described in the program note.

4   Internal Movement: Maintaining Intensity

A very different kind of movement takes place at the climax, a section running
from rehearsal mark R1 to W1. Here, the horizontal contrasts are similar to those
of the preceding section, but, rather than creating trajectory, constitute a formally static plateau (with non-directed internal movement), which maintains
the intensity required of a climax. The section is bounded by two moments
of frozen time in which first an F♯ then a B forms the focal pitch. The music
between these points contains three basic thematic groups: static but dense
chords (such as the F♯ and B); a dance-like theme; and rising gestures. These
coexist rather than interact and, although they undergo a degree of blending, create internal rather than goal-directed motion through their exchanges.
Overall there is a move from a section of dense internal movement with F♯ as
the focal pitch, to a similar texture on B, with other tonal centers acting as step-
pingstones in between (Figure 15.13). In relation to other of Benjamin's works
in a spectral context, Graham Lack points to the high degree of energy within a
"static" sound (2001, 11), furthermore pointing to Manuel Rocha Iturbide (1995)
whose theories support this interpretation.

  A section is created that maintains a singular focus by constantly returning to similar sonorities, but retains movement through its internal activity.
The shifting of pitch centers gives some sense of change – indeed a skeletal
progression from dominant to tonic in B minor could be inferred – but the

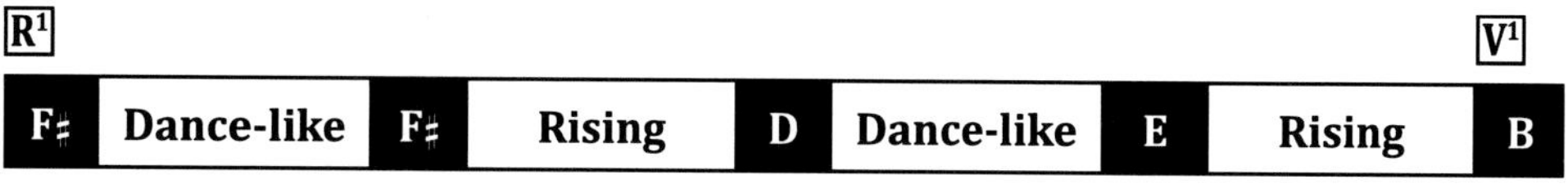

R¹									V¹
F♯	**Dance-like**	F♯	**Rising**	D	**Dance-like**	E	**Rising**	B	

FIGURE 15.13   *Sudden Time*, R1–V1, formal stasis in the climax

rondo-like return of the same texture gives a sense of *formal* stasis. It is certainly as dynamic and exciting as the build-up that precedes it, but its function is to freeze the action in one state, not to progress the work on a formal level. In the section of falling action that follows (rehearsal mark W1–Z1), themes and ideas start to show signs of interaction, and, by the end (Y1–Z1), the rising/falling idea that led into the climax reappears. This transition mirrors the first appearance of the themes and, by reinstating a more teleological interaction, heralds the beginning of the coda.

Although the contrasts and internal movement of the climax are closer to the dense than the sparse end of the continuum outlined in Figure 15.1, a sense of differentiated time is still created. The music does not sound static – there are no frozen sonorities, no held chords, no overtly slow melodies – but, to revisit Rofe's parameters, the localized structure of the section is repetitive. The image of flickering static may be useful again, but, whereas other sections create stasis through non-directed internal movement, here the flickering is at a formal level as we hear material that is seemingly unable to progress. At this moment of formal and narrative significance, time stands still.

It would be difficult to claim that this climax section can be *heard* as static, but given the concentration on matters of time, it might be considered as a conceptual moment of stretched time. Just as the idea of stretched time elsewhere highlights the granular detail of the sound, so the process used here reveals the motion possible within stasis. Returning to the earlier analogy of filming a car at different speeds, we might consider this section as conceptually related to the effect of perceiving detail through extreme slow or fast motion. Whilst the perception of the flow of time is crucial to the work, moments like this demonstrate that concepts that may not be immediately audible have an impact on how we might understand this music.

5   Stratification and Multiple Speeds

The analogy used earlier of parallel trains moving at different rates is a good starting point for looking at stratification in *Sudden Time*. Whilst this analogy was invoked in relation to speed, it may usefully be seen with connections to a variety of parameters, as material types, textures, thematic groups, and implied tempi are all used to create distinct strata. I have already referred to these strata as timezones: by maintaining this logic (that layers of material may simultaneously evoke different temporalities) we may enrich the temporal reading of the work. The effect of layering has already been observed in the rising action of P1–R1, but other vertical relationships can also be examined to understand better the gradual shifting between timezones.

The opening ten bars (up to rehearsal mark A) utilize stratification that immediately evokes a sense of multiple types of time (the melodic lines are shown in Figure 15.14). The bassoon holds the focus with a short octatonic melody,[12] shadowed by two celli which independently follow the line at different paces. This imitative texture creates a melodic haze that gives the impression of the same melody at different speeds, but it is the brass that provide a stretched version that appears to operate in a wholly different timezone. The trumpets' combined line only covers the first two notes but may still be heard as a stretched version of the melody. Crucially, this is strengthened by the timbre of the harmon mutes, the low range, and the slow crescendos over single notes, all of which imbue the sound with a sense of existing in a different temporality.

After its initial statement at the opening, the theme continues to develop until, at rehearsal mark C, it moves beneath the surface, adopting a slower pace and emerging in slow lines in the brass. It appears in a more recognizable (but transposed) version at the end of this section (Figure 15.15). Across this first part of the work, the theme is heard in different timezones, sometimes concurrently, sometimes successively. In its final form, the original four-bar melody is transformed in sonority and stretched over more than twice its original duration. Moreover, unlike the first iteration, which is suspended above an accompanimental bedding in the strings, it is now heard against a fast, rhythmically active timezone in the rest of the orchestra, drawing attention to its stretched-out sound. At the start, the theme is the fast-moving train; now it is the slow carriage being overtaken by more urgent services. This example demonstrates that any one idea can be presented in different temporal contexts, be it in compressed time, stretched time, or somewhere in between. This stratification of types of time is apparent across the work, but in some cases can be explained by more nuanced ideas, as the following example demonstrates.

---

12   An octatonic scale uses eight notes (in contrast to the seven notes of the major or minor scales) alternating intervals of semitones and tones.

FIGURE 15.14   *Sudden Time*, melodies at opening

FIGURE 15.15   *Sudden Time*, transformed opening melody (melodic reduction)

The feeling of a moment of time being stretched is not simply created by events changing less frequently. One of the slowest sounding points in *Sudden Time* is the opening of section two, where a pizzicato-like solo timpani motive is first heard. This figure dictates a slow pulse through its sparsity, creating a relatively undifferentiation continuum, whilst a layer of fast-moving background material is heard in the strings. The quick stratum, however, does not necessarily alter the perception of time: this may not be two distinct timezones, as is used at the opening, but two sides of the same temporal coin. Consider a slowed-down moving image in which we see incredible levels of motion otherwise invisible to the naked eye: a slow-motion video of a cymbal being struck reveals the extremely fast vibrations and the huge warping of the metal that cannot otherwise be observed. Akin to this, the granular details of the sound are orchestrated in this section in a manner that signifies the stretching of time: as time is stretched out we can hear the otherwise hidden details, and, in just the same way, compressing time can reveal new details (as in the sped-up film of a car). Furthermore, the techniques that are employed lend an aura of slowness to the sonority: strings with practice mutes, played on a single up-bow at a low volume produce a scurrying sound that evokes the fast motion of the cymbal in the visual example given above. As the section develops, further timbres enhance this: *legatissimo* phrasing, *flautando* bowing and use of harmonics in the strings; two bowed vibraphones; hand-stopped horn; flutter-tongued and trilled notes in the clarinets. The combination of all of these sounds – which do not individually suggest slow motion – gives a compelling sense of stretched time.[13]

In this section, slow and fast material operate together to maintain the impression of a drawn-out event. This technique has already been observed conceptually in relation to the climax – where stasis is apparent at a formal level – but here it is quite audible. These two events exhibit the tendency for ideas to be presented in both a conceptual manner (the climax, which can be understood, but not necessarily heard, as stretching time) and in more explicit

---

13   A similar effect is found in Benjamin's opera *Written on Skin* (2012) where members of the cast move in slow motion whilst sonorities redolent of slow motion are heard in the score.

manifestations (the opening, in which different temporalities are readily experienced), strengthening the case for time being the modus operandi here.

6 Conclusions

The various theories that informed the introduction to this study orbited around the connections present between speed, distance, and time. Their mathematical relationship was not specifically drawn upon as the ideas of speed and motion are metaphorical, whilst time is manipulated separately. Nevertheless, the three ideas may be useful in framing some conclusions.

6.1 *Time and Illusion*

Time may be seen as constant or malleable; as Newtonian or Einsteinian. Whilst a perception based on each is possible, bringing these ideas together generates the most rounded view. We can perceive the manipulation of time through textures and sonorities redolent of slow-motion, contrasting strata of different timezones and engaging with a morphing between different temporal states. The conception of the work embraces an Einsteinian view – this is apparent from reading the program note alone – but the result can be perceived as either maintaining this or operating in a chronometric framework. The issue of whether we are on the train or observing it from the sidelines remains a matter of perception. It is possible to experience the work as a passive observer (our internal clock ticks regularly), whilst it is also an option to experience our own time as stretching and contracting as a result of the musical devices used (our clock ticks irregularly). The latter takes a certain suspension of disbelief – ignoring the inevitably chronometric aspect of listening – but may help a listener engage more fundamentally with the essence of the work.

6.2 *Speed and Density*

The speed of this music is never simple. Indeed, given the above analysis is framed primarily in terms of stretched and compressed *time*, considering speed can complicate matters. Whilst precise metronome marks appear throughout the score, there are more often than not multiple pulses and implied tempi with which to contend. Simultaneous tempi highlight the use of stratification, as different timezones are negotiated, sometimes changing focus between concurrent strands, whilst elsewhere moving between adjacent sections. Speed is also engaged with at a conceptual level: this has been observed in the conceptually static climax, whilst its audible result can be heard in sonorities which create a sense of slow-motion. Moreover, fast material and a high tempo can sometimes be perceived in terms of slowness or stretched time: in the slowest section we can hear the fast, scurrying detail revealed by stretching out the material in a manner akin to the sped-up film of the slow car. Returning to the context of the introduction we might draw a connection between speed of material and *density* more than speed and time.

These factors come together in a work that might be understood in terms of types of time and motion and the organic flow of material between different timezones. Ideas are presented in both manifest, perceivable versions and more latently. If as listeners we consider these more nuanced versions of speed and motion – alongside those which we perceive quite readily – an interpretation of the work as fundamentally addressing issues of time may become yet more compelling.

6.3   *Distance and Structure*

Quantifying distance relies even more on metaphor, as the temporal nature of music is not directly equatable to physical space. However, the visual analogies used to demonstrate the structural readings of the work show that close attention is given to the space created by architectural scaffolding, the use of frozen time to demarcate sections being the most striking. The huge canvas of the orchestra offers an instrumental space in which Benjamin is able to implement ideas virtuosically. With metaphors of time and speed, the idea of distance is necessarily invoked, but given the framework provided by the program note, the actual distance covered may be quite small: the thunderclap was short but heard over a long time, and so *Sudden Time* may indeed be travelling a short distance, our perception of the journey's speed ever changing. As with the example of two films of a car used in the introduction, the work may be conceived as a single speed, with time slowed down and sped up as part of the compositional process. This indeed explains why at some points we observe the detail revealed through stretched time and strengthens the notion of different timezones that has underpinned various parts of this analysis.

• • •

George Benjamin puts time front and center in this work: the word itself is there in the title, whilst his reference to a multi-temporal thunderclap in the program note suggests an interpretation based on types of time (not simply speeds of music). By taking him at his word and searching for understandings of musical time through reference to the variety of approaches explored above, this study has highlighted both the complexities and usefulness of understanding this music through a temporal lens. And whilst many of the features are revealed through detailed analysis and repeated listening, I am confident that the concepts explored have a bearing on how a listener might experience the piece. These ideas may help us as listeners understand this music more and take these ideas forward in our other listening experiences and indeed our other temporal experiences. That one piece of music opens up so many questions about our experience of time only seems to highlight that much more thinking, listening, and talking on this subject is needed.

References

Adlington, Robert. 2003. "Moving Beyond Motion: Metaphors for Changing Sound." *Journal of the Royal Musical Association* 128(2): 297–318.

Anderson, Julian. 2000. "A Provisional History of Spectral Music." *Contemporary Music Review* 19(2): 7–22.

Benjamin, George. 1991. *A Mind of Winter*. London: Faber Music.

Benjamin, George. 1993. *Sudden Time*. London: Faber Music.

Benjamin, George. 1997. *Three Inventions*. London: Faber Music.

Benjamin Orchestral and Chamber Music. 2005. George Benjamin, conductor, London Philharmonic Orchestra. CD, Nimbus Records, NI5505, track 6.

Clarke, Eric F. 2005. *Ways of Listening: An Ecological Approach to the Perception of Musical Meaning*. Oxford: Oxford University Press.

Epstein, David. 1979. *Beyond Orpheus: Studies in Musical Structure*. Cambridge, Massachusetts: MIT Press.

Fineberg, Joshua. 2000. "Guide to the Basic Concepts and Techniques of Spectral Music." *Contemporary Music Review* 19(2): 81–113.

Grisey, Gérard. 1987. "Tempus ex Machina: A Composer's Reflections on Musical Time." *Contemporary Music Review* 2(1): 239–275.

Grisey, Gérard. 2000. "Did You Say Spectral?" Translated by Joshua Fineberg. *Contemporary Music Review* 19(3): 1–3.

Harvey, Jonathan. 2000. "Spectralism." *Contemporary Music Review* 19(3): 11–14.

Hayden, Sam. 2016. "Complexity, Clarity and Contemporary British Orchestral Music." *Tempo* 70(277): 63–78.

Iturbide, Manuel Rocha. 1995. "Unfolding the Natural Sound Object through Electroacoustic Composition." *Journal of New Music Research* 24(4): 384–391.

Lack, Graham. 2001. "Objects of Contemplation and Artifice of Design: Sonic Structures in the Music of George Benjamin." *Tempo* 215: 10–14.

Nieminen, Risto, ed. 1997. *George Benjamin*. Translated by Julian Anderson and Michael Durnin. London: Faber and Faber.

Rofe, Michael. 2014. "Dualisms of Time." *Contemporary Music Review* 33(4): 341–354.

Schiffer, Brigitte. 1988. "George Benjamin's 'Ringed by the Flat Horizon.'" *Tempo* 133(4): 78–79.

Service, Tom. 2008. "Sudden Time: The Music of George Benjamin." *Roche Commissions: George Benjamin.* http://www.roche.com/dam/jcr:7e30f031-e542-488a-a79a-0e28140d8f28/en/rochecommissions_08_georgebenjamin.pdf (accessed 10 January 2017).

Stockhausen, Karlheinz. 1959. "Structure and Experiential Time." *die Reihe No. 2, Anton Webern,* 64–74.

Sudden Time. On *Benjamin Orchestral and Chamber Music.* 2005. George Benjamin, conductor, London Philharmonic Orchestra. CD, Nimbus Records, NI5505.

Whittall, Arnold. 1983. "Review: Benjamin, George, Piano Sonata; *Flight,* for solo flute; *Ringed by the Flat Horizon,* for orchestra. Score." *Music and Letters* 6(1–2): 137–138.

Index

Note: The letter *f* following a page number denotes a figure; the letter *t* followed by a page number denotes a table.

Printed in the United States
by Baker & Taylor Publisher Services